Withdrawn
University of Waterloo

BIODIVERSITY

BIODIVERSITY
BIOMOLECULAR ASPECTS OF BIODIVERSITY AND INNOVATIVE UTILIZATION

Edited by

Bilge Şener
Gazi University
Ankara, Turkey

Kluwer Academic / Plenum Publishers
New York, Boston, Dordrecht, London, Moscow

Library of Congress Cataloging-in-Publication Data

Biodiversity: biomolecular aspects of biodiversity and innovative utilization/edited by Bilge Sener.
    p.   cm.
   "Proceedings of the 3rd IUPAC International Conference on Biodiversity (ICOB-3), November 3–8, 2001, Antalya, Turkey."
   Includes bibliographical references and index.
   ISBN 0-306-47477-8
     1. Biological products—Congresses. 2. Biological diversity—Congresses. I. Sener, Bilge, 1953– II. IUPAC Conference on Biodiversity (3rd: 2001: Antalya, Turkey).

QH345 .B528 2003
577—dc21

2002037011

"Biodiversity is one of the most important issues of our century"

Proceedings of the 3rd IUPAC International Conference on Biodiversity (ICOB-3), November 3–8, 2001, Antalya, Turkey

ISBN 0-306-47477-8

©2002 Kluwer Academic/Plenum Publishers, New York
233 Spring Street, New York, New York 10013

http://www.wkap.nl/

10 9 8 7 6 5 4 3 2 1

A C.I.P. record for this book is available from the Library of Congress

All rights reserved

No part of this book may be reproduced, stored in a retrieval system, or transmitted in any form or by any means, electronic, mechanical, photocopying, microfilming, recording, or otherwise, without written permission from the Publisher, with the exception of any material supplied specifically for the purpose of being entered and executed on a computer system, for exclusive use by the purchaser of the work

Printed in the United States of America

Contributors

Berhanu M. ABEGAZ, *Prof.Dr.*, University of Botswana, Department of Chemistry, Private Bag 0022, Gaborone-Botswana Fax: +267-355-2836; E-Mail: Babegaz@hotmail.com

Magid ABOU-GHARBIA, *Dr.*, Wyeth-Ayerst Research, Chemical Sciences, CN 8000, Princeton,NJ 08543-8000, USA Fax: +1-732-274 4500; E-Mail: abougam@war.wyeth.com

Sjamsul Arifin ACHMAD, *Prof.Dr.*, Institut Teknologi Bandung, Department of Chemistry, Jalan Ganeca 10, Bandung 40132, Indonesia Fax: +62 22 250 4154; E-Mail: sjamsul@indo.net.id

N. Leyla AÇAN, *Assoc.Prof.Dr.*, Hacettepe University, Faculty of Medicine, Department of Biochemistry, 06100 Ankara-Turkey Fax: +90-312-310 05 80; E-Mail: nla@hacettepe.edu.tr

Khalid AFTAB, *Dr.*, University of Karachi, H.E.J. Research Institute of Chemistry, International Center for Chemical Sciences, 75270, Karachi-Pakistan Fax: +92-21 924 31 90/91; E-Mail: kaftab@cyber.net.pk

Emine AKALIN, *Assist.Prof.Dr.*, İstanbul University, Faculty of Pharmacy, Department of Pharmaceutical Botany, 34452 İstanbul-Turkey Fax: +90-212-519 08 12; E-Mail: adaemine@hotmail.com

Alireza ALIGHANADI(GHANNADI), *Assoc.Prof.Dr.*, Isfahan University of Medical Sciences, Faculty of Pharmacy, Department of Pharmacognosy, Hazar Jerib Av. P.O.Box 81745-359, 81744 Isfahan-Iran
Fax: +98-311-668 00 11; E-Mail: aghannadi@yahoo.com

Filiz AYANOĞLU, *Prof.Dr.*, Mustafa Kemal University, Faculty of Agriculture, Department of Field Crops, 31034, Antakya-Turkey
Fax: +90-326-245 58 32; E-Mail: fayanogl@mku.edu.tr

Feray AYDOĞAN, *Dr.*, Yıldız Technical University, Faculty of Arts & Sciences, Department of Chemistry, Davutpaşa Campus, 34210 İstanbul-Turkey Fax: +90-212-449 18 88; E-Mail: faydogan@yildiz.edu.tr

Eyüp BAĞCI, *Assist.Prof.Dr.*, Fırat University, Faculty of Arts & Sciences, Department of Biology, Elazığ-Turkey Fax: +90-424-233 00 62; E-Mail: ebagci@firat.edu.tr

William S. BOWERS, *Prof.Dr.*, The University of Arizona, College of Agriculture, Department of Entomology, Laboratory of Chemical Ecology, P.O. Box 210036, Forbes Bldg. Room 410, Tucson, Arizona 85721-USA
Fax: +1-520-621-9166; E-Mail: wbowers@ag.arizona.edu

İhsan ÇALIŞ, *Prof.Dr.*, Hacettepe University, Faculty of Pharmacy, Department of Pharmacognosy, 06100 Ankara-Turkey
Fax: +90-312- 311 47 77; E-Mail: icalis@hacettepe.edu.tr

Ermias DAGNE, *Prof.Dr.*, Addis Ababa University, Department of Chemistry, Miazia 27 Square, P.O.Box 30270, Addis Ababa-Ethiopia
Fax: +2511 551244; E-Mail: eda@telecom.net.et

Salvatore DE ROSA, *Dr.*, Universita Cattaloica, ICMIB-CNR, Via Toiono 6, 80072 Arco Felice, Napoli-Italy Fax: +39-81 8041770; E-Mail: sderosa@icmib.na.cnr.it

Nurgün ERDEMOĞLU, *Dr.*, Gazi University, Faculty of Pharmacy, Department of Pharmacognosy, 06330 Ankara-Turkey
Fax: +90-312-223 50 18; E-Mail: nurgun@gazi.edu.tr

Bilgen ERYILMAZ, *M.Sc.,* Gazi University, Faculty of Pharmacy, Department of Pharmacognosy, 06330 Ankara-Turkey
Fax: +90-312-223 50 18; E-Mail: ebilgen@gazi.edu.tr

Mary J. GARSON, *Prof.Dr.*, The University of Queensland, School ofMolecular and Microbial Sciences, Brisbane QLD 4072- Australia
Fax: +61-7-3365 4273; E-Mail: marygarson@hotmail.com

Eugene V. GRISHIN, *Prof.Dr.*, Shemyakin and Ovchinnikov Institute of Bioorganic Chemistry, Russian Academy of Sciences, Ul.Miklukho-Maklaya 16/10, RU-117871 Moscow V-437-Russia Fax: +7-95 310 7007; E-Mail: grev@mail.ibch.ru

Ahmad K. HEGAZY, *Prof.Dr.*, Cairo University, Faculty of Science, Department of Botany, Giza 12613- Egypt Fax: +20-2-572- 7556; E-Mail: akhegazy@main-sec.cairo.eun.eg

Vernon HEYWOOD, *Prof.Dr.*, The University of Reading, School of Plant Sciences, Centre for Plant Diversity and Systematics, P.O. Box 221, Reading RG6 6AS-UK Fax:+44-118-9891745;E-Mail: v.h.heywood@reading.ac.uk

Tatsuo HIGA, *Prof.Dr.*, University of the Ryukyus, Department of Chemistry, Biology and Marine Sciences, Nishihara, 903-0213 Okinawa – Japan Fax: +81- 98 895 8538; E-Mail: thiga@sci.u-ryukyu.ac.jp

Abdel Khalek M. HUSSEIN, *Dr.*, Plant Protection Research Institute, Nadiel-Saied Street, Dokki-Cairo-Egypt Fax: +204 822 69 71; E-Mail: esabeegypt@lycos.com

Zahia KABOUCHE, *Prof.Dr.*, University of Constantine-Mentouri, Faculty of Science, Department of Biology, 25000 Constantine-Algeria
Fax: +213-31 61 43 31; E-Mail: zkabouche@lycos.com

Nutan KAUSHIK, *Dr.*, Tata Energy Research Institute, Bioresources and Biotechnology Division, Darbari Seth Block, Lodhi Road, 110 003 New Delhi-India Fax: +91-11-468 2144; E-Mail: kaushikn@teri.res.in

Nuray KAYA, *M.Sc.*, Akdeniz University, Faculty of Arts & Sciences, Department of Biology, 07070 Antalya-Turkey Fax: +90-242- 227 89 11; E-Mail: n.kaya@sci.akdeniz.edu.tr

Jochen KLEINSCHMIT, *Dr*, Lower Saxony Forest Reseach Institute, Department of Forest Genetic Resources, D-34355 Escherode-Germany Fax: +49-5543 9408 61; E-mail: jochen@kleinschmit.de

Herbert KOLODZIEJ, *Prof.Dr.*, Freie Universität Berlin, Fachbereich Institut für Pharmazie, Pharmazeutische Biologie, Königin-Luise-Strabe 2+4 D-14195 Berlin-Germany Fax: +49-30-838 537 31; E-Mail: kolpharm@zedat.fu-berlin.de

Mehmet KOYUNCU, *Prof.,Dr.*, Ankara University, Faculty of Pharmacy, Department of Pharmaceutical Botany, 06100, Ankara-Turkey
Fax: +90-312-213 10 81; E-Mail: mkoyuncu@yyu.edu.tr

Mahir KÜÇÜK, *Dr.*, Eastern Black Sea Forestry Research Institute, 61040 Trabzon-Turkey Fax: +90-462-230 59 92; E-Mail: dkoae@superonline.com

Bao-Rong LU, *Prof.Dr.*, Fudan University, School of Life Sciences, Institute of Biodiversity Science 220 Handan Road, Shanghai 200433- P.R. China
Fax: +86-21-656 42 468; E-Mail: brlu@fudan.edu.cn

Sabiha MANAV YALÇIN, *Prof.Dr.*, Yıldız Technical University, Faculty of Arts & Sciences, Department of Chemistry, Davutpaşa Campus, 34210 İstanbul-Turkey Fax: +90-212-449 18 88; E-Mail: syalcin@yildiz.edu.tr

Péter MÁTYUS, *Prof.Dr.*, Semmelweis University, Institute of Organic Chemistry, Budapest, Högyes E.U.7.H-1092 Hungary
Fax: +36-1-217 08 51; E-Mail: matypet@szerves.sote.hu

S.Qasim MEHDI, *Dr.*, Biomedical and Genetic Engineering Labs. P.O.Box 2891, 24 Mauve Area, KRL Hospital New Bldg., 3rd Floor, Islamabad-Pakistan Fax: +92-51 926 1144; E-Mail: sqmehdi@isb.comsats.net.pk

Kenan OK, *Assist.Prof.Dr.*, İstanbul University, Faculty of Forestry, Department of Forest Economics, 80895, İstanbul-Turkey
Fax: +90-212-226 11 13; E-Mail: kenanok@istanbul.edu.tr

Emi OKUYAMA, *Prof.Dr.*, Chiba University, Graduate School of Pharmaceutical Sciences, Chiba 263-8522, Japan, Fax: +81-43 290 3021;
E-Mail: emi@p.chiba-u.ac.jp

Nükhet ÖCAL, *Assoc.Prof.Dr.*, Yıldız Technical University, Faculty of Arts & Sciences, Department of Chemistry, Davutpaşa Campus, 34210 İstanbul-Turkey Fax: +90-212-449 18 88; E-Mail: nocal@yildiz.edu.tr

Contributors

Virinder S. PARMAR, *Prof.Dr.*, University of Delhi, Department of Chemistry, 110 007, New Delhi-India, Fax: +91-044 235 28 70; E-Mail: virinder_parmar@uml.edu

Nanjian RAMAN, *Prof.Dr.*, University of Madras, Centre for Advanced Studies in Botany, Guindy Campus, Chennai 600025, India, Fax: +91-044-235 28 70; E-Mail: raman55@vsnl.com

Seyed Ebrahim SAJJADI, *Assoc.Prof.Dr.*, Isfahan University of Medical Sciences, Faculty of Pharmacy, Department of Pharmacognosy, Hazar Jerib Av. P.O. Box 81745-359, 81744 Isfahan-Iran Fax: +98-311-668 00 11; E-Mail: sesajjadi@yahoo.com

M. Koray SAKAR, *Prof.Dr.*, Hacettepe University, Faculty of Pharmacy, Department of Pharmacognosy, 06100 Ankara-Turkey Fax: +90-312-311 47 77; E-Mail: ksakar@ada.net.tr

Ekrem SEZİK, *Prof.Dr.,* Gazi University, Faculty of Pharmacy, Department of Pharmacognosy, 06330 Ankara-Turkey Fax: +90-312-221 06 49; E-Mail: esezik@gazi.edu.tr

Bina S.SIDDIQUI, *Prof.Dr.*, University of Karachi, HEJ Research Institute of Chemistry, International Center for Chemical Sciences,75270 Karachi-Pakistan Fax: +92-21-924 31 90/91; E-Mail: bina@khi.comsats.net.pk

Ren Xiang TAN, *Prof.Dr.*, Nanjing University, School of Life Sciences, Institute of Functional Biomolecules, Nanjing 210093, P.R.China, Fax: +86-25-359 3201; E-Mail: rxtan@netra.nju.edu.cn

Nur TAN, *Assist.Prof.Dr.*, İstanbul University, Faculty of Pharmacy, 34452 İstanbul-Turkey Fax: +90-212-519 08 12; E-Mail: etan@rios.de

Deniz TAŞDEMİR, *Assist.Prof.Dr.*, Hacettepe University, Faculty of Pharmacy, Department of Pharmacognosy, 06100, Ankara-Turkey Fax: +90-312- 311 47 77; E-Mail: dtasdemi@hacettepe.edu.tr

Nilgün YAŞARAKINCI, *Dr.*, Plant Protection Research Institute, 35040 Bornova-İzmir-Turkey Fax: +90-232- 374 16 53; E-Mail: nyasarakinci@yahoo.com

Erdem YEŞİLADA, *Prof.Dr.*, Gazi University, Faculty of Pharmacy, Department of Pharmacognosy, 06330, Ankara-Turkey
Fax: +90 312 223 50 18; E-Mail: yesilada@gazi.edu.tr

Galiya ZHUSUPOVA, *Assoc.Prof.Dr.*, Kazakh State National University, Department of Organic Chemistry, 172 Aimanov Street 137, 480057 Almaty-Kazakhstan Fax:+7-3272-33 99 32;
E-Mail: zhusupova@yahoo.com

Preface

Biodiversity or biological diversity is a complex system involving plants, animals, microorganisms and human beings. So far, about 1.75 million species have been identified, mostly small creatures such as insects. Biodiversity is the combination of life forms and their interactions with each other and environment that has made Earth. It also provides a large number of goods and services that sustain our lives. Genetic diversity also plays an important role in the variability of chemical metabolites and their classes within the same plant species.

One aspect of biodiversity is chemical in origin. The chemistry of bioresources has been offered understanding of complex biological and ecological interactions for future generations. Biomolecular aspects of biological diversity have therefore become an important topic. Dependant on human population growth and economic pressure, the loss of the earth's biodiversity is one of the most pressing environmental and development issue today with each species that disappears, developing countries stewards of most of the planet's biological wealth lose potential for sustainable development. As natural sources are threatened possible cross-sectoral approach can save them. Realizing the global environmental concern, the United Nations Environment Programme established the *Convention on Biological Diversity* in Rio de Janeiro, June 1992. This convention marked a significant step in the global efforts in the conservation and sustainable utilization of biodiversity.

Innovative utilization of bioresources links societies, governments, universities and industrial organizations. The mission of the international programs in the area of biodiversity is to provide information on the status, security, management and utilization of the world's biological diversity to support preservation and sustainable development.

The source of this book is the *3rd IUPAC International Conference on Biodiversity (ICOB-3)*, held in Antalya-Turkey on Nov. 3-8, 2001 under the auspices of the IUPAC organized by the Society of Biological Diversity, Ankara-Turkey. This conference was part of a series of IUPAC-sponsored international conferences that are held every two years initiated by Organic and Biomolecular Chemistry Division. The aim of this series of conferences is to understand the value of bioresources and the need for their conservation in terms of the biomolecular chemistry of naturally occurring molecular systems. The development of pharmaceutical, agricultural and industrial products from bioresources can be used to promote incentives for conservation by providing an economic return to sustainable use of those sources. This conference featured more than 250 scientists from five continents generated novel and fruitful ideas to accelerate the future researches and valuable guidelines regarding the development of activities for sustainable utilization of biodiversity and conservation of the environment for welfare of humans and other living organisms. The ICOB-3 session topics were: Chemical Basis of Biodiversity, Biomolecular Aspects of Biodiversity and Innovative Utilization. In addition, a panel was held as a discussion forum on International Co-operation on Molecular Diversity. A draft report from the IUPAC project on "Molecular Basis of Biodiversity; Conservation and Sustained Innovative Utilization" was discussed as well.

This book includes 54 chapters presented as invited lectures and posters at the conference. This articles update readers on the search for insight into the species and documents how much of live remains to be scientifically identified. They also explore identification strategies and methods along with the implications for protecting biodiversity. As a summary, biomolecular aspects of biodiversity and innovative utilization of bioresources were discussed from very diverse points of view ranging from their botanical, zoological, taxonomic and genomic expressions to their biomolecular, structural, mechanistic and functional aspects.

The plenary lectures have published in an issue of *Pure and Applied Chemistry*, Vol.74, No.4, April 2002; the titles of these presentations are given as an Annex at the end of the book.

As chairperson, I am honoured and delighted to write the preface to the *Proceedings of the ICOB-3* which made the conference a significant occasion, I thank all contributors for the preparation of their presentations.

I am deeply grateful to Professor Z. Sacit Önen, Vice-President of the Society of Biological Diversity, Ankara-Turkey for making this organization possible.

I wish to thank to various organizations and pharmaceutical companies both home and abroad who kindly supported the conference.

Preface

I would like to express my gratitude to Kluwer Academic/Plenum Publishers for taking on the task of publishing the *Proceedings of the 3rd IUPAC International Conference on Biodiversity (ICOB-3)* and I particularly thank Ms. Joanne Duggan for her effort during formatting and pagination procedures.

Prof. Dr. Bilge Şener
Editor

Contents

1. Integration of Conservation and Utilization in Temperate Hardwood Species 1
   Jochen Kleinschmit and Jörg R. G. Kleinschmit

2. The Conservation of Genetic and Chemical Diversity in Medicinal and Aromatic Plants 13
   Vernon H. Heywood

3. Conservation and Sustainable Use of Biodiversity in Wild Relatives of Crop Species 23
   Bao-Rong Lu

4. Perspectives on Human Genome Diversity within Pakistan using Y Chromosomal and Autosomal Microsatellite Markers 35
   S. Qasim Mehdi, Qasim Ayub, Raheel Qamar, A. Mohyuddin, Atika Mansoor, K. Mazhar, A. Hameed, M. Ismail, S. Rahman, Saima Siddiqui, Shagufta Khaliq, M. Papaioannou, Chris Tyler-Smith, and L. L. Cavalli-Sforza

5. Lessons from Nature Show the Way to Safe and Environmentally Pacific Pest Control 49
   William S. Bowers

6. Biodiversity of Soil Fauna in Different Ecosystems in Egypt with Particular References to Insect Predators 59
   Adbel Khalek M. Hussein

7. Optimization of Natural Procedures Leads: Discovery of
   Mylotarg™, CCI-779 and GAR-936 63
   Magid Abou-Gharbia

8. Bioactive Compounds from Some Endangered Plants of Africa   71
   Berhanu M. Abegaz and Joan Mutanyatta

9. New Bioactive Substances Reported from the African Flora 79
   Ermias Dagne

10. Bioactive Components of a Peruvian Herbal Medicine,
    Chucuhuasi (*Maytenus amazonica*) 85
    *Emi Okuyama, K. Shimamura, C. Nagamatsu, H. Fujimoto,
    M. Ishibashi, O. Shirota, S. Sekita, M. Satake, J. Ruiz,
    F. A. Flores, and S. Yuenyongsawad*

11. Discovery of Natural Products from Indonesian Tropical
    Rainforest Plants: Chemodiversity of *Artocarpus* (Moraceae) ... 91
    *Sjamsul Arifin Achmad, Euis Holisotan Hakim,
    Lukman Makmur, Didin Majahidin, Lia Dewi Juliawaty,
    and Yana Maolana Syah*

12. Seminal Findings on a Novel Enzyme: Mechanism of
    Biochemical Action of 4-Methylcoumarins, Constituents of
    Medicinal and Edible Plants 101
    *Virinder S. Parmar, Hanumantharao G. Raj,
    Ashok K. Prasad, Subhash C. Jain, Carl E. Olsen,
    and Arthur C. Watterson*

13. Medicinal Plants—A Source of Potential Chemicals of Diverse
    Structures and Biological Activity 109
    *Bina S. Siddiqui, Farhana Afshan, Munawwer Rasheed,
    Nadeem Kardar, Sabira Begum, and Shaheen Faizi*

14. Biodiversity in Turkish Folk Medicine 119
    Erdem Yeşilada

15. Biodiversity of Phenylethanoids Glycosides 137
    İhsan Çalış

16. The Chemo- and Biodiversity of Endophytes 151
    Ren Xiang Tan and Ren Xin Zou

17. Molecular Diversity and Specificity of Arthropod Toxins 161
    *Eugene V. Grishin, T. M. Volkova, Yu. V. Korolkova,
    and K. A. Pluzhnikov*

18. Chemical Diversity of Coral Reef Organisms 169
    *Tatsuo Higa, Michael C. Roy, Junichi Tanaka,
    and Ikuko I. Ohtani*

19. Chemical Signals from Sponges and their Allelopathic Effects
    on Other Marine Animals 179
    Mary J. Garson

20. Anti-Cancer Metabolites from Marine Sponges 187
    Deniz Taşdemir

21. Altitudinal and Latitudinal Diversity on the Flora on Eastern and
    Western Sides of the Red Sea 197
    Ahmad K. Hegazy and Wafaa M. Amer

22. Biodiversity and Free Market Mechanism 217
    Kenan Ok

23. Domestication and Determination of Yield and Quality Aspects
    of Wild *Mentha* Species Growing in Southern Turkey 227
    Menşure Özgüven, Saliha Kırıcı, and Filiz Ayanoğlu

24. Some Ornamental Geophytes from the East Anatolia 247
    Mehmet Koyuncu

25. Bioactive Molecules from *Cynodon dactylon* of Indian
    Biodiversity 253
    *Nanjian Raman, A. Radha, K. Balasubramanian,
    R. Rughunathan, and R. Priyadarshini*

26. Phenylethanoid Glycosides with Free Radical Scavenging
    Properties from *Verbascum wiedemannianum* 257
    *İhsan Çalış, Hasan Abou Gazar, Erdal Bedir,
    and Ikhlas A. Khan*

27. Antioxidant Activity of *Capsicum annuum* L. Fruit Extracts on
    Acetaminophen Toxicity 261
    Bilgen Eryılmaz, Göknur Aktay, and Funda Bingöl

28. In Vitro Antileishmanial Activity of Proanthocyanidins and
    Related Compounds 265
    Herbert Kolodziej, O. Kayser, A. F. Kiderlen, H. Ito, T.
    Hatano, T. Yoshida, and L. Y. Foo

29. Evaluation of the Antileishmanial Activity of Two New
    Diterpenoids and Extracts from *Salvia cilicica* 269
    Nur Tan, M. Kaloga, O. A. Radtke, and Herbert Kolodziej

30. Antibacterial and Antifungal Activities of *Sedum sartorianum*
    subsp. *sartorianum* 273
    M. Koray Sakar, M. Arısan, M. Özalp, M. Ekizoğlu,
    D. Ercil, and H. Kolodziej

31. Blood Pressure Lowering Activity of Active Principle from
    Ocimum basilicum 279
    Khalid Aftab

32. Chemical Variability in *Azadirachta indica* Growing in Tamil
    Nadu State of India 283
    Nutan Kaushik and B. Gurdev Singh

33. Pesticidal Activity of *Eucalyptus* Leaf Extracts against
    Helicoverpa armigera Larvae 287
    Nutan Kaushik

34. Two New Lignans from *Taxus baccata* L. 291
    Nurgün Erdemoğlu and Bilge Şener

35. Lignans from *Taxus baccata* L. 297
    Nurgün Erdemoğlu and Bilge Şener

36. Heraclenol and Isopimpinellin: Two Rare Furocoumarins from
    Ruta montana 303
    N. Benkiki, M. Benkhaled, Zahia Kabouche,
    and C. Bruneau

37. A Chemotaxonomic Study on the Genus *Ferulago*, Sect.
    Humiles (Umbelliferae) 309
    Emine Akalın, Betül Demirci, and K. Hüsnü Can Başer

38. Aromatic Biodiversity among Three Endemic *Thymus* Species of
    Iran 315
    Seyed Ebrahim Sajjadi

39. Volatile Constituents of the Leaves of *Ziziphus spina-christi* (L.) Willd. from Iran 319
   Alireza Alighanadi (Ghannadi)
   and Mozhgan Mehri-Ardestani

40. Fatty Acid Composition of the Aerial Parts of *Urtica dioica* (Stinging Nettle) L. (Urticaceae) 323
   Eyüp Bağcı

41. Fatty Acid Composition of *Aconitum orientale* Miller and *A. nasutum* Fisch. ex Reichb Seeds, A Chemotaxonomic Approach   329
   Eyüp Bağcı and Hasan Özçelik

42. New Peptide from a Bacterium Associated with Marine Sponge *Ircinia muscarum* 335
   Salvatore de Rosa, Maya Mitova, Salvatore de Caro,
   and Giuseppina Tommonaro

43. Sheep Brain Glutathione Reductase: Purification and Some Properties 341
   N. Leyla Açan and E. Ferhan Tezcan

44. Some Morphological and Phenological Characters of Tobacco (*Nicotiana tabacum* L.) Grown in Hatay Province of Turkey ...   345
   Durmuş Alpaslan Kaya and Filiz Ayanoğlu

45. Genetic Diversity of Two Native Forest Tree Species in Turkey: *Pinus brutia* Ten. and *Cupressus sempervirens* L.   349
   Nuray Kaya and Kani Işık

46. The Vegetation Studies in the Pure Stands of Kürtün (Gümüşhane) Forests in Turkey 355
   Mahir Küçük and Ömer Eyüboğlu

47. 7th Year IPM Implementation: The Biodiversity of Pests and their Beneficial Species in the Protected Vegetable in the Aegean Region of Turkey 363
   Nilgün Yaşarakıncı

48. Phytopreparations from the Species of *Limonium* Mill   367
   Galiya Zhusupova, K. Rachimov, T. Shalakhmetova,
   and Z.H. Abilov

49. The Biological Activities of New Heterocyclic Compounds
    Containing Nitrogen and Sulphur 371
       Feray Aydoğan, Zuhal Turgut, Çiğdem Yolaçan,
       and Nüket Öcal

50. Synthetic Modification of Iridoids to Non-natural Indole
    Alkaloids 375
       Ákos Kocsis and Péter Mátyus

51. The Complexation of New 1,3-Dithiocalix[4]arene Containing
    Oxime Derivative 379
       Naciye Yılmaz Coşkun, Sabiha Manav Yalçın,
       Ayşe Gül Gürek, and Vefa Ahsen

52. Convenient Route to Quinoline-Tetrahydroquinolines from
    Quinoline-Carboxaldehydes 383
       Leonor Y. Vargas Méndeza, Vladimir V. Kouznetsov,
       Nüket Öcal, Zuhal Turgut, and Çiğdem Yolaçan

53. Quantum Chemical Research of Quercetin, Myricetin, their
    Bromo- and Sulpho Derivatives 387
       Galiya Zhusupova and V. Gapdrakipov

54. Destruction and Conservation of Turkish Orchids 391
       Erkem Sezik

Annex 401

Author Index 403

Subject Index 405

Integration of Conservation and Utilization in Temperate Hardwood Species

[1]JOCHEN KLEINSCHMIT and [2]JÖRG R.G. KLEINSCHMIT
[1]*Lower Saxony Forest Research Institute, Department of Forest Genetic Resources, D-34355 Escherode, Germany, 2University of Göttingen, Institute of Forest Genetics and Forest Tree Breeding, Büsgenweg 2, D-37077 Göttingen, Germany*

1. INTRODUCTION

If we compare the present situation of forest area and of tree species composition within forests with the situation which would be the natural pattern of species after the last glaciation, there are drastic differences due to human interference. Clearing of land for agriculture and settlement, overutilization of certain species, preference of species in plantations, replacement of hardwoods by faster growing conifers, intensive forest management, reduction of specific ecological niches like riparian forests, introduction of pests and diseases, seed transfer partly over long distances changed forest area, tree species composition, and genetic structure of the tree species. Conifers extended their relative participation, hardwood species were reduced. Especially minor hardwood species were without economic interest and some were regionally extinct.

Industrialization, mechanization and human population growth speeded up these processes, leading to the fact, that worldwide less forest surface has to supply wood for a steadily increasing human population (Fig.1).

Regional wood production deficits resulted in the extension of plantations with a small number of intensively managed forest tree species.

This development and first scientific results[1], demonstrating the drastic changes and the risks of irretrievable losses of biodiversity, created public awareness and concern. Political actions started and first regional plans of action were developed. The conference in Rio de Janeiro 1992 with the Biodiversity Convention is a markstone in this development and the term sustainability - which was originally developed for wood production in forestry more than 200 years ago – became a central topic in all aspects of nature development. National and international programs for the conservation of forest tree species were developed. A network for conservation of forest genetic resources (EUFORGEN)[2] was established with the aim to coordinate national and international activities within Europe. Scientific studies were intensified as well on the genetic structure and adaptability of forest trees as on conservation strategies and implications of conservation. Questions of conservation became an important topic for the International Union of Forest Research Organizations (IUFRO).

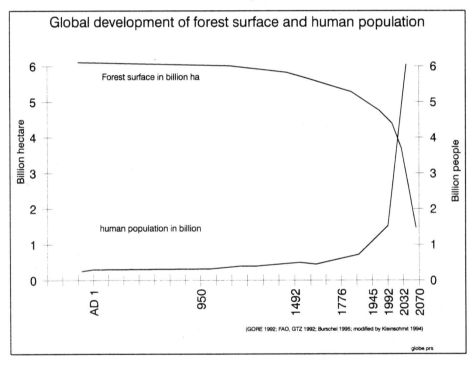

Figure 1. Global development of forest surface and human population

Quite often conservation and utilization are regarded to be mutualy exclusive. The aim of this paper is to give an idea how both objectives can be combined and to discuss the problems arising from this task.

Some explanations for a better understanding of the following chapters:

Conservation is a long term activity with the aim to maintain the adaptability of the species. Since genetic variation is the base for adaptability it means conservation of genetic variation.

Utilization means primarily wood production to serve human needs. Wood is one of the few self regenerating resources. It can be used sustainable.

Temperate hardwood species are broadleved species growing in the temperate climatic zone, where forests disturbed by humans are dominating. Undisturbed relicts are extremely rare in Europe. Forests in the temperate zone originally were dominated by hardwood species. Europe is the main focus of this presentation. Temperate hardwood species are long living, which includes a high environmental variability in time, and the uncertainty of the use and value of wood in the future. They occupy heterogeneous site conditions which means environmental variability in space. They form one of the most important land ecosystems, which is a habitat for many other coevolving species and which has considerable economic importance.

2. CONSERVATION AND UTILIZATION

2.1 Conservation

For efficient conservation strategies we need an inventory, the evaluation and the conservation activities. For conservation, information about the tree species, their present occurrence and their genetic situation is necessary, however financial resources are limited.

2.1.1 Inventory

For the knowledge of the endangerment of tree species the numbers of trees, the ecosystems and sites where they occur, the age structure of the populations, the influences of management on the species, and their economic importance are important. It is obvious, that there are differences between frequent species and rare species. For frequent species it is impossible to count individuals due to technical and economic reasons and unnecessary due to existing extended populations. For very rare species counting may be necessary at least for mature individuals. On the other hand forest inventory data usually are available for the economically most important species but missing for rare species.

Information may come from historical descriptions, aerial pictures, earlier inventories or surveys and local mapping. A problem here is communication between the different organizations involved, to establish comparable data bases and to value these data.

Some of the very rare species may be close to extinction regionally. Here immediate actions are necessary. If financial means are very restricted it may be better in some situations to register intact biotops, which are typical for certain tree species, than to focus on single species directly.

2.1.2 Evaluation

Evaluation is necessary to have a better understanding of the species and its genetic structure. A main interest is the study of the variability of the characters within the species and their genetic control. For this two ways of research are possible: Classical field tests and genetic markers.

The *classical field* testing is done in multiple site provenance-, progeny-, or clonal trials. Such experiments have a long observation time of many decades, they need an institutional organization and long term funding. They allow besides others the study of adaptive traits which are important to judge adaptedness and adaptability of certain genetic units. The long time span between establishment of field tests and the final results as well as the costs are a major problem.

Genetic markers can be evaluated much faster and therefore, with the improvement of the methods, an increasing number of studies uses genetic markers.

Genetic markers may be costly too if sample sizes increase. The interpretation of the results and their importance for adaptability is difficult.

Therefore a synthesis of different methods and characters is necessary. A good example for such a synthesis is the EU funded oak study, which gave a comprehensive view of oak history, genetic structure, and adaptive potential of oak in Europe[3].

For many of the rare hardwood species long term field experiments and for some of the rare species even studies of genetic markers still have to be established to get the fundamental knowledge necessary for efficient conservation measures.

2.1.3 Collection

The aim of a collection is to get a sufficiently representative sample to allow for long term survival and adaptability of the tree species.

Conservation is done *in situ or ex situ*.

In situ collection needs the identification and conservation of representative stands, which cover the ecological conditions of the species range. It is a dynamic conservation strategy, since natural selection processes go an. However risks of loss are not completely excluded due to catastrophic events like forest fires, storms, pests e.g. Therefore ex situ conservation is a necessary complement. For rare species ex situ conservation can be the only option.

Ex situ collection and conservation is done by seed, pollen, scions for cutting propagation or grafting, and by ex situ plantations.

Since the time needed for sufficiently reliable information conflicts with the necessity to act, most forest tree conservation programs started with preliminary information and therefore follow a conservative strategy. Some of the characters can be studied only with ex situ material which needs previous collection.

It is necessary to define priorities by species due to their endangerment. The optimal combination of in situ and ex situ conservation according to the frequency of the species, the management systems in which it is included, and the present knowledge has to be defined. Differences of the fertilization vectors and the reasons for rareness (e.g. human made or naturally) must be considered.

2.2 Utilization

Trees can grow in undisturbed natural forests or in managed forests. Undisturbed forests are existing only in small relicts in temperate regions and they are therefore excluded from utilization. Managed forests can be naturally regenerated or planted. Planting can be done with reproductive material from the same stand or from different origin. The species composition can be close to nature or completely artificial. Legal restrictions exist in some countries for reproductive material to be used in plantations. In managed forests certain species were excluded in the past. Therefore these species are endangered today. Economically important and easy to handle species were preferred and therefore extended.

With the increasing ecological awareness the close to nature management

Table 1. Statistics on countries providing a large proportion of their industrial wood from plantation[4]

Country	Forest area (1000 ha)		Share of plantation (%)	
	Natural	Plantation	Total area	Total production
Argentina	36000	800	2.2	60
Brazil	396000	6500	1.6	60
Chile	6300	1400	22.2	95
New Zealand	6270	1240	19.8	93
Zambia	12900	60	0.4	50
Zimbabwe	28800	117	0.4	50

is favored today. This includes the active integration and favorization of rare species into management decisions and the conversion to a more natural species composition with more broadleaves.

At the same time forestry as a typical primary production is subjected to a strong economic pressure, due to the discrepancy between costs and benefit of wood production. This favors intensively managed highly productive plantations. They can produce more wood on the same area than conventional forests and therefore take utilization pressure from the latter[5] (Table 1).

Today silvicultural management favors systems in agreement with sustainable use of the ecosystem. This considers more aspects than the original concept of sustainable wood production.

2.3 Integration

The biodiversity convention and political declarations are in favor of sustainability. This will last as long as the political priorities are on this line and sufficient money is available.

The forest owner has to live from his forests. For him conservation has no value for its own but it becomes interesting if there is an economic return behind it. Due to the uncertainty of future market development his intention should be to have a differentiated composition of species in his forest providing at any time a big variety of goods for potential markets.

For those who are responsible for conservation it is therefore necessary to show possibilities for utilization even of rare species and to supply reproductive material of high value.

The aim should be to do conservation as efficient and as economic as possible and to integrate conservation into conventional forest management systems. Since managed forests will be the vast majority this enlarges the base for conservation.

In situ conservation has priority in forest tree species due to the longevity of trees, the coevolution with other organisms, and the possibility for adaptational processes. In situ conservation stands include as well completely protected areas as managed stands. Completely protected areas are a good base for major tree species and associated species, they can have restrictions for tree species which are weak competitors. In systems with natural regeneration in situ conservation is possible for all tree species, if the ecological situation offers possibilities for the regeneration of all species and if the silvicultural intentions do not favor certain species only. Boyle[6] and El-Kassaby[7] discussed the problems of human interference on conservation of genetic diversity. The genetic diversity in natural regeneration systems is restricted to the base material which is available more or less accidentally in the stand and which may be far from the optimal potential of the species. Therefore "genetic enrichment planting" can be an interesting addition in such systems for the forest owner, especially for rare species. If new stands for an ecologically oriented silviculture are established, this is the chance to use reproductive material, which optimally combines conservation and economic aspects.

However the management system can restrict the possibility for in situ conservation even more. In plantation forestry with clearcutting in situ conservation ends with the felling of the stand. Therefore conservation and utilization have to be separated on different areas. Here *ex situ* conservation, mostly in combination with tree breeding programs, is usual. The choice of efficient breeding strategies like multiple population breeding systems (MPBS)[8] allows an efficient control and development of genetic variation[9]. Thus breeding provides a means for genetic conservation. It allows at the same time to procure the forest owner with improved reproductive material for all types of management systems, which is not possible from in situ stands.

Ex situ conservation can be static or dynamic as far as the development of genetic diversity is concerned. Storage of seed, grafted clonal archives or seed orchards, and clonal cutting propagation are static, ex situ seedling plantations and the MPBS are dynamic approaches.

Storage of seed is possible only for certain species, in recalcitrant species seedling plantations are an alternative to long term seed storage. Seedling plantations can support the evaluation and utilization, if the experimental design is structured as a progeny test.

Grafted seed orchards are a conventional method for collection and conservation. They open the option for utilization at the same time and are therefore favored in practice. Some questions are open concerning epigenetic effects[10].

Cutting propagation of mature trees is restricted to those species, which can be rejuvenated. Cutting propagation offers a direct access to utilization.

The different *management systems* set limitations how far conservation and utilization can be integrated or have to be separated on the same area. In most cases there will be not just one method (in situ or ex situ) applied to a tree species but different methods parallel.

Further points which have to be taken into account to guarantee the success of integration are:
- International cooperation can help to balance conservation and utilization. If a tree species is abundant under similar ecological conditions in another country, it may be justified to put more emphasis on utilization than on conservation.

- Communication between the different groups involved (forest owners, conservationists, politicians and scientists) is necessary to be able to integrate conservation and utilization efficiently.
- Information and education of the forest owners are of central importance ("Science goes public!").
- Certification of forestry, which is discussed worldwide, and applied in many countries, should integrate aspects of genetic conservation more than today.
- Finally legal instruments are necessary which support the integration of conservation into utilization.

A general conclusion of this chapter is, that utilization is without problem for conservation, if the genetic information is carried on to the following generation and if it is used to establish genetically adaptable stands. Then it is the most secure method since it is not depending on long term funding so much.

2.4 Example Germany

The following gives an example of the integration of genetic conservation into forest management.

Systems of ecologically oriented forest management are discussed since long and locally applied to a limited extent. Since 1972 the state forest services began to apply such systems. 1985 a national program for the conservation of forest genetic resources was developed and applied by the State and Federal Forest Services since then[11-13]. A main emphasis was on in situ conservation for major tree species and on ex situ conservation of rare noble hardwood species. For in situ conservation long term monitoring of genetic diversity is established for some major hardwood species. For 14 noble hardwood species altogether 134 seed orchards and 26 clonal archives have been established in Germany. This work was supported by the fact, that the rare noble hardwood species attended a higher economical interest as a

Table 2. Top prices for quality wood of oak, wild cherry, wild pear tree, birch and wild service tree in Germany

Species	maximal price per m³	
Oak (*Quercus robur* and ssp. *petraea*)	2.000.-	US Dollar
Wild Cherry (*Prunus avium*)	3.700.-	"
Wild pear (*Pyrus pyraster*)	4.800.-	"
Birch (*Betula pendula*)	800.-	"
Wild service tree (*Sorbus torminalis*)	12.000.-	"

substitute for tropical hardwoods during this time. The prices reached unexpected levels (Table 2).

This development and its publication created an immediate interest of forest owners to plant such species. A major problem is the fact that up to now not sufficient valuable reproductive material can be procured from the seed orchards which only start flowering and which open the option for a MPBS. This work is integrated into the EUFORGEN noble hardwood network.

The conservation work is closely integrated into forest management and education of forest owners is a main emphasis.

An increasing amount of scientific studies accompanied this work and continuously improved the base for decisions (Figure 2). There seems to be a slight decrease of interest for the topics "nature conservation" and "sustainability" and an increase of the topic "certification", which may

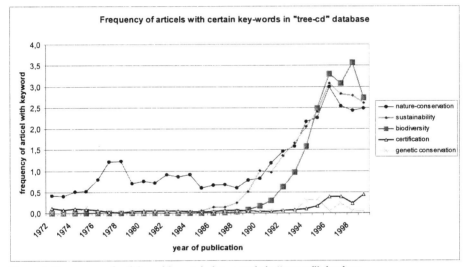

Figure 2. Frequency of articles with certain key-words in "tree-cd" database.

indicate, that after the discussions and studies funding is more restricted and now practical application is sought.

The international political development supported these activities which get additional assistance by the recent process of certification.

3. DISCUSSION

The preconditions for conservation of forest genetic resources improved during the last 20 years. With the increase of economic limitations one of the first decisions of politicians is to cut back funding in fields which have no immediate consequences. Conservation is one of these fields. Therefore all activities which safeguard the long term success are of outstanding importance. This can be reached with the integration of conservation into the regular forest management. The restricted funds than can be allocated to those species which are most endangered.

An important question is when an *integration* of conservation into ecologically oriented forestry should be done and where a *separation* between plantations and ex situ conservation is necessary. This very much depends on the present situation of the forests and on the need for wood in a country. If ecologically oriented management systems are existing or can be developed and if wood production from such systems is in balance with the need for wood, this seems to be a reasonable solution.

However criteria and indicators have to be developed to judge how far the conservation of genetic diversity is ensured. If the genetic diversity is not sufficiently maintained, natural regeneration has to be complemented by genetic enrichment planting, and ex situ conservation in seed orchards of rare species are a potential supplement in this situation. Rotach[14] has discussed the possibilities how minor hardwoods can be promoted by silviculture.

If plantation forestry is necessary for wood supply separation of conservation and utilization must be done. Tree breeding for these plantations has to be done not only for the present economic conditions but has to view the long term adaptability of the species. Namkoong[15] has discussed risks and stability of plantations vs. naturally regenerated forests. He concludes, that it very much depends on the way the breeding is performed and therefore risk and stability in plantations may be higher or lower than in naturally regenerated forests.

Since the natural range of forest tree species does not coincide with national borders conservation programs should be international. EUFORGEN is an excellent example how programs can be coordinated and supplemented without interfering with national interests more than absolutely necessary. For local decisions the knowledge about the programs and research of neighboring countries is essential and can help to set priorities.

The communication of knowledge between the scientific community and the other stake holders concerning forests is a permanent challenge. This can be done by publications, public discussions, education and continued studies at the universities and especially field excursions.

There is an obvious gap of knowledge about the effects of long term dynamics of extended forest ecosystems on the genetic structure of tree species. The metapopulation model[16] allows local extinction of a species if a number of local demes interact through geneflow and migration. If such a model applies, then the demography, the pattern of distribution, and the interaction of the single demes are essential for conservation and not the single deme. There are still many gaps in knowledge to be filled.

REFERENCES

1. National Research Council, Board on Agriculture, 1991, Managing global genetic resources – forest trees, National Academy Press, Washington D.C., p.228.
2. Turok, J., Palmberg-Lerche, C., Skrøppa, T., Ouédraogo, A. S., (eds.), 1998, Conservation of forest genetic resources in Europe. *Proceedings of the European Forest Genetic Resources Workshop* 21 November 1995, Sopron, International Plant Genetic Resources Institute Publ., p.57.
3. Special Issue, 2002, *Forest Ecology and Management*, (in press).
4. Pandley, D., 1992, Assessment of tropical forest plantation resources. Department of Forest Survey, Swedish University of Agricultural Sciences. Unpublished report to Forestry Department FAO, Rome.
5. Bastien, J. Ch., 2001, Importance of intensively managed plantations for wood supply. Meeting Forest Tree Breeding in an Ecologically Oriented Forest Management. Escherode (Germany) June 28 – 29, 2001 (in press).
6. Boyle, T.J.B.,1999, Conserving genetic diversity of forest trees in managed landscapes. In *Forest Genetics and Sustainability* (Matyas, Cs., ed.), Kluwer Academic Publishers, London, pp.131-46.
7. El-Kassaby, Y.A., 2001, Impacts of industrial forestry on genetic diversity of temperate forest trees. In *Forest Genetics and Sustainability* (Matyas, Cs., ed.), Kluwer Academic Publishers, London, pp.155-69.
8. Eriksson, G, Namkoong, G., Roberds, I., 1993, *Forest Ecology and Management*, **62**, 15-37.
9. Karlsson, B., Stahl, P.H., 2001, A Swedish view of tree improvement in an ecologically oriented forest management system. *Meeting Forest Tree Breeding in an Ecologically Oriented Forest Management*. Escherode (Germany) June 28 – 29, 2001 (in press).
10. Johnsen, O., Skrøppa, T., 1996, *Euphytica*, **92**, 1-2.
11. Kleinschmit, J.,1995, *Silvae Genetica*, **44**, 269-74.
12. Kleinschmit, J., 2000, Strategic directions in conserving genetic resources, *Proceedings Canadian Tree Improvement Association*, Sault Ste Marie, Ontario.
13. Behm, A., Becker, A., Dörflinger, H., Franke, A., Kleinschmit, J., Melchior, G. H., Muhs, H.-J., Schmitt, H. P., Stephan, B. R., Tabel, U., Weisgerber, H., Widmaier, T.H., 1997, *Silvae Genetica*, **46**, 24-34.
14. Rotach, P., 1999, In situ conservation and promotion of Noble Hardwoods. In *Silvicultural management strategies* (Turok, J., Jensen, J., Palmberg-Lerche, Ch., Rusanen, M., Russel, K., de Vries, S., Lipman, E., eds.) Compilers, International Plant Genetic Resources Institute, Rome, pp. 39-50.
15. Namkoong, G., 1999, Plantations vs. naturally regenerated forests: Risks and stability in using bred, cloned, or biotech products. In *Forest Genetics and Sustainability* (Matyas, Cs., ed.), Kluwer Academic Publishers, London, pp. 147-54.
16. Levins, R.,1970, *Lectures on Mathematics and the Life Sciences*, **2**, 75-107.

The Conservation of Genetic and Chemical Diversity in Medicinal and Aromatic Plants

VERNON H. HEYWOOD

Centre for Plant Diversity & Systematics, School of Plant Sciences, The University of Reading, Reading RG6 6AS UK

1. INTRODUCTION

In most considerations of plant conservation, scant attention is paid to sampling and conserving the chemical diversity and the underlying genetic diversity found in plant populations. Such diversity is of particular importance when dealing with medicinal and aromatic plants where it is precisely their chemical features, such as alkaloids, essential oils, etc. that are the characteristics for which they are valued. In contrast, a great vast amount of effort has gone into isolating and characterizing these chemical constituents from the limited samples that are traditionally used in phytochemical studies. Conservation of the genetic diversity of medicinal and aromatic plants has generally not received a great deal of attention from the genetic resource agencies or national or regional seedbanks[1,2], although they are recognized in principle as an important group[3,4], nor is there is great deal of evidence from the industry that this has been seriously addressed. Many seed banks do contain samples of some of these plants, but they are not generally the result of deliberate sampling campaigns. Public concern tends to focus on the benefits that may be derived from medicinal plants and on issues of safety and the assumption is made that the plant resources will continue to be available although no concerted efforts have been made to ensure this.

2. GENETIC CONSERVATION OF WILD SPECIES

Genetic conservation of medicinal and aromatic plants involves technical procedures such as ecogeographical surveying, population sampling and analytical techniques for determining genetic variation, and socio-economic issues such as wild harvesting and sustainable use of these resources, bioprospecting agreements and the role of local communities and intellectual property rights. These factors, together with the large numbers of species involved, make the slow progress in developing this topic not altogether surprising.

The genetic conservation of all wild plant species differs from that of cultivars or landraces of domesticated species. Attention has been drawn to these issues by Frankel & al.[5], and by Debouck[6] who points out, that the enormous amount of variation and ecologically highly specific requirements of wild species [and relatives of crops] often makes their *ex situ* conservation difficult. Particular problems are[7,8].

- the often scattered or dispersed patterns of distribution of the populations of wild species, leading to:
- considerable effort is required to find a sufficient number of plants to sample
- variation period of flowering
- prolonged seeding and fruiting periods
- the low yield of seeds
- the shattering of the seed cases leading to loss
- the photoperiodic sensitivity of seeds of many wild species
- Many temperate and tropical trees and shrubs possess recalcitrant seeds that cannot be dried and stored at low temperatures

3. GENETIC CONSERVATION OF MEDICINAL AND AROMATIC PLANTS – SPECIAL ISSUES

The very nature of medicinal and aromatic plants poses additional complications. Unlike agricultural plants where it is the primary products of photosynthesis – the carbohydrates, proteins, triglycerides (fats and oils) – that are the focus of as in food crops such as the staples (wheat, barley, maize, rice, beans, potato, sorghum etc.) on which most human nutrition is based; or cellulose and lignin, in the case of wood and fibre crops, in MAPs it is the secondary products of metabolism we are concerned with – such as the alkaloids, terpenoids, flavonoids – that have evolved as responses of plants to stress, predation and competition and constitute what has been called the vast 'chemical library' of biological systems. It is from these that most of our drugs, herbs, ethnomedicines, essential oils, perfumes, cosmetics derive. Thus it is extracts not the plants themselves (or parts of them such as seeds, grains, fruits, leaves) that are used for medicinals and aromatics.

3.1 Chemical variation – chemotypes

As it is the chemical constituents of MAPS that are the basis of their exploitation, either directly or as precursors for industrial copying or modification, particular attention has to be paid in sampling of populations for genetic conservation, to the variation in the constituents, such as essential oils, that may be affects by many factors. Infra-specific variation can occur as a result of differing soil conditions, altitude, climatic conditions, seasonal factors and other environmental features, leading in some cases to the evolution of different chemical variants or chemotypes.

This significant chemical variation occurring in plants, especially those with prominent chemical components such as terpenoids, alkaloids, etc., has attracted considerable attention and an extensive literature has developed. Penfold & Willis[9] described as the distinctive variants detected in *Eucalyptus citriodora* as 'chemical forms' – 'plants in naturally occurring population which cannot be separated on morphological evidence, but which are readily distinguished by marked differences in their essential oils'. Today the term chemotype is used for these forms and the distinguishing criteria for chemotype identification are the major components *only* of the essential oil from a named specific part of the plant (seeds, leaves etc.). A classic example is *Thymus vulgaris*, thyme oil, that shows considerable genetic variation in its monoterpene production. Although there is little conspicuous morphological differentiation in this species, at least six chemical variants occur within this species, two of which contain principally a phenol, either thymol or carvacrol, while the others have an alcohol as the main component. The alcohol may be linalol (also known as linalool), geraniol, thujanol-4 or – terpineol[10].

Recently, qualitative variation has been reported in southern France in the form of of a new chemotype, namely borneol that occurs primarily in association with a-terpineol. Quantitative variation in the percentage composition of thujanol-4 and other similar molecules makes this chemotype difficult to delimit in field samples.

3.2 Sampling chemical variation

It is clearly important that sampling for genetic resource collections reflects what is known of the patterns of chemical variation such as more or less discrete chemotypes or clinal (more or less continual variation along a gradient) variation. In the majority of medicinal and aromatic plants, such variation is incompletely known if at all. Too often, sample are very limited and do not reflect adequately the total variation within a population.

A recent example of good practice in sampling is reported in a study on genetic diversity within Tea Tree *(Melaleuca alternifolia)*[11]. In this study, plant material is being sampled from the entire known range of the species, which occurs in north-eastern New South Wales and far south-east Queensland. Leaf material is collected for extraction of DNA for genetic analysis and also for oil yield and composition studies. In conjunction with sampling, seed collection is also undertaken in order to establish a germplasm collection. The aim of the project is to develop an understanding of the population genetics of the species and to relate genetic markers to oil quality characteristics of various genotypes.

3.3 Assessing genetic variation

Genetic variation has been largely inferred from morphological patterns of variation. On the other hand, we are beginning to obtain new insights is in understanding patterns of genetic variation through the application of molecular marker techniques. The potential widespread distribution of genetic variation in plants and animals of conservation concern presents a problem if one wishes to conserve the maximum amount of diversity for any particular species.

Until recently this has been very difficult because of the lack of suitable rapid tools for assessing and analyzing this variation and revealing where there are concentrations of genetic diversity that may merit special conservation efforts. Most of the studies that have been carried out so far have used isozyme analysis, long used in population genetics and more recently in conservation biology. More sophisticated DNA-based techniques introduced in the last 10-12 years have the capacity to screen rapidly dozens or even hundreds of loci with several alleles detected at each locus, such as restriction enzyme fragment length polymorphism techniques (RFLP), and PCR (Polymerase Chain Reactions) techniques (which include the random amplified polymorphic (RAPD), microsatellite analysis, and amplified fragment length polymorphism (AFLP)). When we have the luxury of being able to use them – and cost and availability of expertise may be limiting factors – these methods will revolutionize our understanding of the extent, patterns and partitioning of genetic and chemical variation in species and our ability to manage this for conservation purposes.

3.4 Ecogeographical sampling

Apart from the particular problems of chemical diversity, sampling the variation at population level within species is a major challenge. It has not been customary in the pharmaceutical industry to consider plants as resources that grow in ever-changing populations in dynamic ecosystems. Particular problems that have to be addressed are:

- Applying appropriate methods for exploring and sampling the ecogeographical variation in populations of species
- The effectiveness of the different techniques currently available (morphological analysis, isozyme analysis, RAPDs, RFLPs, microsatellites, phenetic observation) in detecting 'essential' genetic variation in factors such as ecophysiological tolerance and particular adaptational features
- Deciding how much of a species should be conserved as a genetic resource
- The purpose for which the resources will be used
- The size of the samples[8]
- Is there a minimum sample[8]

The suite of techniques involved in ecogeographical surveying that were developed for sampling crop genetic resources may be adapted and applied to determine the overall pattern of distribution of the geographical and ecological variation in the populations.

4. CONSERVATION APPROACHES

The traditional approach to the conservation of plant genetic resources by the agricultural sector has been *ex situ* – through their collection and maintenance in field and seed banks. While many seed banks contain samples of some of these plants, they are not generally the result of deliberate sampling campaigns. However, an increasing number of national gene banks or national genetic resource programmes are beginning to include medicinal and aromatic plants in their mandate.

On the other hand, field genebanks of medicinal and aromatic plants are rare. A notable exception is the 50 ha field genebank established at the Tropical Botanic Garden and Research Institute, at Thiruvananthapuram in south India[12], through reconstruction of the original rainforest habitat. In a period of six years 30 000 plants belonging to 1000 angiosperm species, were introduced as part of this restoration of the forest, including 100 rare and endangered medicinal and aromatic plants of the Western Ghats forests. All possible genetic variants, cytotypes and chemotypes of the target species were introduced into the field gene bank. Other field gene banks for medicinal plants are found in some Chinese botanic gardens.

The generally accepted paradigm for effective conservation of biodiversity is through maintenance of the ecosystems in which it occurs, in accord with the principle of the 'ecosystem approach' adopted by the Convention on Biological Diversity. Conservation of ecosystems does not necessarily ensure the maintenance of particular component target species (such as medicinal plants) nor of the genetic and chemical diversity that they contain without specific management intervention to ensure their survival. Such *in situ* conservation of species within ecosystems can be a complex and difficult process and has been attempted so far for only a very small number of medicinal plants. In practice both *in situ* and *ex situ* approaches are complementary. An example is the network of 55 conservation sites that has been established for the conservation of medicinal plant diversity in southern India by the Foundation for the Revitalization of Local Health Traditions (FRLHT). These include 30 *in situ* areas, 15 *ex situ* centres and 10 *in situ* medicinal plant development areas in cooperation with the local community[14].

Much more intensive work needs to be undertaken to establish the effectiveness of this approach.

5. THREATS TO GENETIC CONSERVATION

The loss or extinction of species is a matter of concern to society for a variety of reasons, scientific, economic and ethical, but is especially important to pharmaceutical companies in their search for novel products since some of them will have the ability to produce important but as yet undiscovered molecules[13]. The main threats to species today come from loss of habitat, the fragmentation of habitats, growing demand for resources through population pressure, and wild harvesting.

5.1 Wild harvesting

One of the greatest potential risks to medicinal and aromatic plants is from open access in the wild that may lead to unsustainable levels of harvesting. In South Africa, for example, indigenous medicinal plants are used by more than 60% of South Africans in their health care needs or cultural practices and approximately 3,000 species are used by an estimated 200,000 indigenous traditional healers[15]. The indigenous medicinal plant industry is large, but fully based on harvesting from the wild.

Overharvesting is a common problem in many countries where Government support for and supervision of medicinal plant development are often weak. National legislation is often limited to prohibition or restrictions on the collection of rare or endangered species that may include some used as medicinals or aromatics.

In many countries, little information is available as to the extent of gathering of these species from the wild. Likewise, not enough information is known about the extent of the populations of the species concerned nor about their biology, life-cycles and regeneration times to allow a judgement to be made about the effects of wild gathering on population survival nor as a consequence about what levels are sustainable.

A World Bank report, expresses the view that[16],
'If existing supplies of medicinal plants are to keep up with demand, they will need adequate protection through development of appropriate institutions, policies, and legislation. Local communities need support and encouragement to protect these resources. To complement cultivation of adaptable species, harvesting from the wild must be guided by accurate inventories and knowledge about the species concerned. Above all, overexploitation of rare and endangered species must be avoided'.

The perspective of the pharmaceutical industry has been summarized in a recent review[13]:
'There may be a perception amongst conservationists and the public that 'large quantities' of material are being collected from the bush or the oceans for screening for novel natural products. Whilst small quantities may be used for the initial stage of the drug discovery process, there is a clear within the pharmaceutical industry to conserve the world's biota so that more species can be examined for novel chemical molecules and, that compounds of interest are produced routes that do not involve the destructive and costly

harvesting of samples.'

Unfortunately as Schippmann[17] points out, 'Far too many importers, despite their good intentions, are content to leave issues of environmental[ly] responsible sourcing to local exporters and harvesters and are unaware of the destructive effect their trades are having on some wild plant populations and habitats'. Certainly, clear evidence of serious overcollecting has been reported, as in the case of *Voacanga africana* where Cunningham and Mbenkum[18] reported that 900 tonnes of the seed, used for the industrial production of the alkaloid tabersonine, a depressor of the central nervous system activity in geriatric patients, were exported from Cameroon to France between 1985 and 1991, and 11 537 tonnes of the bark of *Prunus africana* (red stinkwood), used to treat prostatitis, in the same period.

6. LOCAL PARTICIPATION, BIOPROSPECTING AGREEMENTS AND IPR

Although historically, unimproved genetic and biochemical resources were regarded as the 'common heritage of mankind' and freely accessible to all[17], intellectual property right regimes (that had been introduced to improved genetic resources over the past century) began to be applied to them in the 1980s onwards. and the debate over ownership and access to genetic resources gained momentum during the negotiations leading to the Convention on Biological Diversity. One of the three objectives of the Convention, as set out in its Article 1, is the *'fair and equitable sharing of the benefits arising out of the utilization of genetic resources, including by appropriate access to genetic resources and by appropriate transfer of relevant technologies, taking into account all rights over those resources and to technologies, and by appropriate funding'*. Article 8(j) contains provision to encourage the equitable sharing of the benefits arising from the utilization of knowledge, innovations and practices of indigenous and local communities embodying traditional lifestyles relevant for conservation and sustainable use of biological diversity.

One of the manifestations of the need to involve local, people and respect their rights when pharmaceutical or other companies exploit medicinal and aromatic plants and other resources, is the development of the concept of biodiversity prospecting. This may be defined as 'the exploration of biodiversity for commercially valuable genetic and biochemical resources'[19] or more widely as 'the exploration, extraction and screening of biological diversity and indigenous knowledge for commercially valuable genetic and biochemical resources'[20]. Contractual agreements are being entered into by a number of developing countries with pharmaceutical companies for the collection of genetic resources and sharing any benefits that may be derived from them. It is particularly difficult, however, to ensure that the benefits reach local people.

Although bilateral bioprospecting agreements are sanctioned by the Convention on Biological Diversity, according to RAFI (Rural Advancement Foundation International)[20], in the vast majority of cases, commercial bioprospecting agreements cannot be effectively monitored or enforced by source communities, countries, or by the Convention, and amount to little more than what they term 'legalized' bio-piracy. This is a highly controversial area which requires much further attention if these bioprospecting agreeements to command the confidence of all parties concerned. A detailed review is given by ten Kate and Laird[21]

7. CONCLUSION

A considerable amount of work is being undertaken around the world on various aspects of the conservation of genetic and chemical diversity of medicinal plants. This work needs to be carefully coordinated and a strategic approach adopted so that it can be undertaken in the most effective manner possible.

REFERENCES

1. Heywood, V.H., 1991, Botanic gardens and the conservation of medicinal plants. In *The Conservation of Medicinal Plants* (Akerele, O., Heywood, V.H., Synge, H., eds.), Cambridge University Press, Cambridge, pp. 213–28.
2. Heywood, V.H., 1999, *Acta Horticulturae,* **500**, 21–9.
3. Arora, R.K., Engels, J.M.M., 1993, *Acta Horticulturae,* **330**, 21-38.
4. Guarino, L., Padulosi, S., Fassil, H., Ouedrago, A-S., Arora, R. , 1999, *Acta Horticulturae,* **500**, 151–9.
5. Frankel, O.H., Brown, A.H.D., Burdon, J.J. ,1995, *The Conservation of Plant Biodiversity,* Cambridge University Press, Cambridge.
6. Debouck, D.G., 2000, Perspectives about *in situ* conservation of wild relatives of crops in Latin Americsa. In *In situ Conservation Research, Proceedings of an International Conference held from 13th to 15th October 1999,* National Institute of Agrobiological Resources, Tsukuba, Japan.
7. Heywood, V.H., 1992, Conservation of germplasm of wild species. In *Conservation of Biological Diversity for Sustainable Development* (Sandlund,O.T., Hindar, K. and Brown, A.H.D., eds.), Scandinavian Press, Oslo, pp. 189-203.

8. Williams, J.T., 1993, Scientific issues affecting gene conservation and exploitation of some tropical perennials. In *Gene Conservation and Exploitation* (Gustafson, J.P., Appels. R., Raven, P., eds.), Plenum Press, New York, pp.15-28,
9. Penfold, A.R., Willis, J.L., 1953, *Nature*, **171,** 883-4.
10. Thompson, J.D., Ehlers B., Chalchat, J.-C., 2001, Qualitative and quantitative variation of monoterpene production in thyme: linking biochemistry, genetics and ecology, *Volatile organic compounds (VOC) produced by plants. PhytoVOC 2001*, An International Workshop Montpellier, France, 22-23 March 2001.
11. Rossetto, M., Slade, R.W., Baverstock, P.R., Henry, R.J, Lee, L. S.,1999, *Molecular Ecology,* **8,** 633-43.
12. Pushpangadan, P., Thomas J., 1999, *Acta Horticulturae,* **500**, 177-82.
13. Wildman, H.G.,1999, Pharmaceutical bioprospecting and its relationship to the conservation and utilization of bioresources. Invited lecture presented at the *International Conference on Biodiversity and Bioresources: Conservation and Utilization.* IUPAC.
14. http:// www.iupac.org/symposia/proceedings/phuket97/wildman.html
15. Tandon, V., 1996, Medicinal plant diversity and its conservation in southern India. In *Floristic Characteristics and Diversity of East Asian Plants* (Zhang Aoluo, Wu Sugong, eds.), China Higher Education Press, Beijing and Springer-Verlag, Berlin, pp. 461-71.
16. Coetzee, C., Jefthas, E., Reinten, E., 1999, Indigenous plant genetic resources of South Africa. In *Perspectives on new crops and new uses* (Janick, J., ed.), ASHS Press, Alexandria, VA, pp. 160–3.
17. World Bank, *Medicinal Plants Local Heritage with Global Importance.* The World Bank Group. http://www.worldbank.org/html/extdr/offrep/sas/ruralbrf/medplant.htm
18. Schippmann,U.,1999, Summary remarks and conclusions. In *Medicinal Plant Trade in Europe: Conservation and supply. Proceedings, First International Symposium on the conservation of medicinal plants in trade in Europe.* TRAFFIC Europe, pp.173-8.
19. Cunningham, A.B., Mbenkum, F.T., 1993, Sustainability of harvesting Prunus africana bark in Cameroon: a medicinal plant in international trade. People and Plants Working Paper 2, UNESCO, Paris.
20. Reid, W.V., Laird, S., Gámez., R., Sittenfeld,A., Janzen, D.H., Gollin, M.A., Juma, C., 1993, 1. A New Lease on Life. In *Biodiversity Prospecting: using genetic resources for sustainable development* (Reid, W.V., Laird, S., Gámez, R., Sittenfeld,A., Janzen, D.H., Gollin, M.A., Juma. C., eds.), World Resouces Institute, Washington DC., pp.1-52.
21. RAFI (Rural Advancement Foundation International), *Bioprospecting/Biopiracy and Indigenous Peoples..* http://www.latinsynergy.org/bioprospecting.htm.
22. Ten Kate, K., Laird, S.A., 1999, The Commercial Use of Biodiversity – Access to genetic Resources and Benefit-Sharing, Earthscan, London.

Conservation and Sustainable Use of Biodiversity in Wild Relatives of Crop Species

BAO-RONG LU
The Ministry of Education Key Laboratory for Biodiversity Science and Ecological Engineering, Fudan University, Institute of Biodiversity Science, Shanghai 200433, People's Republic of China

1. INTRODUCTION

The continuous increases in the global population, the declining availability of farming land, the increasing shortage of water, and the loss of rural labour to the urban centers profoundly challenge the world's food security. To meet the increasing demand for food supply humanity has to significantly enhance crop productivity, for which fuller exploitation and utilization of genetic resources in crop species, particularly those in the gene pool of wild relatives of crop species will provide many more opportunities. Examples of the successful use of wild relative species are hybrid rice, where the male sterility (MS) gene was introduced from the common wild rice (*Oryza rufipogon* L.), and grassy stunt virus resistant rice varieties, in which the virus resistance gene was incorporated from the annual wild rice (*O. nivara* Sharma et Shastry). Many disease resistant genes in wheat, barley and rye, as well as soybean varieties were also transferred from their wild relative species in the tribe *Triticeae* and the genus *Glycine*. Conserving biodiversity of wild relative species is therefore essential for the continued availability and sustainable use of these valuable genetic resources, and essential for the sustainable world's food security.

2. THE NEED FOR AND THREAT TO BIODIVERSITY OF WILD RELATIVES

Wild relatives of crop species (abbreviated as WRCS throughout this paper) are very important sources of material for broadening the genetic background of crop varieties. These include the ancestral species from which the crop species were domesticated, genetically closely related wild species, the weedy types of crops, and hybrid swarms between wild and crop species. Usually, many WRCS occur in agricultural ecosystem, and co-evolution and genetic introgression between WRCS and the cultivated species have occurred frequently since the domestication of the crop species. Therefore, these WRCS remain in a close relationship with the crop species. Through millions of years of evolution and genetic adaptation to environments, the WRCS have accumulated abundant biodiversity. Many traits are unique to the WRSC and could be beneficial to the cultivated species. Serving as a vast genetic reservoir, WRCS provide elite germplasm to improve crop varieties by transferring beneficial genes to the crops. It is apparent that WRCS have played an increasingly important role in continued enhancement of crop production and sustainable development of crop varieties, particularly under the circumstances that modern breeding practices and change of crop management have accelerated the great loss of biodiversity in crop varieties[1].

Genetic erosion or loss of biodiversity of crop species has been recognized as a problem since the early 1960s, owing to the factors such as the adoption of high-yielding crop varieties, farmer's increased integration into the markets, change of farming systems, industrialization, population increase, and cultural change[2]. Similar situation has also been observed for the WRCS. In many places of Asia, populations of wild relatives of crop species are becoming extinct or are threatened because their natural habitats are seriously damaged by the extension of cultivation areas, expansion of communication systems such as road construction, and urban pressures. Global weather change and air pollution have also considerably, in some cases, deteriorated the habitats of many wild relative species. As a consequence, the loss of and habitat change for WRCS have caused the quick diminishing of populations of WRCS and irreversible losses of the valuable germplasm in their gene pool.

According to the unpublished data collected by the Chinese Academy of Sciences (CAS) in 1994 (D. Y. Hong personal comm.), nearly 80% of the common wild rice (*Oryza rufipogon*) sites recorded during 1970s, when Chinese Wild Rice Expedition Team conducted a 5-year comprehensive survey and collection of wild rice species in China, were found to no longer exist. The size of some surviving *O. rufipogon* populations was also found to be significantly reduced. Based on our own survey, some populations of *Triticeae* species and wild soybean, recorded in Chinese Flora and local Floras, and other literatures, were not found when collecting teams revisited these sites (B. R. Lu, unpublished data). A similar situation was reported for many other relative species in China, as well as in other countries.

3. CONSERVATION AND SUSTAINABLE USE OF WILD RELATIVES

Strategic conservation is essential for maintaining the continued availability of biodiversity of WRCS and has become increasingly important for their sustainable use in crop improvement, particularly given that genetic erosion has occurred and continues to take place in agro-ecosystem. This has been well recognized by biologists, plant breeders and conservationists for many decades. Depending on the objectives and scopes of the activity, there are two basic approaches for germplasm conservation of WRCS, i.e., *ex situ* and *in situ* conservation.

3.1 *Ex situ* conservation

Ex situ conservation is an approach that entails the actual removal of genetic resources (seeds, pollen, sperm, individual organisms) from the original habitat or natural environment. *Ex situ* conservation involves activities of collecting seed samples of WRCS from the original sites and then storing the samples in gene bank or planting them in conservation nurseries. This method of conservation has so far been the principle strategy for preserving crop genetic resources and their wild relative species. It provides efficient means for germplasm preservation, utilization, exchange, and information generation, through effective management and value added research of WRCS. However, seed samples placed under *ex situ* conservation in a genebank become isolated from the environment where they originated and grow. The expected microevolution of these genetic resources in the original environment is stopped. Therefore, in evolutionary terms, *ex situ* conservation is static[2]. Concerns have been raised regarding the observation that static conservation may reduce the adaptive potential of the wild species and their populations in the future. Thus, *ex situ* conservation cannot be considered as the only approach for conserving biodiversity of the WRSC. Complementary dynamic approaches such as *in situ* conservation are also necessary.

3.2 *In situ* conservation

In situ conservation is a method that attempts to preserve the integrity of genetic resources by conserving them within the evolutionary dynamic ecosystems of their original habitat or natural environment. In contrast to the *ex situ* approach, *in situ* conservation of WRSC aims to preserve wild species or populations in a dynamic way, because the evolutionary processes of the *in situ* conserved WRSC is continued in their original habitats. As a consequence the new inheritable adaptation to the environmental changes will be selected during the conservation process. For the long term and dynamic conservation, the *in situ* approach has its great value. However, for some reasons, *in situ* conservation of WRCS has received the least attention. Limited scientific input has passed into *in situ* conservation and its design and management for WRCS. For example in the wild rice *in situ* conservation, only 10 species have been reported from 18 reserves in Africa and South and Southeast Asia[3]. In addition, urgent needs to conduct researches, such as taxonomy, biosystematics, genetic diversity patterns, eco-geographic distribution, population dynamics, ecology and species relationships in their community should be met to facilitate more effective *in situ* conservation of WRSC.

4. CONSERVATION OF WILD RELATIVES IN CHINA – CASE STUDIES

4.1 Conservation of wild rice

Rice is one of the world's most important crops that provide staple food for nearly one half of the global population. More than 90% of rice is grown and consumed in Asia where about 55% of the world's population lives[1]. The Asian cultivated rice (*Oryza sativa*) is classified in the genus *Oryza* that includes about 26 species widely distributed in pan-tropics and subtropics. The wild species in the genus *Oryza* and in the related genera in the tribe *Oryzeae* constitute valuable gene pool for the rice improvement[1]. Therefore, efficiently conserving biodiversity of the rice gene pool including the wild relative species is essential for the sustainable rice production and development.

The conservation of wild rice in China can be traced back to as early as 1910s when Dr. E. D. Merrill first found perennial common wild rice (*O. rufipogon*) in Lofu Mountain of Guangdong province. Later in 1926, Prof. Ding Ying collected *O. rufipogon* at more sites in Guangdong, and in 1930's many Chinese scientists found other two wild rice species, i.e. granulated wild rice (*O. granulata*) and medicinal wild rice (*O. officinalis*), in addition to *O. rufipogon* at more localities in southern China[4]. Efforts at collecting and conserving wild rice germplasm were attempted at small scales during the 1950's to 1960's by different agencies. As a consequence, a large number of wild rice seed samples were conserved and used by agricultural research stations and universities for breeding and research. The well known Chinese rice variety "Zhongshan No. 1" tolerant to cold and other abiotic stresses was bred by Prof. Ding Ying in 1931 through wide hybridization with *O. rufipogon*. The famous hybrid rice has also benefited from the finding and introduction to rice varieties of the male sterility trait from *O. rufipogon* in the early 1970's. There are more successful examples of utilizing wild rice genetic resources in rice breeding[4,5].

During 1978-1982, in close cooperation with provincial agricultural agencies, the Chinese Academy of Agricultural Sciences organized a nation-wide extensive survey and collection for wild rice species. More than 5,000 wild rice samples, representing four species in *Oryza* and seven species in the related genera of the *Oryzeae* (Table 1), were collected in different provinces. These collected wild rice samples are deposited in the Chinese National Genebank (Beijing), Genebank of Chinese National Rice Research Institute (Hangzhou), and genebanks and seed storages of provincial and local agricultural agencies, respectively, as the *ex situ* conservation of wild rice germplasm. In addition, more than 2,000 accessions of wild rice samples are also conserved as living stocks in the National Wild Rice Nurseries in Guangdong and Guangxi provinces.

A few *in situ* conservation sites for *O. rufipogon* have also been established in China, such as in Dongxiang (Jiangxi Province), Caling (Hunan Province) and Zhengcheng (Guangdong Province)[5,6]. The central and provincial governments have paid substantial attention to the significance of *in situ* conservation of wild rice species, given that wild rice relatives are important genetic resources, but their habitats – aquatics and wetlands – are very fragile. More conservation sites will be set up in cooperation with national and provincial conservation programs. The conserved biodiversity of wild rice germplasm has not only provided valuable resources for rice breeding[4], but also important materials for scientific research[5]. The conserved wild rice biodiversity has become increasingly important for the sustainable use of wild species resources in rice improvement and development, as well as the generation of scientific information.

4.2 Conservation of wild relative species in the tribe *Triticeae*

The tribe *Triticeae* encompasses more than 500 annual and perennial species in the world, including three of the world's most economically important cereal crops, i.e. wheat (*Triticum aestivum* L.), barley (*Hordeum vulgare* L.) and rye (*Secale cereale* L.), as well as many agronomically valuable forage grasses. Some of the *Triticeae* species such as *Leymus chinensis* (Trin.) Nevski and *L. secalinus* (Georgi) Tzvelve are also important component of grassland maintenance and development, apart from their forage value. Based on the records of latest Chinese flora, there are about 130 wild *Triticeae* species, many of which are widely distributed over China[7]. These *Triticeae* species are taxonomically classified in 11 genera, and majority of them are endemic to China (Table 2). The large number of the wild relative species in the *Triticeae* serves as a vast gene pool that provides important and valuable sources of new genetic variation for the improvement of the wheat crops and forage grasses, apart from the biological interest of these materials *per se* for studies of cytogenetics, molecular biology, and evolution.

Many wild relatives in the *Triticeae* possess agronomically useful traits, such as pathogen and insect resistance, salt and alkalinity tolerance, winter hardiness, high protein content, and perennial habit. A number of such traits have been successfully incorporated into wheat varieties[8-12]. More examples of intensive genetic use of the *Triticeae* wild relative species were focused on the germplasm in the genus *Elytrigia* (*E. intermedia* and *E. elongata*), *Aegilops*, *Agropyron*, *Secale*, and *Elymus*. Noticeably, the Australian *Elymus scabrus* contains apomictic genotypes that provide a method of cloning plants through seeds, potential for fixing hybrid vigour in hybrid breeding, if such genotypes can be transferred into cultivated crops[13].

Conservation and utilization of *Triticeae* wild relative species have always been important components in wheat breeding programs in China[14]. The early collection activities of *Triticeae* germplasm were on relatively small scales, and much effort was focused on exploiting genetic resources in wheat breeding. Major and systematic collections of *Triticeae* species were not conducted until 1985 when the International Board of Plant Genetic Resources (IBPGR), FAO, funded a five-year collection project for *Triticeae* germplasm in Sichuan, Tibet, Shanxi, Gansu, Ningxia, Qinghai, Xinjiang, Inner Mongolia, and the northeast parts of China. Two leading institutions were actively involved in coordination and organization of the *Triticeae* Expedition trips in China, these were the Institute of Crop Germplasm Resources, Chinese Academy of Agricultural Sciences (Beijing), and

Table 1. Wild relative species in the tribe *Oryzeae* (wild rice) in China

Genus	Species	2n=	Genome	Distribution
Chikusichloa Koidz.	*C. aquatica* Koidz.	24	?	Guangdong, Guangxi, Hainan
Hygroryza Nees	*H. aristata* (Retz.) Nees ex Wright & Arn.	24	?	Fujian, Jiangxi, Guangdong
Oryza Linn.	*O. rufipogon* Griff.	24	A	Guangdong, Guangxi, Hainan, Yunnan, Hunan, Jiangxi
	O. nivara Sharma et Shastry	24	A	Hainan, Taiwan
	O. officinalis Wall ex Watt	24	C	Yunnan, Guangxi
	O. granulata Nees et Arn. ex Watt	24	G	Yunnan
Leersia Soland.	*L. hexandra* Swartz	48	?	Southern China, Taiwan
	L. oryzoides (L.) Swartz	48	?	Jiangsu, Anhui, Zejiang, Jiangxi, Guangdong
Porteresia Tateoka	*P. coarctata* (Roxb.) Tateoka	48	?	*
Zizania Linn.	*Z. aquatica* Linn.	30,	?	Over China
	Z. latifolia (Griseb.) Turcz. ex Stapf	30	?	Over China

* Only recorded in literatures

Triticeae Research Institute of Sichuan Agricultural University (Dujiangyan). Foreign scientists from The Swedish University of Agricultural Sciences, the Danish Royal Veterinary and Agricultural University, France Agriculture Research Station, Canadian Central Experimental Farm, the Kihara Memory Foundation, and Kyoto University, also participated in the Expedition trips. A great number of collections as seed samples, representing all available genus of *Triticeae* found in China, were collected during the five-year Expedition. All the collected *Triticeae* samples were deposited in the Chinese National Genebank and the Genebank in the Triticeae Research Institute (Dujiangyan), in addition to other local genebanks. This conserved *Triticeae* germplasm has been intensively used for breeding and scientific researches[15-18]. Owing to the limited understanding of basic knowledge, such as biodiversity pattern, habitat change, and the degree of threats to the *Triticeae* wild relative species, no specific *in situ* conservation activities for *Triticeae* have been initiated.

4.3 Conservation of wild soybean

China is the origin center of cultivated soybean *Glycine max*, and is also the diversity center of the wild soybean *G. soja*, from which the cultivated soybean was domesticated. The intermediate or weedy type of soybean, named by some taxonomists as *G. gracilis*, is considered to be the natural hybrids or introgression types between the wild and cultivated soybeans. Together with the cultivated soybean, *G. soja* and *G. gracilis* are classified in the subgenus *Soja*, indicating their high affinity in terms of evolution. Biodiversity of the wild and weedy soybeans, *G. soja* and *G. gracilis*, is abundant in many parts of China (Table 3). They are found from northern part of Heilongjiang province (ca. 53° N) down to Guangxi province (ca. 24° N), and from east coast of China to Tibet (97° E). Distribution of the wild and weedy soybeans considerably overlaps with that of cultivated *G. max* in China. In addition, two other wild soybean species, *G. tabacina* Benth. and *G. tomentella* Hayata, which are classified in the subgenus *Glycine* and have a relatively low affinity with the cultivated soybean are also found occasionally in south part of China, although their main diversity centers are in Australia or Southeast Asia (Table 3).

Glycine soja and *G. gracilis* are known to hybridize with the cultivated *G. max* under natural conditions. Interspecific hybrids from artificial crosses are usually normal and have relatively high fertility[19]. Cytogenetic studies revealed high frequencies of chromosome translocations (83%) in some cultivated and wild soybeans. Molecular studies indicated identical rDNA

Table 2. Wild relatives of the tribe *Triticeae* (wheatgrass) in China (including introduced but commonly used in wheat breeding)

Genus	No. of species	2n=	Genome	Distribution
Aegilops Linn.	8	14, 28, 42	D, S, CD, DM, MU	Xinjiang, Shanxi, Qinghai, Hebei[2]
Agropyron Gaertn.	5	14, 28, 42	P	Northeast and North China, Gansu, Qinghai, Xinjiang, Shanxi, Inner Mongolia
Elymus Linn. (*sensu lato*)[1]	>80	28, 42	StH, StY, StHY, StPY, StStY,	Allover China
Elytrigia Desv.	5	28, 42	StStH, E^e, $E^b E^e St$	Xinjiang, Gansu, Qinghai, Tibet[2]
Erymopyrum (Ledeb.) Jaub. et Spach	3	14, 28	F, Xe	Inner Mongolia, Xinjiang
Hordeum Linn.	11	14, 28, 42	H, I	Heilongjiang, Inner Mongolia, Qinghai, Gansu, Xinjiang, Shanxi, Ningxia, Tibet, Sichuan, Hebei
Leymus Hochst	9	14-84	NsXm	Xinjiang, Gansu, Qinghai, Liaoning, Jiling, Heilongjiang, Inner Mongolia, Hebei, Shanxi, Sichuan
Psathyrostachys Nevski	5	14, 28	Ns	Shanxi, Xinjiang, Gansu, Inner Mongolia
Pseudoroegneria (Nevski) Löve	4	14, 28	St	Tibet, Ningxia, Gansu
Scale Linn.	2	14	R	North part of China[2]
Triticum Linn.	4	14, 28, 42	A, AB, ABD, AAG	Xinjiang[2]

1. Elymus sensu lato includes Roegneria, Hystrix, and Kengyilia
2. Some introduced wild relative species are cultivated in many provinces in China

Table 3. Wild relative species of the genus *Glycine* Willd (wild soybean) in China and other countries

Species	2n=	Genome	Distribution
G. soya Sieb.et Zucc.	40	G	China (most provinces), Korea, Japan, Russia (Far East)
G. gracilis Skvort.	40	G	China, in the regions where the cultivated soybean and *O. soja* grow sympatrically
G. tabacina Benth.	40, 80	B, AB	China (Guangdong, Fujian, Guangxi, Taiwan), Southeast Asia, Australia
G. tomentella Hayata	38, 40, 78, 80	E, D, DE	China (Guangdong, Fujian, Taiwan), Philippines, Papua New Ginuea, Australia

ITS1 sequences of *G. soja*, *G. gracilis*, and *G. max*, and RAPD analysis also suggested a close genetic relationship of the three species[19]. Many scientists therefore believe that *G. soja*, *G. gracilis*, and *G. max* belong to the same biological species. The wild relatives of soybean have been significantly utilized as genetic resources in soybean breeding programs by introducing agronomically beneficial traits, such as high-yielding, high protein content, disease and insect resistant, abiotic tolerant, and male sterility[19]. The new soybean variety Jilingxiaoli-1 was bred by using the small grain character of *O. soja*[20].

Conservation of the wild relatives of soybean becomes essential for remaining availability of soybean biodiversity in China. Studies on geographic distribution and the associated ecological environment, abundance and its regulating factors, and the biodiversity patterns of the soybean wild relative species are important for the effective conservation and management of the wild soybean germplasm.

The major survey and collection activities for wild soybean diversity were initiated in 1970's, although smaller scales of conservation were attempted much earlier[19]. In 1978, the Jiling Agricultural Academy of Sciences launched a collection project for wild soybean in China. In this expedition, 873 accessions of wild soybean samples were collected and many new eco-types and genetic variations were found in these collected samples. During 1979-1985, Chinese Academy of Agricultural Sciences and Jiling Academy of Agricultural Sciences co-organized a nation-wide comprehensive exploration and collection of wild soybean species, covering 1,189 cities and counties in China. As a result, more than 6,200 seed samples and 4,000 herbarium specimens of *O. soja* and *O. gracilis* were collected, and even more new genetic variations were recorded among these collections[19,20]. The collected seed samples were extensively utilized in breeding and research, and deposited in the China National Genebank, Genebank of Soybean Research Center, Jiling Academy of Agricultural Sciences, and other provincial genebanks. Due to technical difficulties and limited knowledge on related scientific basics, *in situ* conservation of wild soybean has not been initiated in China so far.

Through these expeditions and collections, Scientists in China have a much better understanding of the distribution and diversity patterns of wild soybean species, and have accumulated a lot of wild soybean genetic resources available for breeding and research programs. In fact, a great number of scientific research has been reported based on this conserved wild soybean germplasm.

REFERENCES

1. Lu, B. R., 1996, Diversity of the rice genepool and its sustainable utilization. In *Proceedings of the Intern. Sysmp. on Florestic Characteristics and Diversity of East Asian Plants* (Zhang, A., Wu, S., eds), China Higher Education Press-Berlin, Springer-Verlag. Beijing, pp. 454-60.
2. Bellon, M. R., Brar, D. S., Lu, B. R., Pham, J. L., 1998, Rice Genetic Resources. In *Sustainability of rice in the global food system* (Dwoling, N. G., Greenfield, S. M., Fischer, K. S. eds.) Chapter 16, Davis, Calif. (USA), Pacific Basin Study Center and IRRI, Manila, pp. 251-83.
3. Vaughan, D. A., Chang, T. T., 1992, *Econ. Bot.*, **46**, 368-82.
4. Wu, M. S. (ed.), 1990, Collection of Research Articles on Wild Rice Germplasm, **Chinese Science Techn. Press**, Beijing.
5. Wang, X. K., Sun, C. Q., (eds.), 1996, The Origin and Differentiation of Chinese Cultivated Rice, China Agric. Univ. Press, Beijing.
6. Lu, B. R., 1999, *Intern. Rice Res. Notes*, **24** (2), 41.
7. Kuo, P. C. (ed.), 1987, Flora Reipublicae Popularis Sinicae, Tomus 9 (3), Gramineae, Pooideae, Science Press, Beijing.
8. Sharma, H. C., Gill, B. S.,1983, *Euphytica*, **32**, 17-31.
9. Sears, E. R., 1983, The transfer to wheat of interstitial segment of alien chromosomes. In *Proceedings of the 6th Intern. Wheat Genet. Symp., Japan Kyoto,* (Sakamoto, S. ed.), Kyoto Univ. Press, Kyoto, pp. 5-12.
10. Sharma, H. C., Ohm, H. Y., Lister, R. M., Forster, J. E., Shukle, R. H., 1989, *Theor. Appl. Genet.*, **77**, 369-74.
11. Tosa, Y., Sasai, K., 1991, *Theor. Appl. Genet.*, **81**, 735-9.
12. Li, L. H., Yang, X. M., Li, X. Q., Dong Y. S., 1998, *Scientia Agricultura Sinica,* **31** (6), 1-5.
13. Hanna, W., Bashaw, E. C., 1987, *Crop Science,* **27**, 1136-9.

14. Lu, B. R., 1995, *Chinese Biodiversity*, **3** (2), 63-8.
15. Lu, B. R., Bothmer, R. von, 1989, *Hereditas*, **111**, 231-8.
16. Lu, B. R., Bothmer, R. von, 1990, *Hereditas*, **112**, 109-16.
17. Lu, B. R., 1991, *Nord. J. Bot.*, **11**, 27-32.
18. Lu, B. R., Jensen, K. B., Salomon, B., 1993, *Genome*, **36**, 1157-68.
19. Zhuang, B. C., (ed.), 1999, Biological Studies of Chinese Wild Soybean, Science Press, Beijing.
20. Li, F. S. (ed.), 1990, A Catalog of Chinese Wild Soybean Germplasm, Agric. Press, Beijing.

Perspectives on Human Genome Diversity within Pakistan using Y Chromosomal and Autosomal Microsatellite Markers

S. QASIM MEHDI[1], QASIM AYUB[1], RAHEEL QAMAR[1], A. MOHYUDDIN[1], ATIKA MANSOOR[1], K. MAZHAR[1], A. HAMEED[1], M. ISMAIL[1], S. RAHMAN[1], SAIMA SIDDIQUI[1], SHAGUFTA KHALIQ[1], M. PAPAIOANNOU[2], CHRIS TYLER-SMITH[3] and L. L. CAVALLI-SFORZA[4]

[1] Biomedical and Genetic Engineering Division, Dr. A. Q. Khan Research Laboratories, Islamabad, Pakistan. [2] Unit of Prenatal Diagnosis, Center for Thalassemia, Laiko General Hospital, Athens, Greece. [3] University of Oxford, Department of Biochemistry, Oxford, UK. [4] Department of Genetics, Stanford University, Stanford, California, USA.

1. INTRODUCTION

Current genetic and fossil data has consistently pointed to a recent common origin of man in Africa, less than 200,000 years ago, with subsequent migrations and dispersals of modern humans throughout the rest of the world[1]. Although fossil evidence is lacking, it is postulated that humans arrived in Pakistan, that lies on the postulated coastal route out of Africa to Australia, probably 60,000 to 70,000 years ago. Evidence of neolithic settlements has been found at Mehrgarh, in southern Pakistan and extensive sites have been excavated indicating that the agrarian Harappan culture flourished in the fertile Indus Valley around 2400-2000 BC[2]. Since then the Indo-Pak sub-continent has seen repeated invasions and migrations. Invaders included the nomadic Indo-European tribes from central and west Asia (the Aryans), Alexander's army, Arabs, Afghans and Mongols. All have contributed to the ethnic variety of extant Pakistani populations[3].

The present population of Pakistan exceeds 140 million and is represented by numerous ethnic and linguistic groups[4]. Prominent northern ethnic groups include the Pathan, Punjabi, Hazara, Burusho, Kashmiri, Kalash and Balti. Southern ethnic groups are the Baloch, Brahui, Makrani and Sindhi. The majority of the ethnic groups speak Indo-European languages with the exception of the language isolate, Burushaski. The Baltis speak a Sino-Tibetan language and the Brahuis speak a Dravidian language (Table 1).

Table 1. The Pakistani ethnic groups studied for Y chromosomal polymorphisms. Sample size (n) is given for each population.

Ethnic Group	Language Family	n	Suggested origin(s)
Baloch	Indo-European	59	Syria[5]
Balti	Sino-Tibetan	13	Tibet[6]
Brahui	Dravidian	110	West Asia[7]
Burusho	Language isolate	94	Alexander's army[8]
Hazara	Indo-European	23	Genghis Khan's soldiers[9]
Kalash	Indo-European	44	Greeks[10] or 'Tsyam'[11]
Kashmiri	Indo-European	12	Semetic[12]
Makrani Baloch	Indo-European	25	West Asia[7]
Negroid Makran	Indo-European	33	Africa?
Parsi	Indo-European	90	Iran, via India[13]
Pathan	Indo-European	93	Jewish[14], Greek or Rajput[15,16]
Sindhi	Indo-European	122	Arab, Indo-European, Dravidian[17]

The Punjabis form the majority population of Pakistan. However, this linguistic group encompasses a complex admixture of castes and tribes[18] and will be the subject of a separate study.

In recent years DNA based markers have gained acceptance as the markers of choice for elucidating questions of human evolution and migration[19]. An informative DNA marker should be both highly polymorphic and selectively neutral. Studies of such markers on the non-recombinant portion of the human Y chromosome have shed light on human evolutionary history from the male perspective[20-22]. Unlike the other 22 pairs of human chromosomes (autosomes) in males, the Y chromosome does not undergo "shuffling", or recombination, between both parents before being passed on to the male progeny. Therefore, mutations in the non-recombinant portion of the human Y chromosome are stable and specific for populations in different geographical regions. They provide anthropologists and geneticists with an extremely powerful tool for tracing the patrilineal descent of man[23]. Similarly the mitochondrial DNA is passed only by the mother to children of both sexes and provides a female perspective on human evolution[24]. Microsatellites are highly variable, short, tandemly repeated sequences of two to six nucleotides [25,26]. Genetic variation at these loci has provided additional insights and lent support to the African origins of modern *Homo sapiens*[27,28].

The origin of present day Pakistani populations was investigated by typing sixteen unique event polymorphisms mapping on the non-recombining portion of the human Y chromosome and autosomal microsatellites were used to compare the Pakistani populations with other diverse world populations. These included populations that had been studied earlier with a set of 29 different dinucletide repeats[27].

2. MATERIALS AND METHODS

2.1 Samples

Blood samples were obtained with the informed consent of unrelated donors. The DNA used in the present study was extracted from the lymphoblastoid cell lines. Males (n=718) from twelve Pakistani ethnic groups were examined for Y chromosomal polymorphisms (Table 1). These groups included the Baloch, Balti, Brahui, Hunza Burusho, Hazara, Kalash, Kashmiri, Makrani Baloch, Makrani Negroid, Parsis, Pathans and Sindhis. Autosomal microsatellite variation was studied in individuals belonging to various world populations. These included three populations from sub-Saharan Africa, the Zaire Pygmies (n=9), Central African Republic (CAR) Pygmies (n=9), and Lisongo (n=9). Four populations were from Europe and included North European (n=10), North Italian (n=9), Greek (n=40) and Basque (n=10). East Asian populations were of the Chinese (n=10), Japanese (n=10) and Cambodians (n=10). Oceanic populations included Melanesian (n=10), New Guinean (n=10) and Australian Aborigine (n=10). South American populations were the Mayan Indians from the Yucatan peninsula (n=10) and the Brazilian Indians (n=16) (Surui and Karitiana from the Amazon basin). In addition, autosomal microsatellites were analyzed in 128 Pakistani individuals that were Brahui (n=5), Burusho (n=39), Kalash (n=39) Pathan (n=40) and Sindhi (n=5) individuals.

2.2 Y chromosome Biallelic Polymorphism Typing

Twelve SNPs, an Alu insertion[29], the 12f2 deletion[30] and the polymorphic L1Y retroposon insertion[31] on the Y chromosome were analyzed in Pakistani males. The Y-SNPs studied included 92R7 C to T[32]; SRY-2627 C to T[33]; SRY-1532 A to G to A[34,35]; sY81 (DYS271) A to G[36]; SRY-8299 G to A polymorphism[37] Apt G to A[38]; LLy22g C to A and TAT T to C transition[39]. Other markers included the M9 C to G mutation, the M17 single base deletion, the M20 A to G mutation[40] and the RPS4Y C to T mutation[41].

2.3 Autosomal Microsatellite Analysis

One hundred and thirteen tri- and tetra-nucleotide microsatellite loci were typed in 20 extant human populations (310 individuals). These loci were approximately 7 cM apart and were presumed to have no detectable linkage disequilibrium. The loci were located on human chromosomes 1-6,9-11,14,16,17 and 18. Each sample was PCR amplified in a multiplex reaction consisting of 5 to 7 primer pairs (Human Genome MAPPAIRS MULTIPLEX kit, Research Genetics) labeled either with TET, HEX or FAM. The multiplex PCR assay was performed in a 5 µl final volume. Briefly, the reaction consisted of the following: 1X PCR Buffer (10 mM Tris-HCl, pH 8.3, 50 mM KCl, 0.1 mg/ml gelatin and 1.5 mM $MgCl_2$), 200 µM dNTPs, 0.25 U of AmpliTaq DNA polymerase (Perkin Elmer), 0.083 µg TaqStart Antibody (ClonTech) and 0.05 µM of each primer. $MgCl_2$ was supplemented to give a final concentration of 3 mM. The TaqStart Antibody was pre-incubated with AmpliTaq for 5 to 7 minutes after which this mixture was added to the master mix and dispensed into tubes containing 50 ng genomic DNA. PCR was performed by a "touchdown" protocol as described previously[42]. After amplification, 0.3 µl of each sample was mixed with 0.3 µl of either TAMRA 350 or TAMRA 500 internal lane standard and 2.4 µl of a loading dye consisting of formamide and Dextran Blue. The samples were denatured at 98°C for 2 minutes and 1 µl of this mix was loaded on a 12 cm long, 4% denaturing polyacrylamide gel that was electro-phoresed for 1.5 hours on the ABI 377 DNA sequencer following the manufacturer's instruction. Data were collected using software version 1.1. The fragment sizes were estimated based upon known internal lane size standards using the software GeneScan (Ver 2.1) and the genotypes were called using the software Genotyper (Ver 2.0). Subsequent analyses were carried out on loci for which genotypes could be scored confidently in all human populations.

2.4 Data Analysis

Allelic frequencies were used to construct phylogenetic trees by using DISPAN software (http://bio.psu.edu/People/Faculty/Nei/Lab /Programs) containing the programs GNKDST for calculating genetic distances and TREEVIEW for drawing phylogenetic trees by neighbour joining and UPGMA methods[43,44]. Principal component analysis was carried out by using the ViSta (Visual Statistics) system software version 5.0.2 (http://forrest.psych.unc.edu)[45]. The first and second principal components were plotted using the Microsoft Office Suite Excel 2000 package on Windows 98(Microsoft, WA).

3. RESULTS AND DISCUSSION

The combination of allelic states of the Y chromosomal biallelic (base substitutions, insertions and deletions) examined in 718 Pakistani males enabled the identification of 11 lineages of Y chromosomes referred to as "haplogroups"[23,46] at varying frequencies (Table 2). Based upon the Y haplogroup frequencies, the Hazara population was unique. The Burusho which speak Burushaski, a language isolate, and the Dravidian language speaking Brahuis were not remarkable.

Table 2. Frequency of 8 Y haplogroups in twelve Pakistani ethnic groups.

Population	Y haplogroups							
	1	2	3	8	9	10	21	28
Baloch	0.186	0.000	0.288	0.017	0.119	0.000	0.085	0.288
Balti	0.154	0.077	0.462	0.000	0.154	0.000	0.000	0.154
Brahui	0.082	0.100	0.391	0.027	0.282	0.018	0.000	0.073
Burusho	0.277	0.074	0.277	0.000	0.074	0.085	0.000	0.170
Hazara	0.609	0.043	0.000	0.000	0.043	0.304	0.000	0.000
Kalash	0.091	0.386	0.182	0.000	0.091	0.000	0.000	0.250
Kashmiri	0.250	0.000	0.583	0.000	0.167	0.000	0.000	0.000
Makrani Baloch	0.240	0.000	0.280	0.040	0.240	0.000	0.040	0.160
Makrani Negroid	0.182	0.121	0.303	0.091	0.182	0.000	0.030	0.091
Parsi	0.267	0.033	0.078	0.000	0.389	0.000	0.056	0.178
Pathan	0.108	0.161	0.452	0.000	0.065	0.000	0.022	0.129
Sindhi	0.123	0.090	0.492	0.000	0.205	0.000	0.025	0.066

The most frequent haplogroup in Pakistan was haplogroup 3 (32.5%) which is common in north eastern Europe and Asia[47]. The same haplogroup has been designated as 1D[48]. It is an example of a rare, recurrent G to A transition of SRY-1532 marker and is also identified by the M17 deletion[4,40]. It is found in all Pakistani populations with the exception of the Hazara. This haplogroup is widespread in Europe and Asia but absent in Africa and the New World chromosomes. It has been postulated that this haplogroup coincided with the arrival of Indo-European nomadic tribes from West and Central Asia[49]. In Pakistan the highest frequency of this haplogroup is found in Baltis (46%) Kashmiris (58%), Pathans (45%) and Sindhis (49%).

Haplogroup 1 was also found in all the Pakistani populations (18% of total surveyed) but at a very high frequency in the Hazaras (61%). Its frequency ranged from 8-28% in the other Pakistani populations. It was present at a very low frequency in the Parsis (8%). The presence of this haplogroup in Pakistan points towards genetic influences from north western Europe and West Asia.

Haplogroup 9 identified by the 12f2 deletion was distributed across all Pakistani populations being higher in the southern Parsi (38.9%), Brahui (28.2%), and Makrani Baloch (24.0%) populations. There appears to be a decrease in the frequency of this haplogroup as one moves from the south west to the north east of Pakistan. Its presence in the Pakistani population indicates a Persian and Mediterranean gene flow and is supported by the high frequency of this haplogroup in the Parsis. This population arrived in the Indo-Pak sub-continent from Iran[49].

The YAP insertion was found in 3.5% of the Pakistani Y chromosomes examined and these belonged to haplogroup 21 (2.4%) and haplogroup 8 (1.1%). Haplogroup 21 has been reported earlier to have a north African origin and is rarely found in northern Pakistan[50]. Haplogroup 8 is sub-Saharan in origin and is found in southern Pakistani populations (Baloch, Brahuis and Makranis) only. The highest frequency of haplogroup 8 is found in Makrani Negroid poulation (9.1%) who are reported to have an African origin, but the timing of their migration is uncertain. These could represent the genetic legacy of the African slaves that were brought to the Indo-Pakistan sub-continent by the Arabs and European invaders. There is also the possibility that these may be the genetic clues left behind by the earliest humans that arrived in Pakistan along the postulated coastal route out of Africa.

Haplogroup 10 (identified by the C allele for RPS4Y) was only found in three populations being highest among the Hazara (30.4%) followed by the Burusho (8.5%) and Brahui (1.8%). It is fairly common in Central Asia and Mongolia and points towards the Mongol origins of the Hazara population which is supported by historical records[9].

Haplogroup 28 identified by the M20 marker[40] was also common in the Pakistani populations, with the exception of the Hazara and Kashmiri groups and has not been reported at such a high frequency elsewhere in the world. It may be associated with the spread of populations during the height of the Indus Valley Civilization.

The Pakistani populations were compared with other world population using published data[48]. The latter were limited in the available number of examined markers. Using published[48] frequency data, the Pakistani populations cluster together in the principal component analysis (Figure 1). Eleven of the twelve ethnic groups from Pakistan cluster among themselves and are close to the European (Greek and Italian) and Indian populations. The Kashmiris are the notable exception and they lie close to the Inuit Eskimos. In addition to the small number of Kashmiri samples (n=12) this may reflect the fact that the markers used in this study do not differentiate between Y haplogroups 1C and 1G defined by the DYS257 (92R7) and the DYS199 T allele respectively. These haplogroups are present at an equal frequency (42-44%) in the Inuit Eskimos[48].

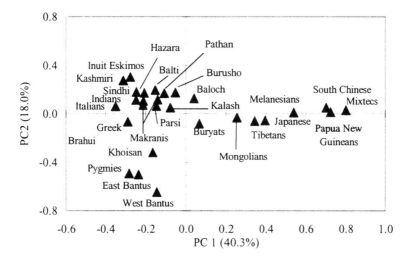

Figure 1. Principal component analysis between Pakistani and selected world populations based on Y chromosomal haplogroup frequencies.

Based upon the analysis of Y chromosomal markers, the Pakistani populations are well separated from populations in Africa, South East Asia, Oceania and America. The Chinese are separate from the Pakistani populations which indicates that the Karakoram Mountains form a formidable barrier to gene flow from the north. The Europeans are closer to the Pakistani population-cluster which is supported by historical records of the Aryan invasion of the sub-continent by tribes from West Asia and Caucasus[3]. The same tribes are supposed to have expanded to the West towards Europe.

Genotypes from 113 autosomal microsatellite loci (tri- and tetra-nucleotide repeats) were scored unambiguously by multiplex PCR amplification and PAGE analysis. A total of 1145 alleles were found in the various human populations. The average number of alleles for each microsatellite locus was 11 with a range of between 7 (D4S2417 and D6S503) and 20 (D3S2406 and D3S2465). The heterozygosity values ranged between 0.56 (*D2S1777*) and 0.91 (D3*S2406*).

All the world populations were equally heterozygous (0.71-0.73) for these autosomal microsatellites. Among the five Pakistani populations, the Sindhis were the most heterozygous (0.78) followed by the Kalash (0.74), Brahui (0.72) and Pathan (0.71). The Burusho were the least heterozygous (0.66).

Three different methods were used for genetic data analysis: (1) Phylogenetic analysis based upon genetic distance between populations[43,44]. (2) Principal component (PC) analysis of the allelic frequency data[1]. (3) Structure analysis of the multilocus genotype data[51]. A phylogenetic tree based upon genetic distances between two populations using the proportion of shared alleles resolved the individual populations into their regional and continental groups (data not shown). The first branch in the phylogenetic tree separates sub-Saharan Africans (Lisongo and Pygmies) from the non-Africans, consistent with the 'out-of-Africa' evolutionary model[1,27]. Subsequent branching separates East Asians (Cambodian, Chinese and Japanese) from the Pakistani (Brahui, Burusho, Kalash, Pathan and Sindhi), European (Basque, Greek, Northern European and North Italian) South American (Brazilian and Mayan Indian) and Oceanic (Australian aborigine, Melanesian and New Guinean) populations.

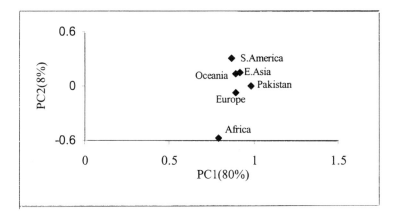

Figure 2. Principal component analysis based upon allele frequencies of 113 autosomal microsatellites in the Pakistani and world populations.

The PC analysis based upon the allelic frequencies of 113 autosomal microsatellite repeats is shown in figure 2. The Pakistani populations are compared with other world populations grouped according to their continental origins. The first two principal components account for 88% of the allele frequency data. The African cluster is separated from all the other world populations. The European and Pakistani populations are close to each other. The East Asians are closer to the Oceanic populations rather than the South Americans.

Structure, a model based clustering method was used to infer population structure[51]. Five Pakistani populations were compared with 15 world populations grouped according to their continental origins (as described above) using the multilocus autosomal microsatellite data. The best fit model separated human populations into three distinct clusters (figure 3). The African and non-African populations separate into well defined clusters. The first cluster (Cluster 1; Figure 3) contains only the sub-Saharan Africans, consistent with an African origin for modern humans. This is also supported by the fact that there is an African component in the second and third population cluster (Cluster 2 & 3; Figure 3), but not vice-versa. The second cluster contains the European and Pakistani populations. The Pakistani populations show considerable admixture with the Europeans, reflecting the common Indo-European origins of these populations from nomadic tribes from West Asia and Caucasus. There is some admixture between the Sindhis and Brahuis and the populations from East Asia, Oceania and South America, mainly represented in cluster 3. Although the Y chromosome shows very little genetic exchange with populations from East Asia, the autosomal data suggests that some exchange did occur between the Pakistani and East Asians. These genetic exchanges may reflect recent admixtures or they may be remnants of ancient migrations out of Africa to East Asia along the postulated coastal route. This is supported by the presence of shared autosomal microsatellite alleles between populations from East Asia and southern Pakistan. The northern Pakistani populations (Kalash, Pathans and Burusho) show no admixture with Eastern Asian populations.

The results presented here suggest that genetic relationships on a local geographic level are dictated primarily by geographic proximity rather than by linguistics. No evidence of genetic affinity among the two language isolate groups was detected. The isolation of the Hunza Burusho and Basque in the mountains of northern Pakistan and Europe respectively may have helped preserve the unique nature of their languages, Burushaski (Burusho), and Euskera (Basque), but the two groups are genetically more closely related to their geographic neighbours than to each other.

The fact that there is gene flow across linguistic boundaries is also borne by the observation that the Dravidian speaking Brahui population of Pakistan are genetically similar to their eastern Indo-European speaking Sindhi neighbours.

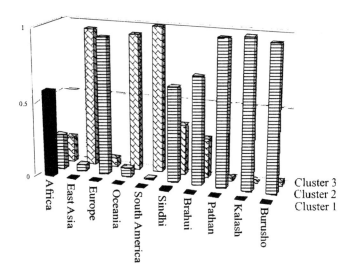

Figure 3. Summary of clustering results using 113 autosomal microsatellite allele frequencies in five Pakistani and various world populations grouped according to their continental origins.

The major clusters identified in this study largely agree with the currently accepted view of human migration and evolution as deduced from archaeological and genetic evidence. Previous studies have typed between 3 to 200 different microsatellite loci (primarily CA repeats) across worldwide populations to generate sufficient resolving power to distinguish among various world populations[27,52,53,54]. The data presented here supports the "out of Africa" theory which proposes an African origin of modern humans and which has also been supported by data from classical markers [1], mitochondrial DNA sequence data[24], Y chromosomal SNPs [22] and primarily di-nucleotide microsatellite repeats[27,28,55].

The present study provides a perspective into the evolutionary history of world populations in general and Pakistan in particular. A positive correlation exists between genetic affinity and linguistic classification only at the continental level. Within continents language isolate groups (such as Burusho and Basque) are closely related to their geographic neighbors. Among the various ethnic groups of Pakistan the Hazara Y chromosomes stand out. Other ethnic groups show genetic admixture with the European populations. This may reflect the arrival of nomadic Indo-European settlers from the pastoral highlands of West and Central Asia. These may have pushed the Dravidian speaking inhabitants further eastwards and southwards into India and Sri-Lanka.

ACKNOWLEDGMENTS

This work was supported by the Wellcome Trust (SQM). Thanks are due to the donors who made this study possible. We are grateful to the Department of Health, Government of Baluchistan, the Baloch Student Federation, M. Imran, A. Talaat, I. Kazmi and F. Sethna for help with the collection of Baloch, Brahui, Pathan, Burusho and Parsi samples. Samples from populations outside Pakistan were from the collection of MP (Greek) and L. L. C-S.

REFERENCES

1. Cavalli-Sforza L.L., Menozzi P., Piazza A.,1994, The history and geography of human genes, Princeton University Press, Princeton, New Jersey.
2. Jarrige J-F., 1991, Mehrgarh: Its place in the development of ancient cultures in Pakistan. In *Forgotten Cities on the Indus. Early Civilization in Pakistan from the 8th to the 2nd Millennium BC* (Jansen, M., Mulloy, M., Urban, G., eds.), Verlag Philipp von Zabern, Mainz, pp. 34-50.
3. Wolpert S., 2000, A new history of India, 6th ed. Oxford University Press, New York.
4. Mehdi S.Q., Qamar R., Ayub Q., Khaliq S., Mansoor A., Ismail M., Hammer M.F., Underhill P.A., Cavalli-Sforza L.L., 1999, The origins of Pakistani populations: Evidence from Y chromosome markers. In *Genomic diversity: Applications in Human Population Genetics* (Papiha, S.S., Deka, R., Chakraborty, R., eds.), Kluwer Academic/Plenum Publishers, New York, pp. 83-90.
5. Quddus S.A.,1990, The tribal Baluchistan, Ferozsons (Pvt.) Ltd., Lahore.
6. Backstrom P.C., 1992, Balti. In *Sociolinguistic Survey of Northern Pakistan. Vol. 2, Languages of Northern Areas* (Backstrom ,P.C., Radloff, C.F., eds.), National Institute of Pakistan Studies, Islamabad, Summer Institute of Linguistics, Dallas, pp. 3-27.
7. Hughes-Buller R., 1991, Imperial Gazetteer of India: Provincial series, Baluchistan. Sang-e-Meel Publications, Lahore.
8. Biddulph J., 1977, Tribes of the Hindoo Koosh. Indus Publications, Karachi.
9. Bellew H.W., 1979, The Races of Afghanistan, Sang-e-Meel Publications, Lahore.
10. Robertson G.S., 1974, The Kafirs of the Hindu-Kush. Oxford University Press, Karachi.
11. Decker K.D., 1992, Sociolinguistic survey of northern Pakistan Vol. 5 Languages of Chitral. National Institute of Pakistan Studies, Islamabad, Summer Institute of Linguistics, Dallas.
12. Ahmad A.K.N., 1952, Jesus in heaven on earth, The Civil and Military Gazette Ltd., Lahore.
13. Nanavutty, P., 1977, The Parsis. National Book Trust, New Delhi.
14. Dorn B., 1999, History of the Afghans, Vanguard Book (Pvt.) Limited, Lahore.
15. Caroe O., 1992, The Pathans. Oxford University Press, Karachi.
16. Bellew H.W., 1998, An inquiry into the ethnography of Afghanistan, Vanguard Books Limited, Lahore.
17. Burton R.F., 1851, Sindh and the races that inhabit the valley of the Indus, WH Allen and Co. Ltd., London.
18. Ibbetson D., 1994, Panjab castes, Sang-e-Meel Publications, Lahore.
19. Cavalli-Sforza L.L., 1998, *Trends Genet.*, **14**, 60-5.

20. Hammer, M.F., Spurdle, A.B., Karafet, T., Bonner, M.R., Wood, E.T., Novelletto, A., Malaspina, P., Mitchell, R.J., Horai, S., Jenkins, T., Zegura, S.L., 1997, *Genetics*, **145**, 785-805.
21. Tyler-Smith, C., 1999, Y-chromosomal DNA markers. In *Genomic Diversity: Applications in Human Population Genetics* Papiha, S.S., Deka, R., Chakraborty, R., eds.), Kluwer Academic/Plenum Publishers, New York, pp. 65-73.
22. Underhill, P.A., Shen, P., Lin, A.A., Jin, L., Passarino, G, Yang,WH, Kauffman, E, Bonne-Tamir, B, Bertranpetit, J, Francalacci, P, Ibrahim, M., Jenkins, T., Kidd, J.R., Mehdi, S.Q., Seielstad, M.T., Wells, R.S., Piaza, A., Davis, R.W., Feldman, M.W., Cavalli-Sforza, L.L., Oefner, P.J., 2000, *Nat. Genet.*, **26**, 358-61.
23. Jobling M.A., Tyler-Smith C., 1995, *Trends Genet.*, **11**, 449-56.
24. Quintana-Murci, L., Semino, O., Bandelt, H-J., Passarino, G., McElreavey, K., Santachiara-Benerecetti, A.S., 1999, *Nat. Genet.*, **23**, 437-41.
25. Weber, J.L., May, P.E., 1989, *Am. J. Hum. Genet.*, **44**, 388-96.
26. Tautz, D., 1989, *Nucleic Acids Res.*, **17**, 6463-71.
27. Bowcock, A. M., Ruiz-Linares, A., Tomfohrde, J., Minch, E., Kidd, J.R., Cavalli-Sforza, L.L., 1994, *Nature*, **368**, 455-7.
28. Jorde, L.B., Watkins, W.S., Bamshad, M.J., Dixon, M.E., Ricker, C.E., Seielstad, M.E., Batzer, M.A., 2000, *Am. J. Hum. Genet.*, **66**, 979-88.
29. Hammer, M.F., 1994, *Mol. Biol. Evol.*, **11**, 749-61.
30. Casanova, M., Leroy, P., Boucekkine, C., Weissenbach, J., Bishop, C., Fellous, M., Purrello, M., Fiori, G., Siniscalco, M., 1985, *Science*, **230**, 1403-6.
31. Santos, F.R., Pandya, A., Kayser, M., Mitchell, R.J., Liu, A., Singh, L., Destro-Bisol, G., Novelletto, A., Qamar, R., Mehdi, S.Q., Adhikari ,R., de Knijff, P., Tyler-Smith, C., 2000, *Hum. Mol. Genet.*, **9**, 421-30.
32. Mathias, N., Bayés, M., Tyler-Smith, C., 1994, *Hum. Mol. Genet.*, **3**, 115-23.
33. Bianchi, N.O., Bailliet, G., Bravi, C.M., Carnese, R.F., Rothhammer, F., Martínez-Marignac, V.L., Pena, S.D.J., 1997, *Am. J. Phys. Anthropol.*, **102**, 79-89.
34. Santos, F.R., Pandya, A., Tyler-Smith, C., Pena, S.D.J., Schanfield, M., Leonard, W.R., Osipova, L., Crawford, M.H., Mitchell, R.J., 1999, *Am. J. Hum. Genet.*, **64**, 619-28.
35. Whitfield, L.S., Sulston, J.E., Goodfellow, P.N., 1995, *Nature*, **378**, 379-80.
36. Seielstad, M.T., Hebert, J.M., Lin, A.A., Underhill, P.A., Ibrahim, M., Vollrath, D., Cavalli-Sforza, L.L., 1994, *Hum. Mol. Genet.*, **3**, 2159-61.
37. Santos, F.R., Carvalho-Silva, D.R., Pena, S.D.J., 1999, PCR-based DNA Profiling of Human Y Chromosomes. In *Methods and Tools in Biosciences and Medicine* (Epplen, J.T, Lubjuhn, T., eds.), Birkhauser Verlag, Basel, pp. 133-52.
38. Pandya, A., King, T.E., Santos, F.R., Taylor, P.G., Thangaraj, K., Singh, L., Jobling, M.A., Tyler-Smith, C., 1998, *Ind. J. Hum. Genet.*, **4**, 52-61.
39. Zerjal, T., Dashnyam, B., Pandya, A., Kayser, M., Roewer, L., Santos, F.R., Schiefenhövel, W., Fretwell, N., Jobling, M.A., Harihara, S., Shimizu, K., Semjidmaa, D., Sajantila, A., Salo, P., Crawford, M.H., Ginter, E.K., Evgrafov, O.V., Tyler-Smith, C., 1997, *Am. J. Hum. Genet.*, **60**, 1174-83.
40. Underhill, P.A., Jin, L., Lin, A.A., Mehdi, S.Q., Jenkins, T., Vollrath, D., Davis, R.W., Cavalli-Sforza, L.L., Oefner, P.J., 1997, *Genome Res.*, **7**, 996-1005.
41. Bergen, A.W., Wang, C.Y., Tsai, J., Jefferson, K., Dev, C., Smity, K.D., Park, S.C., Tsai, S.J., Goldman, D., 1999, *Ann. Hum. Genet.*, **63**, 63-80.
42. Jin, L., Underhill, P. A., Buoncristiani, M., Robertson, J. M., 1997, *J. Forensic Sci.*, **42**, 496-9.
43. Nei, M., 1973, Analysis of gene diversity in subdivided populations, *Proc. Natl. Acad. Sci. USA* **70**, 3321-3.

44. Nei, M., Tajima, F., Tateno, Y., 1983, *J. Mol. Evol.,* **19,** 153-70.
45. Young, F.W., Bann, C.M., 1996, A visual statistics system. In *Statistical Computing Environments for Social Researches* (Stine, R.A., Fox, J., eds.), Sage Publications, New York, pp. 207-36.
46. Jobling, M.A., Tyler-Smith, C., 2000, *Trends Genet.,* **16,** 356-62.
47. Zerjal, T., Pandya, A., Santos, F.R., Adhikari, R., Tarazona, E., Kayser, M., Evgrafov, O., Singh, L., Thangaraj, K., Destro-Bisol, G., Thomas, M.G., Qamar, R., Mehdi, S.Q., Rosser, Z.H., Hurles, M.E., Jobling, M.A., Tyler-Smith, C., 1999, The use of Y-chromosomal DNA variation to investigate population history: recent male spread in Asia and Europe. In *Genomic Diversity: Applications in Human Population Genetics* (Papiha, S.S., Deka, R., Chakraborty, R., eds.), Kluwer Academic/Plenum Publishers, New York, pp. 91-101.
48. Karafet, T., Zegura, S.L., Posukh, O., Osipova, L., Bergen, A., Long, J., Goldman, D., Klitz, W., Harihara, H., de Knijff, P., Wiebe, V., Griffiths, R.C., Templeton, A.R., Hammer, M.F., 1999, *Am. J. Hum. Genet.,* **64,** 817-31.
49. Quintana-Murci, L., Krausz, C., Zerjal, T., Sayar, S.H., Hammer, M.F., Mehdi, S.Q., Ayub, Q., Qamar, R., Mohyuddin, A., Radhakrishna, U., Jobling, M.A., Tyler-Smith, C., McElreavey, K., 2001, *Am. J. Hum. Genet.,* **68,** 537-42.
50. Qamar, R., Ayub, Q., Khaliq, S., Mansoor, A., Karafet, T., Mehdi, S.Q., Hammer, M.F., 1999, *Hum. Biol.,* **71,** 745-55.
51. Pritchard, J.K., Stephens, M., Donnelly, P., 2000, *Genetics,* **155,** 945-59.
52. Cooper, G., Amos, W., Bellamy, R., Siddiqui, M. R., Frodsham, A., Hill, A.V.S., Rubinsztein, D.C., 1999, *Am. J. Hum. Genet.,* **65,** 1125-33.
53. Goldstein, D.B., Linares, A.R., Cavalli-Sforza, L.L., Feldman, M.W., 1995, *Proc. Natl. Acad. Sci. USA,* **92,** 6723-7.
54. Pérez-Lezaun, A. Calafell, F., Mateu, E., Comas, D., Ruiz-Pacheco, R., Bertranpetit, J., 1997, *Hum. Genet.,* **99,** 1-7.
55. Mountain, J.L., Cavalli-Sforza, L.L., 1997, *Am. J. Hum. Genet.,* **61,** 705-18.

Lessons from Nature Show the Way to Safe and Environmentally Pacific Pest Control

WILLIAM S. BOWERS
The University of Arizona, Laboratory of Chemical Ecology, Department of Entomology, Forbes Bldg., Tucson, AZ 85721, USA

1. INTRODUCTION

We apply great effort to the propagation and growth of a variety of crops, domesticated animals and beneficial microorganisms because of their obvious contribution to our welfare. We are, on the other hand, often inclined to relegate the remaining biota to the status of weeds or pests overlooking their actual necessity in maintaining the planetary biological web. Misunderstanding the need for a diverse biota has finally led to the realization of the ecological damage that now prompts us to seek to restore the complexity we have previously denied.

Plants and animals, including man, have been shaped over eons by the competitive forces of nature which ever refine organisms prowess for survival. Thus, the antibiotic penicillin produced by common bread mold destroys competitive bacteria. The origin of the mold's defensive chemistry depended on the need for protection against bacteria and a timeless evolutionary process favoring the eventual appearance and retention of a unique protective chemical. Since the discovery of penicillin investigation of other microorganism has yielded additional antibiotics e.g., streptomycin etc. evidencing the university of the evolution of defensive responses between competing organisms. Thus, conflicts over space and resources among and between organisms has generated over time a host of unique chemicals from which we have developed important products. From studies of the chemical ecology of biological organisms we have derived many important resources including drugs, foods and pesticides. We now recognize the imperative to maintain the rich diversity of living organisms, for out of their continuing competition energes the special biological chemistry to sustain our own survival.

1.1 Historical Perspectives

It must be apparent to everyone that the excellent food and public health enjoyed in most of the world is due directly to the use of insecticides for the control of insect pests of agriculture and vectors of disease. At the same time we recognize that the use of certain insecticides has damaged the environment and promoted the development of resistance in many target pests. Our expanding population requires ever increasing food production and our public health status cannot return to the devastating plagues of yesteryear. Since we are increasingly dependent on pesticides we must clearly define our target and develop pest management strategies that are inherently rational in concept and design as well as environmentally benign.

Historically, early pesticides were composed of plant poisons that could be extracted and applied to food and fiber crops, ie., pyrethrins, carbamates, rotenoids, alkaloids etc. Nature can seldom provide the quantity of natural chemicals to support extensive pest control and plantation methods of production for any but the most efficacious pesticides is impractical. Post WWII the world awakened to an explosion in population and an enhanced realization of the cost of vector borne diseases. Industrial screening of toxic chemicals and war gases yielded a host of pesticidal chemistry eg., chlordane, benzene hexachloride, DDT etc. Their great efficacy depended upon their rapid acting neurotoxicity, increased stability and cost effectiveness compared to plant toxins. Although cheaply produced, the ultimate cost to the environment has become unacceptable. Clearly, future pest management strategies require the discovery and development of chemistry selective to pest management and benign to the environment.

1.2 Insect Growth Regulators

Vertebrates and invertebrates are separated by 600 million years of evolution allowing the appearance of numerous biochemical and physiological differences that suggest opportunities for the development of selective methods of pest management. The discovery of the juvenile hormone (JH) of insects by Wigglesworth[1] and Williams[2] and its unique role in regulating insect metamorphosis and reproduction offered the first possibility to develop a specific method to disrupt of insect growth processes. Its characterization by Bowers et al[3]. provided the basis for the first selective approach to insect control. Synthetic optimisation of juvenile hormon III[4] yielded the first commercial products (Fig.1) called insect groth regulators (IGRs).

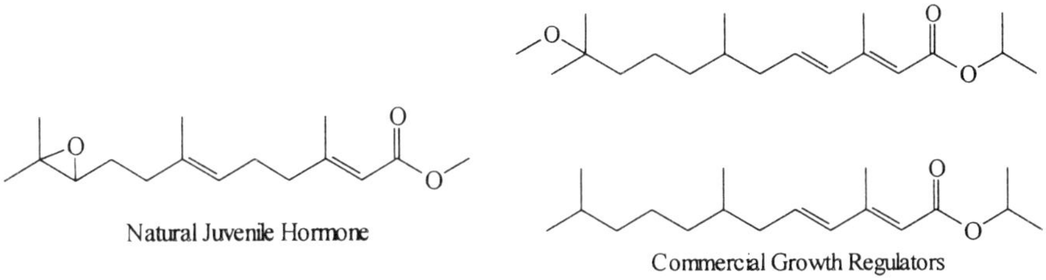

Figure 1. Natural juvenile hormone of insects provided the lead to the first commercial insect growth regulator products.

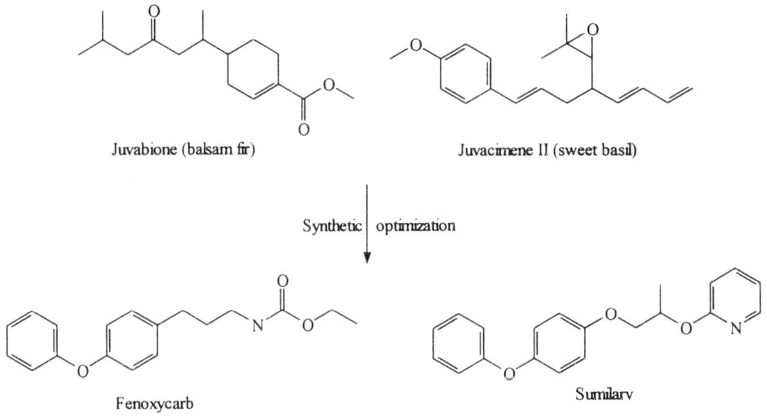

Figure 2. Optimization of phytojuvenoids yielded a second generation of IGRs with improved activity. Fenoxycarb (Syngenta), Sumilarv (Sumitomo).

Since the IGRs were developed from a natural growth regulator intrinsic to the biology of the target pests the discovery process was termed "biorational" a term emphasizing the process of defining a specific biological target for disruption in the pest species which lacking any counterpart in higher animals may ensure safety.

Subsequently, the discovery of novel phytochemistry with JH activity by Bowers et al.,[5,6] lent new chemical models for optimisation birthing a second generation of insect growth regulator products with increased efficacy and continuing environmental safety (Fig.2).

2. BEHAVIOUR MODIFYING CHEMISTRY

Repellent chemistry first line defense against insect vectors of disease.
All of the five major tropical diseases are transmitted by insect vectors including malaria, filariasis, African and American trypanosomiasis and leishmaniasis are responsible for over 1 million deaths and cause of morbidity in hundreds of millions more annually.

2.1 Pheromones

Insects utilize a variety of volatile chemicals for communication. Called pheromones they are used for sexual receptivity, courtship, resource marking, colonization, alarm, social recognition, orientation[7] etc. After a 30 year effort the first pheromone identified was of the commercial silkworm *Bombyx mori* by Butenandt et al.,[8].However, the greatest stimulus to pheromone research followed the development of the electroantennegram method for identifying the sex pheromones of *Lepidoptera*[9]. Highly species specific, the sex pheromones are used to monitor the presence of reproductively active insects in the vicinity of susceptible crops. Successful application of pheromones in monitoring for pests in apple orchards can reduce insecticide and miticide applications by up to 50 %[10]. Pheromones are also used to permeate the atmosphere confusing males and preventing mating of certain pests including the cabbage looper and pink bollworm[10]. Before the sex pheromones of the American cockroach were isolated and identified[11,12] phytochemicals in *Verbena* and spruce oil were found to stimulate the identical attraction and mating display cockroaches and were identified as (+)-bornyl acetate, (+)-E-verbenyl acetate and germacrene D[13-16].This discovery was the first to find arthropod pheromone mimics in plants.

2.2 Repellents

As the principal vectors of all of the ancient plagues insects seem to conspire constantly to keep man in an endless state of ill health. While human reproductive potential appears to be a constant, malignant diseases like malaria, yellow fever typhus and trypanosomiasis have historically, savagely checked population growth. The use of plant extractives to repel biting arthropods is an ancient technology universal to all human cultures. Equally, the widespread presence of mono- and sesquiterpenes in plants, many with potent insect repellency, suggests that repellents constitute a plant defensive modality as well. Early repellents were generally derived from plant volatiles and included plant secondary chemicals such as geraniol, citronellal, camphor and menthol. However, the need for a long-lasting repellent during WW II supported a massive screening effort of industrial chemicals that led to the adoption of the synthetic chemical diethyl m-toluamide (DEET).

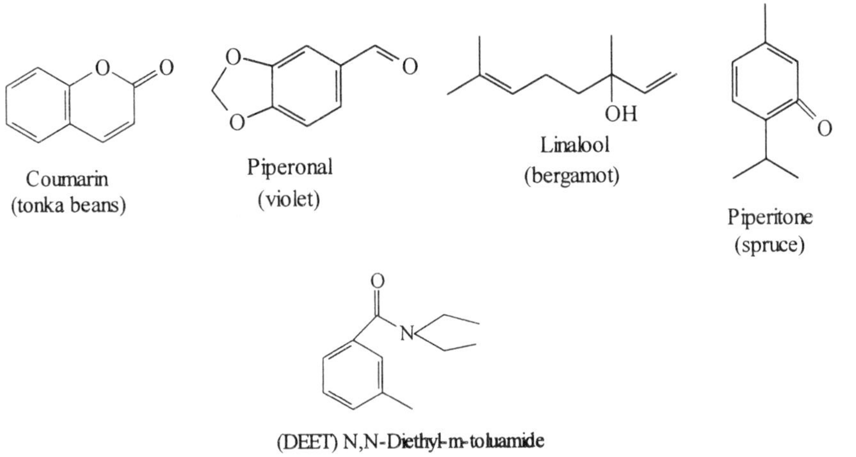

Coumarin (tonka beans)

Piperonal (violet)

Linalool (bergamot)

Piperitone (spruce)

(DEET) N,N-Diethyl-m-toluamide

Figure 3. Phytochemical repellents and commercial product DEET.

DEET can be credited with the protection of millions of soldiers from vector borne disease. Postwar, DEET entered the civilian market and largely displaced natural repellents. DEET a patented repellent commercial preparation could be sold for a premium price whereas natural repellents lacking patent advantage were largely eliminated from the market. However, DEET, like the aforementioned synthetic insecticides was revealed to possesses significant toxicity to humans[17].

Our improved understanding of the natural chemical defenses of plants opens the possibility of finding new, natural repellents lacking the toxicity of DEET. We have discovered numerous plants that possess potentially useful natural repellents. We find among plant secondary chemicals numerous repellents like coumarin, piperonal, piperitone and linalool are equal in repellency activity to DEET[18-22]. These are highly effective space repellents and modern controlled-release formulation technology dramatically improves the persistence and efficacy of the more volatile compounds.

Repellents are also used by animals to protect against parasites and predators. Insect defensive secretions especially contain a striking array of repellent chemistry[23]. Benzoquinones, naphtoquinones and phenolics are common repellent components of arthropods[24]. Following the isolation and identification of the alarm pheromones of aphids, (E)-β-farnesene[25] and (-)-germacrene A[26,27] (Fig.5) as well known plant secondary chemicals it was shown that, acting as natural repellents they could be used to drive aphids from their host plants. Jowever, due to their extreme lability in air and light no useful commercial applications might be expected. Gibson and Pickett[28] found that (E)-β-farnesene is a constituent released from the trichomes of an aphid resistant potato cultivar. This demonstrated, for the first time, the use by a plant of a natural insect pheromone for defense. Slow release formulations for (E)-β-farnesene have been developed[29], although full field efficacy remains to be demonstrated. Numerous investigations currently focus on the application of natural repellents for the direct protection of crops.

3. INDUCIBLE PLANT DEFENSES

One of the most exciting discoveries to impact future plant protection strategies is the discovery that the natural chemical defenses in plants are inducible. These chemical counteradaptations occur in response to damage by herbivores and pathogens. The pioneering works[30,31] have stimulated world-wide investigations of these phenomena.

Studies reveal that insect feeding damage and salivary components can provoke the synthesis and release of volatiles from plants that attract parasitoids to infect their herbivores[32,33]. An elicitor in caterpillar oral secretions that induces corn seedlings to emit chemical signals attractive to parasite wasps has been isolated and characterized[34].

The classical work of Ryan[35], demonstrated that jasmonic acid a signalling compound that occurs as an intermediate in the octadecanoid cycle part of the wound hormone response in plants can be used to artificially induce some plant defenses.

3.1 Ecdysone: An Inducible Defense against Insects and Nematodes

The presence of phytoecdysteroids in plants has been known for over 25 years yet attempts to demonstrate for them a role in defense against insect attack has met with little success. Indeed some insects have been shown to successfully detoxify high levels of ingested ecdysteroids. This led many to dismiss them as a serious defensive tactic. We decided to re-investigate the role of ecdysones in plants focusing on the plant response to stress. The herb spinach, *Spinacia oleracea*, is well known to produce and accumulate high levels of the ecdysteroids 20-hydroxyecdysone (90 %) and polypodine B (20 %) in the leaves. We found that removal of portions of the roots and/or shoot damage induced by mechanical means or by insect feeding, caused an elevation of the ecdysteroids in the plant[36].

We examined effect of treatment with methyl jasmonate on spinach plants and realized a dosage dependent increase of the ecdysteroids in plants[37]. On analysis of separated parts of the induced plants for their ecdysteroid content we were astonished to find that only the roots possessed elevated titers. These results revealed the response to stress or treatment with methyl jasmonate was root specific. We reasoned that the enhanced titer of ecdysteroids in roots must imply a defense targeted to root pests and tested this by exposing induced and naïve plants to the root pest Bradysia impatients. We found the Bradysia larvae died as prepupae seeming to undergo apolysis without ecdsis[38]. The specific nature of the spinach root induction opened the possibility that other root pests might also be affected.

The presence of ecdysones has been long suspected to occur in nematodes and even postulated to be important to their growth and molting. To examine this possibility we infested jasmonate induced and un-induced control spinach plants with infective juveniles of the cyst forming nematode *Heterodera schachtii*. After seven weeks the cysts were washed from the roots and soil, counted and measured. There was a 50% decrease in the cysts produced on the roots of induced plants and the cysts were much smaller that those from control plants. Moreover, on dissection nearly all of the juveniles present in the cysts of the induced plants were dead[39].

I believe our investigations, support a role for ecdysones in nematode endocrinology particularly for the regulation of their growth and morphogenetic development similar to their role in insects. Equally, our studies establish that the phytoecdysones compose a powerful natural defense against both insects and nematodes.

4. SUMMARY

Our flora and fauna contain a host of natural treasures that have evolved during eons of competitive interactions. To the natural product chemist this vast reservoir of biologically active chemistry can be tapped to provide useful drugs and biorational pest controls and represents another step in the discovery of just how precious is our biological diversity. The emerging insights into the induction of plant defensive chemistry is another exciting frontier for the chemical ecologist to plumb and an opportunity to discover natural, biorational methods with which to deal with those few pests that persistently conspire to steal our calories and corpuscles.

REFERENCES

1. Williams, C.M., 1956, *Nature,* **178**, 212-3.
2. Wigglesworth, V.B., 1934, *Quart. J. Mic. Sci.,***77**, 191-222.
3. Bowers, W.S., Thompson, M.C., Uebel, E.C., 1965, *Life Sci.,***4**, 2323-31.
4. Henrick, C.A., Staal, G.B., Siddall, J.B., 1973, *Agr. Food Chem.,* **21**, 354-9.
5. Bowers, W.S., Fales, H.M., Thompson, M.J., Uebel, E.C., 1966, *Science,* **54**, 1020-22.
6. Bowers, W.S., Nishida, R., 1980, *Science,* **209**, 1030-32.
7. Bell, W.J., Carde, R.T., 1984, Chemical Ecology of Insects, Sinauer Associates, Inc., Sunderland, MA.
8. Butenandt, A.R., Beckmann, C., Stamm, D., Hecker, E., 1959, *Naturforschung,* **146**, 283-4.
9. Roelofs, W.L., 1979, *Chemtech,* **9**, 222-7.
10. Roelofs, W.L., 1981, Semiochemicals. In *Their Role in Pest Control* (Nordlund, D.A., Jones, R.L., Lewis, W.J., eds.), John Wiley & Sons Pub., New York.
11. Persoons, C. J., Verwiel, P.E., Ritter, J., Talman, E., Nooijen, P.J.F., Nooijen, W.J., 1976, *Tetrahedron Lett.,* 2055-8.
12. Nishino, C., Kobayashi, K., Fukushima, K., Imanari, M., Nojima, K., Kohno, S., 1988, *Chemistry Letters,* 517-20.
13. Bowers, W.S., Bodenstein, W.G., 1971, *Nature,* **232**, 59-61.
14. Nishino, C., Tobin, T.R., Bowers, W.S., 1977, *Appl.Ent. Zool.,* **12** (3), 287-90.
15. Tahara, S., Yoshida, M., Mizutani, J., Kitamura, C., Takahashi, S., 1975, *Agr. Biol. Chem.,* **39**, 1517-8.
16. Kitamura, C., Takahashi, S., Tahara, S., Mizutani, J., 1976, *J. Agr. Biol. Chem.,* **40**, 1965-9.
17. Knowles, C.O., 1990, Encyclopedia of Pesticide Toxicology In *Missouri Agricultural Station Journal* (Hayes, W. J., Laws, E. R., eds.), 10308, Columbia, MO.
18. Bowers, W.S., Ortego, F., Xioquing, Y., Evans, P.H., 1993, *J. Nat. Prod.,* **56**, 935-8.
19. Rusell, G.B., Hunt, M.B., Bowers, W.S., Blunt, J.W., 1994, *Phytochemistry,* **35**, 1455-6.
20. Bowers, W.S., 1996, *The IPM Practitioner,* **18** (8), 13.
21. Bowers, W.S., 1997, *Acta Botanica Gallica,* **144** (4), 383-90.
22. Bowers, W.S., Evans, P.H., Spence, P., 2000, *J.Herbs, Spices & Medicinal Plants,* 7 (4), 85-9.
23. Blum, M.S., 1981, Chemical Defenses of Arthropods, Academic Press, Orlando.
24. Tschinkel, W.R., 1969, *J. Insect Physiol.,* **15**, 191-5.
25. Bowers, W.S., Nault, L.R., Webb, R.E., Dutky, S.R., 1972, *Science,* **177**, 1121-2.
26. Bowers, W.S., Nishino, C., Montgomery, M.E., Nault, L.R., Nielsen, M.W., 1977, *Science,* **196**, 680-1.

27. Nishino, C., Bowers, W.S., Montgomery, M.E., Nault, L.R., 1976, *Agr. Biol. Chem.*, **40**, 2303-4.
28. Gibson, R.W., Pickett, J.A., 1983, *Nature*, **302**, 608-9.
29. Dawson, G.W., Griffiths, D.C., Pickett, J.A., Smith, M.C., Woodcock, C.M., 1982, *J.Chem.Ecology*, **2**, 1111-7.
30. Ryan, C.A., 1992, *Plant Mol. Biol.*, **19**, 123-33.
31. Baldwin, I.R., Schmelz, E.A., Zhang, Z. P., 1996, *J.Chem.Ecology*, **22**, 61-74.
32. Turlings, T.C.J., McCall, P.J., Alborn, H.T., Tumlinson, J.H., 1993, *J.Chem.Ecology*, **19**, 411-25.
33. Takabayashi, S., Dicke, M., Posthumus, M.A., 1994, *J.Chem.Ecology*, **20**(6), 1329-54.
34. Alborn, H.T., Turlings, T.C.,J., Jones, T.H., Stenhagen, G., Loughrin, J.H., Tumlinson, J.H., 1997, *Science*, **276**, 945-9.
35. Ryan, C.A., Jagendorf, A., 1995, Self defense by plants, *Proc. Natl. Acad. Sci.*, **92**, 4075.
36. Schmelz, E.A., Grebenok, R.J., Galbraith, D.W., Bowers, W.S., 1998, *J. Chemical Ecology*, **24**(2), 339-60.
37. Schmelz, E.A., Bowers, W.S., 1998, Induction of 20-hyroxyecdysone in spinach *(Spinacia oleracea)*, Vol.1, SixthAustralasian Applied Entomological Research Conference, 29 Sept-2 Oct., Brisbane, Australia,
38. Schmelz, E.A., Grebenok, R.J., Galbraith, D.W., Bowers, W.S., 1999, *J. Chemical Ecology*, **8**, 1739- 57.
39. Davies, K.A., Fisher, J.M., 1994, *Int. J. for Parasitology*, **24**(5), 649-55.

Biodiversity of Soil Fauna in Different Ecosystems in Egypt with Particular References to Insect Predators

ABDEL KHALEK M. HUSSEIN
Plant Protection Research Institute, Nadiel-Saied Street, Dokki, Cairo, Egypt

1. INTRODUCTION

Three different ecosystems were chosen for this study. The first : The desert one, in the western desert of Egypt of Omayed 83 km west of Alexandria and about 10 km south of the sea shore (29° 12` 15`` E, 30°C 45` 45``) the soil is made of lagoonal deposits composed of gypsum intermixed with sand. Surface soil layer are loose and subject to active erosion and deposition creating micro-topographic variations. Soils are often compact, surface layers includes large amounts of snail shell fragments increasing the compactness of soil[1]. The area has been classified as part of the "Attenuated xero-Mediterranean climatic province of Egypt[2]. Rainfall varying between 50 - 110 mm / year. The mean air temperature varied from 13.6 to 27°C and the mean air relative humidity varied from 56 - 87 %. The vegetation is dominant by *Asphodelus microcarpus, Plantago albicans, Anabasis articulata*[3] and *Thymeleae hirsuta* and *Gymnocarpos decandrum* [4].

Semi-desert ecosystem was represented in Tahrir province (30° 25` N, 30°C 30` E) 100 km south of the Mediterranean coast and 40 West of the Nile river) is characterized by natable reclamation projects and agricultural extension using different vegetables and field crops and horticultural plantations depend on both Nile and under ground water vegetation cover is mixed of these cultivated crops and wild shrubs herps[5].

Agro-ecosystem lies in Menoufiya area the southern part of Nile Delta. The soil is completely clay fertile soil with heavy texture. Field crops and vegetables covers > 95 % of the whole land area with permanent irrigation supply from Nile and with special intensified crop rotation system[6].

2. MATERIAL AND METHODS

Sampling of soil fauna in the first ecosystem, the desert one was carried out using seiving method according to Ghabbour *et al.*[7], Sampling of these fauna in the both semi - desert (reclaimed area), and agro-ecosystems in the clay fertile soil in the Nile delta area was conducted by pit-fall traps according to Southwood[8]. The obtained fauna was kept in 70 % ethyl alcohol, and directly subjected to identification and classified to 3 groups (herbivores or pests which feed on the vegetation, carnivores or predators which feed on other insects and deteritivores feeding on organic matter as essential diet and dead insects).

3. RESULTS AND DISCUSSION

Survey of the fauna for a whole year in the soil under the main vegetation species *Thymealae hirsuta* showed there was 22 of fauna species and / or higher taxa. *Anabasis articulata* was the second species of important plants consisting the vegetation with 26 species of fauna. Investigation of the fauna under the four dominant species of plants showed 36 species of invertebrates. So, the number of fauna species and their densities increase with the number of vegetation cover. Investigation in leeward and widword direction, showed that densities and species of fauna recorded higher values in leeword direction, this may due to the ecological local protection aginst winds in such open climatic conditions.

Depending on the fauna associated with the dominant plants consisting the vegetation cover in the area, the 36 species consists of 8 herbivores species, 11 carnivore species, species and 17 deteritivore species. Total density represented as 0.85 %, 4.83 % and 94.01 % for the three prementioned categories respectively. The main herbivores represented in *Messor* sp. *Capsodes* sp. (Pentatonidae : Coleoptera), and *Brachycerus spinicollis*. The main carnivores are Antlion and spider with 2.71 % and 1.33 % of the total population density respectively. The main deteritivore species was *Heterogamia syriaca* with 69.71 % of the whole population density.

Semi desert ecosystem, the reclaimed land in Tahrir province showed 45 species of invertebrates, devided as 14, 18 and 13 species of herbivores, Carnivores and detritivores respectively. Population densities of these three prementioned categories wee 73.18 %, 11.98 % and 15.78 %, respectively. The main herbivores were *Monomorium* sp., *Messor* sp. (Formicidae : Hymenoptera), *Zophosis* sp. (Tenebrionidae : Coleoptera) and *Aphis* sp. (Aphidae : Homoptera) with 62.14 %, 5.70 %, 1.61 % and 1.22 % of the density of the whole population, respectively. The main carnivores (natural enemies) were : spiders (Arachnidae), *Labidura riparia* (Labiduridae : Dermaptera) and (Pompilidae : Hymenoptera) with 4.55 %, 2.02 % and 1.07 % of the whole population density respectively.

Agro-ecosystem in Nile Delta area was characterized with 52 species of arthropods in 18 vegetable and field crops within a whole year. Number of herbivores, carnivores and detritivores recorded 22, 17 and 13 species, respectively. The population densities for the 3 categories were 30.83 %, 41.07 % and 28.06 %, respectively. The main herbivores (insects and pests) were : *Monomorrium* sp., Formicidae, *Aiolopus* sp. (Acridiidae) and *Tropinota squalida* (Scarabidae : Coleoptera) with 13.61 %, 7.10 %, 0.95 % and 0.63 % of the total population density of the whole arthropods. The most important carnivores (Natural enemies) were spiders, *L. reparia, Medon ochracen* (staphylinidae : coleoptera) and *Philanthus* sp (sphecidae : Hymenoptera) with 27.8 %, 3.06 %, 0.88 % and 0.85 % of the whole population density, respectively.

4. CONCLUSION

Pit-fall traps technique according to Southwood[8] was used to survey the soil fauna, especially (macro-invertebrates). Three different ecological systems (e.g. desert, semidesert and agro-ecosystems) in Egypt, were subjected to the investigation. Soil chemical and physical properties gradually changed from pure sandy soil to moderate loamy soil then heavy fertile clay texture in the 3 ecosystems respectively.

Biodiversity of soil fauna was expressed as number of species and population density per certain area and/or per trap in addition to index of species abundance. The main ecological function groups were presented in each ecosystem as percentage of the total number of species and whole populations. Results showed that the main herbivores in desert ecosystem was *Messor* sp. (Formicidae : Hymenoptera) and *Capsodes* sp. (Pentatomiae : Hemiptera) with 0.36% and 0.20 % of the total population of soil fauna. The main Carnivores were represented by the antlion and the spiders with 2.71 and 1.33 % respectively. The main deteitivores in the desert was *Heterogamia syriaca* 69.71 % (Blattidae : Dictyoptera) followed by *Pssamobius porcicollis* and isopoda 8.03 % and 3.24 % respectively. In the semi-desert ecosystem (newly reclaimed area), the main herbivores was

Monomorrium sp., *Messor* sp. and *Zophosis* sp. (Tenbrionidae : Coleoptera) with 62.1 %, 5.7 % and 1.6 % of the whole fauna respectively. The most important carnivores are the spiders and *Labidura riparia* (Labiduridae : Dermoptera) with 4.55 % and 2.02 % respectively, while the main dominant detritivores were Tephritidae, Muscaidae and isopoda with 7.28 %, 2.45 % and 1.5 % respectively.

Agro-ecosystem, in Nile Delta was characterized with *Monomorrium* sp., Formicidae and *Drasterius biomaculatus* (Elateridae : Copleoptera) 13.6 %, 7.1 % and 1.9 % respectively. The main carnivores were the spiders and carabidae 27.8 and 4.22 % of the whole fauna, while the main species of detritivores were collembola and isopoda with 9.2 and 8.9 % respectively. In the 3 different ecosystems : desert, semi-desert and agro-ecosystem, the herbivores (insects for crops) represented with 8, 14 and 22 species respectively, while their natural enemies (carnivores) reported 11, 18 and 17 species and the detritivores were 17, 13 and 13 species respectively.

REFERENCES

1. El-Kadi, H.F.,1989, Effect of grazing pressure and certain ecological parameters on some fodder plants of the Mediterranean cosat of Egypt, M.Sc. Thesis, Univ. of Tanta, Tanta.
2. Ayyad, M.A.,El-Kadi, H.F. ,1982, REMDENE Progress Report, No 3.
3. Abd-El-Razik, M.S., 1976 , A study on the Vegetation Composition, Productivity and Phenology in a Mediterranean Desert Ecosystems at Omayed, Egypt, M.Sc. Thesis, Univ. of Alexandria, Alexandria.
4. Hussein, A.M.,1985, Studies on populations of some soil animals- (Macro-invertebrates) in an African Desert Ecosystem. M.Sc. Thesis, Cairo Univ., Cairo.
5. Hussein *et al.*,1997, *Bull. Fac. Agric., Univ. Carieo*, **49**. 597- 610.
6. Hussein, A.M., 1993, Ecological evaluation of some technologies for biologically improving sandy soil fertility and their effect on some soil-borne pests, Ph.D. Thesis, Ain Shams University, Cairo.
7. Ghabbour, S.I., Mikhail, W.Z.A., Rizk, M.A.,1977, *Rev. Eco. Biol. Sol.*, **14**, 429- 59.
8. Southwood, T.R.E., 1978, Ecological methods with particular references to the study of inssect populations, Chapman and Hall, London , p. 524.

Optimization of Natural Procedures Leads: Discovery of Mylotarg™, CCI-779 and GAR-936

MAGID ABOU-GHARBIA
Wyeth-Ayerst Research, Chemcial Sciences, CN 8000, Princeton, NJ 08543-8000, USA

1. INTRODUCTION

Drug Discovery and Development is a challenging and complex process that involves the dedicated multidisciplinary efforts of many R&D functions. Breakthroughs in innovation and process refinements have dominated drug discovery during the last decade, which have been aimed at increasing efficiencies and, thus, reducing cycle time. Despite these technological advances, the number of new chemical entities (NCEs) approved for human use has declined from an average of 53 NCEs during the 1980s to 41 NCEs per year in the 1990s[1,2]. At the turn of the century, the deciphering of the human genome will lead to an explosion in genomic technologies and, subsequently, to the identification and characterization of new targets for new small molecule ligands. Meeting these demands will require a well-integrated discovery effort and the application of many developmental functions at the earlier stages of the R&D process to increase the "hit rate" for identifying clinical candidates that reach NDA registration (Figure 1).

Optimization of the early leads into developmental drug candidates is the cornerstone of drug discovery and development. Several medicinal chemistry approaches have been utilized successfully to optimize initial leads identified via screening of natural products, compound libraries, rational and/or structure-based drug design[3,4]. This report will highlight some of our approaches that were successfully utilized in optimizing early natural product leads into drug candidates. Modification of calicheamicin, rapamycin and minocyclin provided the anticancer agents Mylotarg™ and CCI-779 and the broad-spectrum antibiotic GAR-936, respectively.

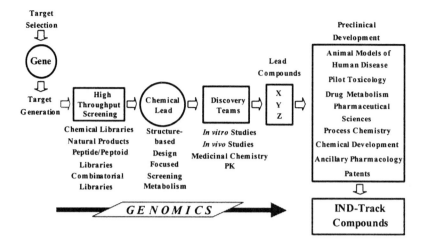

Figure 1. Drug Discovery Process

2. THE DISCOVERY OF MYLOTARG™

Calicheamicin γ^I_1(Gamma) is the parent of a family of extremely potent cytotoxic natural products. It was first isolated in 1982 from caliche soil samples and the structure was fully elucidated in 1987[5]. Calicheamicin γ^I_1 is more than 10,000 times as cytotoxic as adriamycin. This is apparently due to the remarkable DNA cleaving ability of this agent. Bioreductive cleavage of the enediyne moiety when bound to the minor groove of DNA leads to a transient p-phenylene-diradical which in turn initiates strand scission by abstracting proximal hydrogen atoms from the targeted deoxyribose sugars. This highly cytotoxic activity limits its use as a stand-alone anticancer agent and prompted us to explore an alternate approach to deliver it to the target tumor cells. The use of an antibody-targeted chemotherapy approach to target cancer cells expressing a specific antigen on their cell surface, while not affecting most normal cells and tissues, seemed to be a reasonable strategy. This approach required the identification and selection of an appropriate antigen, development of the corresponding humanized monoclonal antibody, and linking of the antibody to calicheamicin. This strategy was successfully applied[6] and resulted in the discovery and development of Mylotarg™ (1), as the first marketed antibody-targeted chemotherapeutic agent (Figure 2).

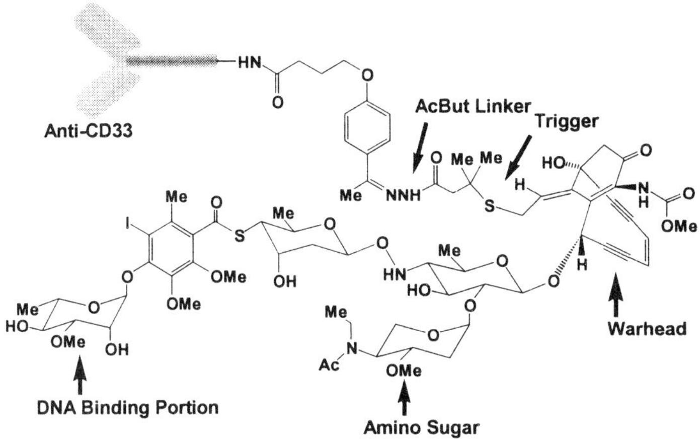

Stategies for Optimization of Natural Procedures Leads 65

Figure 2. Calicheamicin Conjugates

Mylotarg™ is the optimized antibody conjugate containing engineered human CD33 antibody attached bifunctional "AcBut" hybrid linker to N-acetyl calicheamicin γ_1^1 dimethylhydrazide (DMH). The antibody targets the DC33 antigen commonly found on the surface of acute myelogeneous leukemia (AML) cells. The resulting calicheamcin conjugate[6]/CD33 complex then becomes internalized and once inside the cell the calicheamicin derivative is released by hydrolysis inside acidic lysosomes and migrates, at least in part, to the nucleus. Calicheamicin binds to the DNA minor groove causing double-stranded DNA breaks and, ultimately, cell death. Although nuclear DNA appears to be the primary target it is possible that mitchondrial DNA or even certain cellular proteins may also be targeted. Mylotarg™ was successfully marketed in 2000 as an effective anti-tumor agent for the treatment of AML.

3. DISCOVERY OF CCI-779

Rapamycin (2) is a novel immunosuppressant natural product with unique mechanisms of action[7,8]. It binds to FKBP and forms a complex that binds with m-TOR (mammalian target of rapamycin) and inhibits cell cycle progression. Rapamycin was marketed in 1999 as Rapamune® for the treatment of transplantation rejection. A great deal of effort has been devqoted by us and many other researchers in order to determine the therapeutic potential of rapamycin, to identify novel analogs as back-ups for transplantation and to expand the therapeutic utility of this interesting class of natural products. The complex molecular structure of rapamycin has numerous sites of reactivity and instability (Figure 3). The molecule includes a 31-membered macrocyclic lactone, a hemiketal masked tricarbonyl unit, an all trans triene, an allylic alcohol, a β,γ-unsaturated ketone, 15 chiral centers, a retro-aldol site (C_{31}-C_{32}) and a segment susceptible to β-elimination (C_{25}-C_{26}). These unique structural features lend the rapamycin molecule to a multitude of chemical manipulations[9] which vary from alcohol functionalization, ketone manipulation, ring opening, contraction and expansion. These are just a few strategies as outlined in Figure 3. The C-42 alteration proved to be the most fruitful approach because of biological tolerability and synthetic feasibility. Such modifications represented no alteration in the core macrocycle and offer easy access for derivation of the C-42 alcohol, as it is more accessible than C-31 alcohol. Over 700 rapamycin derivatives were synthesized (Figure 4) and evaluated for their potential biological activity. The hindered ester (Cell Cycle Inhibitor; CCI-779) was selected from a short list of 31 novel rapamycin derivatives for further evaluation.

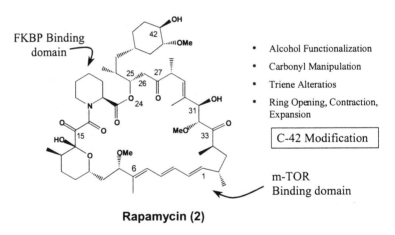

Figure 3. Rapamycin Chemical Manipulations

In pre-clinical pharmacological studies[10-13] CCI-779 (4) inhibited the growth of human tumor cells *in vitro* and in nude mouse zenografts *in vivo*. Phase II clinical evaluation is in progress. CCI-779 was readily synthesized from rapamycin a 4-step synthesis shown in Figure 5. This process involved acylation of rapamycin with the tri-chlorophenyl mixed anhydride (3), in presence of DMAP to afford a mixture of C-42 monoacylated and C-42/C-31 diacylated derivatives which were separated by chromatography. Deprotection of the C-42 monoacylated product diol intermediate afforded CCI-779 (4) in a 30% overall yield[13].

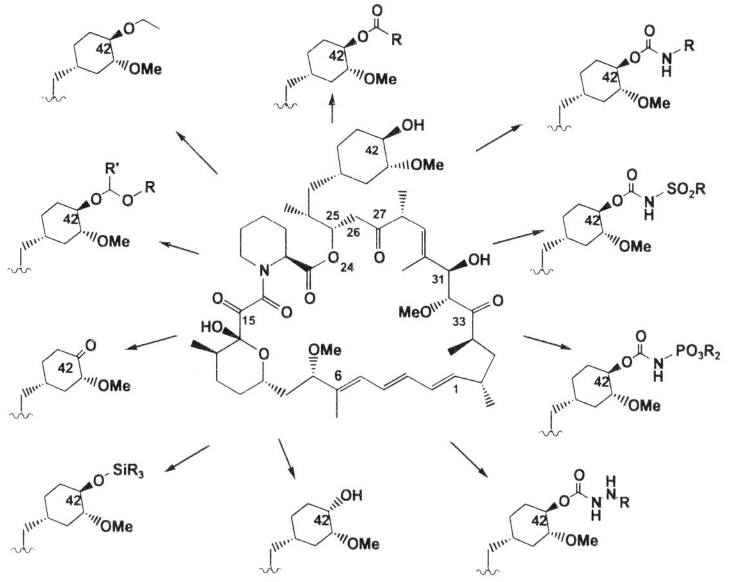

Stategies for Optimization of Natural Procedures Leads 67

Figure 4. Representation of Rapamycin C-42 Functionalization

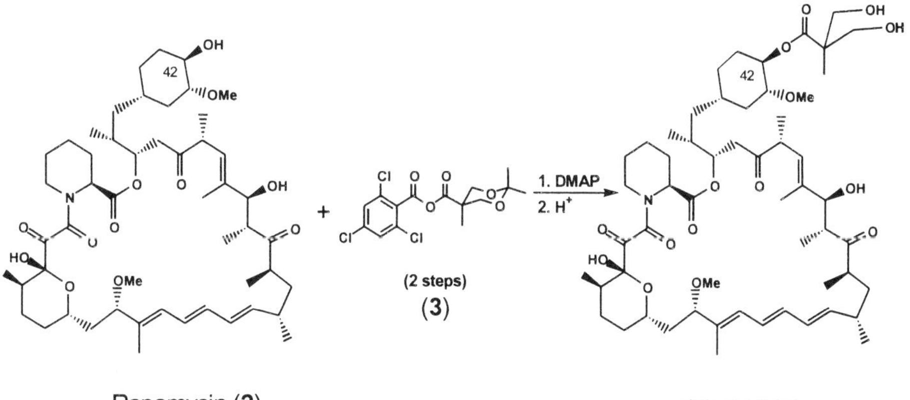

Figure 5. Synthesis of CCI-779

4. THE DISCOVERY OF GAR-936

The broad-spectrum tetracycline antibiotics, chlortetracycline (Aureomycin-Lederle), oxytetracycline (Terramycin-Pfizer) and tetracycline itself (Achromycin-Lederle), were discovered in the late 1940's and early

1950's.[14, 15]. However, the subsequent occurrence of organisms resistant to these agents led to a flurry of research activities in search of a new generation of natural and semisynthetic tetracyclines. However, no new clinically useful agents in this class have been reported since Doxycycline (in 1967) and Minocycline (in 1972, Figure 6).

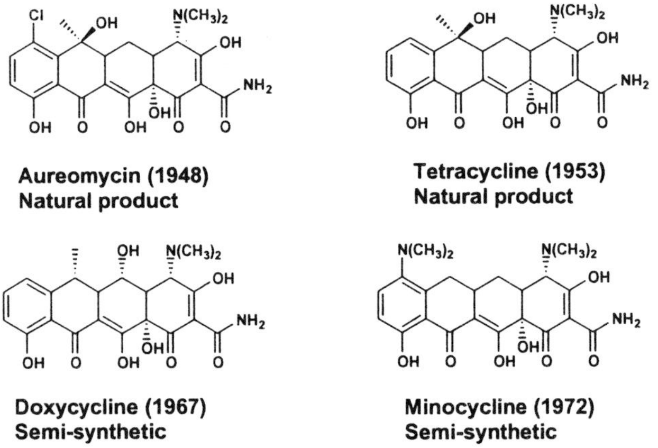

Aureomycin (1948)
Natural product

Tetracycline (1953)
Natural product

Doxycycline (1967)
Semi-synthetic

Minocycline (1972)
Semi-synthetic

Figure 6. The Tetracyclines

The structural complexity of the tetracycline nucleus, where many functionalities are required for antibacterial activity, has limited progress in this area (Figure 7). These structural requirements include β-carbonyl systems in the A and B/C rings, the 4-α-dimethylamine moiety, the D-ring phenol and the carboxamide. All these collectively are essential for the maximum spectrum and potency of antibacterial activity and present a semisynthetic challenge for medicinal chemists[16, 17].

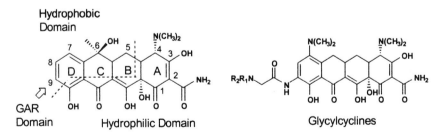

Figure 7. SAR in Tetracyclines

Figure 8. GAR-936
R_1 = H, R_2 = tBu (Phase III)

We embarked on semisynthetic efforts to design and synthesize a new generation of tetracyclines with an improved antibacterial spectrum including activity against bacteria resistant to the currently marketed tetracyclines and other antibiotics. Minocycline, the most active tetracycline antibiotic discovered, developed and marketed by our R&D group was used as a model for designing a new generation of tetracyclines. Our efforts were focused on modifying C-7 and C-9 positions of the 6-demethyl-6-deoxytetracycline nucleus[18,19]. These efforts led to a new class of tetracyclines known as "glycylcyclines", one of which, GAR-936 (Figure 8) was demonstrated to have broad spectrum antibacterial activity and represented a new class of tetracycline antibiotics against highly resistant pathogens. GAR-936 is currently under Phase III clinical investigations.

ACKNOWLEDGEMENTS

The author is indebted to many Wyeth researchers who are responsible for the discovery of these three breakthroughs, first-in-class drug candidates. Sincere thanks to my colleagues, Jerry Skotnicki, P-E Sum, Phil Hamman, Janis Upeslacis, Guy Carter, Hassan Elokdah, Boyd Harrison and George Ellestad for providing constructive feedback and suggestions to the content of the presentation; sincere thanks to Ms. Marie Arcaro for her tireless efforts during the preparation of the presentation and the manuscript.

REFERENCES

1. Gaudilliére, B., Bernz, P., 2000, *Ann. Report. Med. Chem.*, **35**, 331-52.
2. Graul, A.I., 1999, *Drug News & Perspectives*, **12**, 27-43.
3. Ashton, M.J., Jay, M. C., Mason, J. S., 1996, *Drug Discovery & Develop.*, **2**, 71.
4. Hugh, D., 1996, *Drug Discovery & Develop.*, **1**, 1.
5. Lee, M. D., Dunne, T. S., Siegel, M. M., Chang, C. C., Morton, G. O., Borders, D. B. 1987, *J. Am. Chem. Soc.*, 109, 3464-6 ; Lee, M. D., Dunne, T. S., Chang, C. C., Ellestad, G. A., Siegel, M. M., Morton, G. O., McGahren, W. J., Borders, D. B., 1987, *J. Am. Chem. Soc.*, 109, 3466-8.
6. Hinman, L. M., Hamann, P. R., Menedez, A. T., Wallace, R., Durr, F.E., Upeslacis, J., 1993, *Cancer Res.*, **53**, 3336-42.
7. Sehgal, S.N., Molnar-Kimber, K., Ocain, T.D., Weichman, B.M., 1994, *Med. Res. Rev.*, **14**, 1-22.
8. Molnar-Kimber, K.L., 1996, *Transplantation Proceedings*, **28**, 964-9.
9. For representative examples from Wyeth-Ayerst see:
   (a) Caufield, C.E., 1995, *Curr. Pharm. Design*, **1**, 145-60.
   (b) Nelson, F. C., Stachel, S. J., Mattes, J. F., 1994, *Tetrahedron Lett.*, **35**, 7557-60.
   (c) Nelson, F. C., Stachel, S. J., Eng, C. P., Sehgal, S. N., 1999, *Bioorg Med. Chem. Lett.*, **9**, 295-300.

(d) Steffan, R. J., Kearney, R. M., Hu, D. C., Failli, A. A., Skotnicki, J. S., Schniksnis, R. A., Mattes, J. F., Chan, K. W., Caufield, C.E., 1993, *Tetrahedron Lett.,* **34,** 3669-72.

(e) Chen, Y., Nakanishi, K., Merrill, D., Eng, C.P., Molnar-Kimber, K.L., Failli, A., Caggiano, T.J., 1995, *Bioorg. Med. Chem. Lett.,* **5,** 1355-8.

(f) Grinfeld, A.A., Caufield, C.E., Schiksnis, R.A., Mattes, J.F., Chan, K.W., 1994, *Tetrahedron Lett.,* **35,** 6835-8.

(g) Skotnicki, J.S., Kearney, R.M., 1994, *Tetrahedron Lett.,* **35,** 201-2; Skotnicki, J.S., Kearney, R.M., 1994, *Tetrahedron Lett.,* **35,** 201-2

(h) Skotnicki, J.S., Kearney, R.M., Smith, A.L., 1994, *Tetrahedron Lett.,* **35,** 197-200.

10. Yu, K., Toral-Barza, L., Discafani, C., Zhang, W. G., Skotnicki, J. S., Frost, P., Gibbons, J. J., 2001, *Endocrine-Related Cancers,* **8,** 249-58.

11. Yu, K., Zhang, W., Lucas, J., Toral-Barza, L., Petersen, R., Skotnicki, J., Frost, P., Gibbons, J., 2001, *Proc. Amer. Assoc. Cancer Res.,* **42,** 4305.

12. Gibbons, J. J., Discafani, C., Petersen, R., Hernandez, R., Skotnicki, J., Frost, P., 1999, *Proc. Amer. Assoc. Cancer Res.,* **40,** 2000.

13. Skotnicki, J. S., Leone, C. L., Smith, A. L., Palmer, Y. L., Yu, K., Discafani, C. M., Gibbons, J. J., Frost, P., Abou-Gharbia, M. A., 2001, *Clin. Cancer Res.* 7 (November 2001 Suppl): 37492; *Proc. AACR-NCI-EORTC International Conference*: Abstr. 477.

14. Duggan, B. M., 1948, *Ann. N.Y. Acad. Sci.,* **51,** 177.

15. Blackwood, R. K., Beereboom, J. J., Rennhard, H. H., Schach Von Wittenau, M., Stephens, C. R., 1963, *J. Am. Chem. Soc.,* **85,** 3943-53.

16. Hlavka, J. J., Ellestad, G. A., Chopra, I. T., 1992, *Encylopedia of Chemical Technology, 4th Edition,* John Wiley and Sons, NY, ,331-46.

17. Sum, P-E., Lee, V. J., Testa, R. T., Hlavka, J. J., Ellestad, G. A., Bloom, J. D., Gluzman, Y., Tally, F. P., 1994, *J. Med. Chem.,* **37,** 184-8.

18. Barden, T.C., Buckwalter, B. L., Testa, R. T., Petersen, P. J., Lee, V. J., 1994, *J. Med. Chem.,* **37,** 3205-11.

19. Sum, P-E., Sum, F-W., Projan, S. J., 1998, *Current Pharm. Design,* **4**(2), 119-32.

Bioactive Compounds from Some Endangered Plants of Africa

BERHANU M. ABEGAZ and JOAN MUTANYATTA
Department of Chemistry, University of Botswana, Private Bag 0022, Gaborone, Botswana

1. INTRODUCTION

Surveys conducted in a number of African countries, particularly in Ethiopia, Kenya, Uganda, Tanzania, and Botswana reveal the existence of thriving businesses based on the sale and utilization of unprocessed and semi-processed plant parts[1]. These businesses are generally located in traditional markets, in bus and railway terminals, near churches, mosques, etc. A number of these plants have become endangered because they are utilized in a non-sustainable fashion. The three most important features that contribute to this danger are: first, the plants are not cultivated and thus have to be collected from the wild; second, the underground parts are required and hence collection means total uprooting and, thirdly the rate of utilization of these plants is higher than their natural regeneration process.

Several years ago we became interested in the phytochemical investigation of marketed plants and we observed that the beneficial parts of the plants were sold, sometimes in powder form, and that the plants are generally referred to by their vernacular names. The immediate problem that was recognized at the outset was establishing the botanical names of these plants. For some, it was possible to refer to scientific monographs that provided the Latin binomials against the vernacular names. This approach, we soon discovered, could only be used with caution because some plants have local names that describe their traditional uses and this often leads to confusion since more than one plant can have similar uses. In other cases, we found that the same plant may have many vernacular names, and this heightened the need for more reliable method of identification. For some marketed plants, it was possible to obtain botanical specimens by asking the vendors to help collect the plants from their natural habitats. In other cases we decided to use the market samples for propagation and obtaining identifiable specimens from an experimental garden. More recently we were able to work together with colleagues in the Department of Biological Sciences of the University of Botswana to initiate micro-propagation techniques. This has opened up not only the possibility of generating botanical specimens but also to potentially contribute to conservation efforts through the provision of seedlings to nurseries and communities. This report will deal with studies of marketed plants belonging to the genera: *Bulbine, Scilla, Ledebouria,* and *Dorstenia*. The first three plants were obtained from a vendor in Gaborone while *Dorstenia* was sourced from the Central African country of Cameroon.

2. BOTANICAL DESCRIPTIONS ETHNOMEDIC USES

2.1 *Scilla nervosa* (Burch.) Jessop ssp. *rigidifolia*

S. nervosa (Hyacinthaceae) is a robust, smooth herbaceous geophyte, with a large bulb, globose and covered with a dense layer of bristles and old leaf bases. The leaves, usually 6-12, are rigid, erect, bright green and slightly hairy. The inflorescence is solitary, many-flowered raceme with many greenish-yellow flowers. There are 40 species in the genus *Scilla*, and these occur mainly in Europe, Asia, Southern and Eastern Africa, Middle East and some parts of the Northwest Africa[2]. *S. plumbea, S. flomifolia, S. natalensis* and *S. nervosa* are found in Southern Africa. A subspecies of the latter, *S. nervosa* sub. *rigidifolia* occurs in Botswana[3]. The Setswana and Kalanga names for this plant are *Jajane* and *Chinyame chikulu,* respectively. A vendor in Gaborone claims that *Jajane* can be used to treat infertility in women and the cold-water extract of the bulb has utility as a multi-infection remedy.

2.2 *Ledebouria graminifolia* (Bak.) Jessop (Hyacinthaceae)

This geophyte has a flowering stem arising from an ovoid bulb with a brown tunic. It has six light green leaves with dark green markings. The raceme is often found to have up to 30 crimson-maroon to purple flowers, which are paler at the edges and on the lower surfaces. It occurs in rocky areas. *Ledebouria* species are mainly found in Africa and in particular Southern Africa. The genus consists of 16 species, six of which are known to occur in Botswana. These are: *L. cooperi, L. hyacinthina, L. luteola, L. revoluta, L. undulata* and *L. graminifolia*[4]. The latter is an important medicinal plant used for cleansing of urinary tracts and for the treatment of heart diseases. A cup of the water infusion of the finely cut bulbs is drunk three times a day.

2.3 *Bulbine* species

These are succulent, caulescent, much branched, rhizomatous and caespitose or solitary geophytes[5]. They have hard roots, fused to form a tuberous, lobed organ with spreading attenuated ends. Stems are adorned with circular leaf scars, branching freely. Leaves are cauline and alternate, flesh with tubular basal sheath. With few exceptions, all species have yellow flowers and the filaments are bearded with yellow pointed or clavate hairs[6]. The genus *Bulbine* (Asphodelaceae) consists of 80 species found in Australia and Africa. Seven species are known to occur in Botswana *B. capitata* (V. Pollen), *B. abyssinica* (A. Rich) *B. frutescens* (L.) Willd, *B. asphodeloides* (L.) Willd, *B. angustifolia* (V. Pollen), *B. narcissifolia* Salm-Dyck, *B. tortifolia* Verdoorn. *Bulbine* species are used to treat various ailments that probably arise from bacterial and fungal infections. The leaf juice of *B. asphodeloides* is applied to wounds, and is also drunk as treatment for venereal diseases. The sap of *B. narcissifolia* is used for wart and corn remedy. Root preparations of *B. frutescens* are administered as enemas to reduce fever and the leaf sap is used for mosquito bites, cuts, grazes and burns.

2.4 *Dorstenia* species

Dorstenia (Moraceae) is a large genus (170 taxa) of herbs and small shrubs found mostly in wet forest environments[7]. The receptacles are usually plate-shaped, boat-shaped or linear and with or without elongate lobes at the plate margins. Male and female flowers are found on the same receptacle. The genus occurs in Asia, Africa and Central and South America[8]. *Dorstenia* plants are widely used in traditional medicine, for treatment of skin diseases (*D. barnimiana*), cough, headache and stomach pains (*D. kameruniana*), rheumatism and stomach disorders (*D. mannii*), rheumatism, headaches, stomach disorders and as snakebite remedy (*D. psilurus*).

3. CHEMISTRY AND BIOLOGICAL ACTIVITIES

3.1 Homoisoflavonoids

Homoisoflavonoids belong to a small class of C-16 flavonoids and are believed to be biosynthesised from chalcones and a one-carbon unit derived from methionine. Their occurrence is restricted to a few genera, namely:

	R^1	R^2	R^3
1	HO	MeO	HO
2	HO	MeO	MeO
3	HO	H	HO
4	MeO	H	HO

Eucomis, Scilla, Veltheimia, Drimiopsis, Ledebouria and *Muscari*. Our studies on marketed plants *S. nervosa* subsp. *rigidifolia*[9] and *L. graminifolia* led to the identification of several novel homoisoflavonoids. The former yielded thirteen homoisoflavonoids (**1-13**), of which nine, **2, 4-6, 8, 10** and **11-13** were previously unreported.

	R^1	R^2	R^3	R^4	R^5	R^6
5	H	MeO	H	MeO	H	MeO
6	H	MeO	H	HO	MeO	HO
7	H	HO	H	HO	HO	MeO
8	H	MeO	MeO	HO	MeO	HO
9	H	HO	MeO	HO	H	HO
10	H	HO	H	HO	MeO	MeO
11	H	MeO	HO	MeO	H	MeO
12	H	MeO	MeO	MeO	H	HO
13	HO	MeO	H	MeO	H	MeO
14	H	HO	H	HO	H	HO
15	H	OMe	H	HO	H	HO
16	H	HO	H	HO	H	MeO
17	H	MeO	MeO	HO	H	HO
18	MeO	MeO	H	MeO	H	HO
19	H	HO	H	HO	HO	HO

20 R = H
21 R = Me

22 R^1 = OMe R^2 = H
23 R^1 = OH R^2 = OH

This species is identified as a rich source of the 3-benzylidene-4-chromanone **1**, and the 3-benzyl-4-chromanone derivatives **5, 7,** and **9**. *Ledebouria graminifolia* also yielded 13 homoisoflavonoids (**9, 14-23**) of which five (**15, 17, 18, 22** and **23**) have not been reported previously (unpublished results). The bulbs of this plant were not as rich as the *Scilla* described earlier but were found to contain two representatives of the very rare scillascillin type homoisoflavonoids **22** and **23**. Nine of the homoisoflavonoids (**1, 5, 7, 9,15,17,18, 22** and **23**) were sent to the NCI, USA for anti-tumor evaluations and following the initial screen on three cell lines were subjected to the full assessment on 60 cell lines. Although the overall activity of all the compounds were not very high, compound **7** was found to be the most active with significant effects on a few colon and breast cell lines. Compounds **1, 5, 17, 18** and **23** had intermediate activity and **9, 15** and **22** were the least active.

3.2 Phenyl anthraquinones and isofuranonaphthoquinones

Knipholone (**31**) was first isolated from *Kniphofia foliosa* by Dagne and Steglich in 1984[13]. Subsequently derivatives of this novel compound, knipholone anthrone, isoknipholone (**32**) and foliosone (**36**), were reported in several species of the Asphodlelaceae[14]. Our investigations[15,16,17] on *Bulbine* have led to these and other new phenyl anthraquinones and have also revealed interesting biological properties for them[18,19]. *B. abyssinica*[15,18] yielded four isofuranonaphthoquinones (**24-27**) and six phenylanthraquinones **31, 32**, knipholone-6'-O-methylether (**33**), 4'O-demethylknipholone (**34**), bulbine knipholone (**35**) and foliosone (**36**). *B. capitata* was found to be a rich source of isofuranonaphthoquinones (**24-30**)[16]. The roots of *B. frutescens* yielded **31, 34**, the glycoside **37** and the novel phenyl aloe-emodin derivatives **38** and **39**. These phenyl anthraquinones display significant antiplasmodial activities with knipholone showing the highest activity, which is only slightly lower than that of chloroquine (IC_{50} 0.38 µM, cf chloroquine 0.09 µM)[18]. It is interesting to note that neither the upper anthraquinone part (chrysophanol) nor the lower half (phloroacetophenone derivative) show significant antiplasmodial activity, an observation that leads us to attribute the biological activity to the axially chiral molecular architecture of the phenylanthraquinones[18]. Recently Bringmann *et al* have reported the absolute configuration and the total synthesis of knipholone and derivatives[20].

	R¹	R²
24	H	H
25	OH	H
26	OMe	H
27	OH	OH
28	OH	OTigl
29	H	OAc
30	OH	OAc

	R¹	R²	R³	R⁴
31	Me	H	H	H
32	H	Me	H	H
33	Me	H	Me	H
34	H	H	H	H
35	H	H	Me	H
37	Gluc	H	H	H
38	Me	H	H	OH
39	H	Me	H	OH

36

3.3 Prenylated flavonoids

The genus *Dorstenia* yielded a variety of coumarins, mono-, di-, and tri-prenylated flavonoids. The chemistry of this genus has recently been reviewed. Our own work stems from investigations of *D. barnimiana* from Ethiopia, *D. kameruniana*, *D. manni*, *D. poinsetifolia*, *D. proropens*, *D. psilurus* and *D. zenkeri*, from Cameroon in Central Africa. *D. barnimiana* yielded styrene and benzofuran derivatives, which did not show significant antimicrobial activity against a number of common microorganisms[21]. *D. kameruniana* was found to be rich in mono- and di-prenylated chalcones and flavones. Some of the bichalcones have moderate activity against HL-60 promyelocytic leukaemia cells. *D. mannii* is characterized by the presence of a wide array of di-prenylated flavanones with the substituent prenyl groups having undergone modifications, hydroxylation, and cyclizations. Many of these substances have more potent antioxidant properties than quercetin[22]. *D. poinsettifolia* yielded the novel and rare group of dihydrocoumarins having a C-4 phenyl group (**40**)[23]. *D. psilurus* is typically characterized by benzofurans and tri-prenylated flavones with considerable structural diversity resulting from modifications of the prenyl groups. Of particular interest was dorsilurin E (**41**) with three cyclized prenyl substituents and ring B modified to a dienone functionality[24]. *D. zenkeri* yielded a number prenylated flavonoids including compound **42** which is believed to arise by an initial Diels-Alder reaction of a diene and dienophile followed by further cyclization to furnish the bicyclic ether. *D. proropensin* also yielded several prenylated compounds of which the most novel is the first example of a natural bisgeranylated chalcone (**43**)[25].

ACKNOWLEDGEMENTS

Financial support from the Research Office of the University of Botswana and from the International Programs in Chemical Sciences (IPICS) is gratefully acknowledged.

REFERENCES

1. Abegaz, B.M., Demissew, S., 1998, Field Survey of Indigenous African Food and Useful plants with special emphasis on medicinal plants in Eastern Africa, UNU/INRA, Ghana.
2. Mathew, B., Baytop, T., 1984, *Bulbous Plants of Turkey*, p.10, B.T. Batsford Ltd., London.
3. Arnold, T.H., De Well, B.C., 1983, Plants of Southern Africa: Names and Distribution, Natal Botanical Institute, Pretoria, p.88.
4. Hutchings, A., 1997, Zulu Medicinal Plants: An Inventory, University of Natal Press, Pietermaritzburg, pp. 501-2.
5. Barnes, J. E., Turton, L., 1994, A List of Flowering Plants of Botswana. In *the Herbarium at the National Museum*, Sebele and University of Botswana, The Botswana Society and the National Museum, Monuments and Art Gallery, p.47.
6. Hall, L. I., 1984, *South African J. Botany*, 3, 356-8.
7. Olorode, O., 1984, Taxonomy of the West African Flowering Plants, Longman, London, p. 45,
8. Satabie, B., 1985, *Flore du Cameroun Moraceae-Cecropiaceae*, p.74.
9. Silayo, A., Ngadjui, B.T., Abegaz, B.M., 1999, *Phytochemistry*, 52, 947-55.
10. Heller, W. Tamm, C., 1981, *Fortsch. Chem. Org. Natur.*, 40, 105-52.
11. Crouch, N. R., Bangani, V., Mulholland, D. A., 1999, *Phytochemistry*, 5, 943-6.
12. Pohl, T.S., Crouch, N.R., Mulholland, D. A., 2000, *Current Organic Chemistry*, 4, 1287-324.
13. Dagne, E., Steglich, W., 1984, *Phytochemistry*, 23, 1729-31.
14. van Wyk, S.E., Yenesew, A., Dagne, E., 1995, *Biochem. Syst. Ecol.*, 23, 277-81.
15. Bezabih, M-T., Motlhagodi, S., Abegaz, B. M., 1997, *Phytochemistry*, 46, 1063-7.
16. Bezabih M-T., Abegaz, B. M., 1998, *Phytochemistry*, 48, 1071-3.
17. Bezabih M-T., 1999, PhD Dissertation, University of Botswana, Botswana.
18. Bringmann, G., Menche, D., Bezabih, M-T., Abegaz, B.M., Kaminsky, R., 1999, *Planta Medica*, 65, 757-8.
19. Bezabih, M-T., Abegaz, B. M., Dufall, K., Croft, K, Skinner-Adams, T., Davis, T. M. E., 2001, *Planta Medica*, 67, 297-390.
20. Bringmann, G., Menche, D., 2001, *Angew. Chem. Int. Ed.*, 40, 1687-90.
21. Abegaz, B. M., Ngadjui, B. T., Dongo, E., Bezabih. M.-T., 2000, *Current Contents in Organic Chemistry*, 4, 1079-90.
22. Woldu, Y., Abegaz, B., Botta, B., Delle Monache, G., Delle Monache, F., 1988, *Phytochemistry*, 27, 1227.
23. Ngadjui, B.T., Kouam, S.F., Dongo, E., Kapache, G. W. F., Tamboue, H., Abegaz, B.M., 2000, *Phytochemistry*, 55, 915-9.
24. Ngadjui, B.T., Kapache, G.W.F., Tamboue, H., Abegaz, B.M., Connolly, J.D., 1999, *Phytochemistry*, 51, 119-23.

25. Ngadjui, B.T., Tabopda, T., Dongo, E., Kapche, G.W.F., Sandor, P., Abegaz, B.M., 1999, *Phytochemistry*, **52**, 731-5.
26. Abegaz, B.M., Ngadjui, B.T., Dongo, E., Ngameni, B., Nindi, M., Bezabih, M-T., 2001, *Phytochemistry*, (in press).

New Bioactive Substances Reported from the African Flora

ERMIAS DAGNE
African Laboratory for Natural Products (ALNAP), Department of Chemistry, Addis Ababa University, P.O. Box 30270, Addis Ababa, Ethiopia

1. INTRODUCTION

Africa is a continent rich in biodiversity. It is endowed with vast resources of plants used by the people since time immemorial as medicines, flavors and fragrances. Of the 150,000 species of higher plants believed to occur in tropical countries 30,000 species divided into about 2,500 genera are found in Africa.

Throughout Africa medicinal plants play an important role in the daily lives of millions of people and are of immense value in healthcare. These plants constitute important sources of novel compounds with diverse biological activities such as anti-cancer, anti-HIV, anti-fungal, anti-malarial, etc. The leading natural-product journals, namely *Journal of Natural Products, Phytochemistry, Planta Medica, Journal of Ethnopharmacology, Phytotherapy Research* etc as well as African scientific journals, dissertations, monographs, proceedings etc regularly report structure elucidation and biological activities of crude extracts and purified substances from African plants. As part of our activity to build a database on natural products, we regularly compile information on compounds reported from African plants[1]. We present in this paper the most outstanding novel bioactive compounds reported from the African flora in the years 2000 and 2001. Lead compounds of previous years were subject of a previous report[2].

2. ANTICANCER AND CYTOTOXIC COMPOUNDS FROM AFRICAN PLANTS

2.1 *Combretum. erythrophyllum* (Combretaceae) from South Africa

The South African willow tree, *Combretum caffrum*, is well known for its anti-neoplastic constituents prominent of which are combretastatin A-1 (**1**) and B-1 (**2**). A clinical trial of an analogue of the combretastatins was launched in 1998. Recently C. *erythrophyllum* was also found to yield compounds **3** and **4**, equally active glucosides of combretastatin A-1 and B-1 respectively[3].

1, R = H, Combretastatin A-1
3. R = Gluc, Combretastatin A-1 2' β-D- glucoside

2. R = H, Combretastatin B-1
4. R = Gluc, Combretastatin B-1 2' β-D- glucoside

2.2 *Commiphora erlangeriana* (Burseraceae) from Ethiopia

Novel cytotoxic (**5**) and cytostatic (**6**) named as erlangerin A and erlangerin D respectively were isolated by our research group from the resin of *Commiphora erlangeriana*, a plant occurring in Ethiopia and Somalia[4]. The structures of these compounds were elucidated on the basis of spectral evidence, chemical data, and X-ray crystallographic analysis. The latter compound is an analogue of the well known anti-cancer compound podophyllotoxin.

5, Erlangerin A   6, Erlangerin D

3. ANTI-HIV COMPOUNDS

3.1 *Monotes africanus* (Dipterocarpaceae) from Tanzania

The extract of *Monotes africanus* collected from Tanzania was found to be active in an anti-HIV *in vitro* primary screen[5]. Bioassay guided fractionation of the extract led to the isolation of the active compound, 6,8-diprenylaromadendrin (**7**). The presence of prenyl groups on the flavonoid core was found to be important for anti-HIV activity.

7, 6,8-Diprenylaromadendrin

3.2 *Croton tiglium* (Euphorbiaceae) from Egypt

The phorbol diesters 12-O-acetylphorbol-13-decanoate (**8**) and 12-O-tetradecanoylphorbol-13-acetate (**9**) isolated from the seeds of *C. tiglium* from Egypt were found to completely inhibit replication of HIV-1 on MT-4 cells at concentration (IC_{100}) of 0.48 and 7.6 ng/mL, respectively. These compounds exhibited minimal cytotoxic properties[6].

8, $R_1 = Ac,$ $R_2 = C_{10}H_{19}O$
9, $R_1 = C_{14}H_{27}O,$ $R_2 = Ac$

4. ANTI-LEISHMANIAL COMPOUNDS

4.1 *Stephania dinklagei* (Menispermaceae) from Ghana

Several oxoaporphine alkaloids isolated from the aerial parts of this plant exhibited significant activity against *Leishmania donovani* amastigotes[7]. N-methylliriodendronine (**10**) was found to be the most active substance. Leishmania is a debilitating disease prevalent in many countries of Africa.

10, N-Methylliriodendronine

4.2 *Corynanthe pachyceras* (Rubiaceae) from Ghana

Five indole alkaloids, corynantheidine, corynantheine, dihydrocorynantheine, α-yohimbine and corynanthine, isolated from the bark of this plant exhibited IC$_{50}$ values against *Leishmania major* promastigotes of 1 μM or below, thus placing them among the most potent leishmanicidal natural products known[8].

5. ANTI-MALARIAL COMPOUNDS

Malaria is a serious worldwide problem causing at least one million deaths out of 400 million cases reported each year.

5.1 Kigelia pinnata (Bignoniaceae) from Zimbabwe

The two naphtoquinones **11** and **12**, isolated from root bark of *K. pinnata* showed good anti-malarial activity[9].

11

12, Isopinnatal

5.2 *Strychnos icaja* (Loganiaceae) from Congo-Zaire

The above plant was a subject a recent report[10] dealing with the *in vitro* anti-malarial properties of the bisindole alkaloids strychnogucines A and B and 18-hydroxyisosungucine.

5.3 *Cryptolepis sanguinolenta* (Asclepiadaceae) from Guinea Bissau

During the isolation and anti-malarial evaluation of two new compounds cryptolepinoic acid (**13**) and methyl cryptolepinoate (**14**) from the above plant, the known alkaloid cryptolepine (**15**) was found to be the most active alkaloid present in the plant with IC$_{50}$ values comparable to that of chloroquine[11].

13, R= COOH, Cryptolepinoic acid
14, R= COOCH₃, Methyl cryptolepinoate
15, R= H, Cryptolepine

16, Knipholone

5.4 *Kniphofia foliosa* (Asphodelaceae) from Ethiopia

Knipholone (**16**), which has interesting structure and good anti-malarial activity with little cytotoxicity, was synthesised recently by Bringmann and Menche[12].

REFERENCES

1. In the ALNAP (African Laboratory for Natural Products) Database relevant published and unpublished documents, monographs, dissertations, conference proceedings etc, dealing with African plants are regularly entered with title, authors, keywords, abstract, biological activity, geographic area etc. The entries in the Database now exceed 12,000. As of September 2001, the ALNAP Database is on the Internet managed by the National Information Services Corporation (NISC) and can be accessed along with 13 other databases such as CAB Health Database, Medline etc. at www.nisc.com.
2. Dagne, E., 2000, New Biologically Active Compounds from African Plants. In *2000 Years of Natural Products Research Past, Present and Future* (Luijendijk, T.J.C., ed.), Congress Proceedings, Phytoconsult, Netherlands, pp. 1-13.
3. Schwikkard, S., Zhou, B-N., Glass, T.E., Sharp, J.L., Mattern, M.R., Johnson, R.K., Kingston, D.G.I., 2000, *J. Nat. Prod.*, **63**, 457-60.
4. Dekebo, A., Lang, M., Polborn, K., Dagne, E., Steglich, W., 2001, *J. Nat. Prod.*, (submitted).
5. Meragelman, K.M., McKee, T.C., Boyd, M.R., 2001, *J. Nat. Prod.*, **64**, 546-8.
6. El-Mekkawy., S., Meselhy, M.R., Nakamura, N., Hattori, M., Kawahata, T., Otake, T., 2000, *Phytochemistry*, **53**, 457-64.
7. Camacho, M.R., Kirby, G.C., Warhurst, D.C., Croft, S.l., Phillipson, J.D., 2000, *Planta Med.*, **66**, 478-80.
8. Staerk, D., Lemmich, E., Christensen, J., Kharazmi, A., Olsen,C.E., Jaroszewski, J.W., 2000, *Planta Med.*, **66**, 531-6.
9. Weiss, C.R., Moideen, S.V.K., Croft, S.L., Houghton, P.J., 2000, *J.Nat.Prod.*, **63**, 1306-9.
10. Frederich, M., De Pauw, M-C., Prosperi, C., Tits, M., Brandt, V., Penelle, J., Hayette, M-P., De Mol, P., Angenot, L., 2001, *J. Nat. Prod.*, **64**, 12-6.
11. Paulo, A., Gomes, E.T., Steele, J., Warhurst, D.C., Houghton, P.J., 2000, *Planta Med.*, **66**, 30-4.
12. Bringmann, G., Menche, D., 2001, *Angew. Chem.*, **40**, 1687-90.

Bioactive Components of a Peruvian Herbal Medicine, Chucuhuasi (*Maytenus amazonica*)

EMI OKUYAMA[1], K. SHIMAMURA[1], C. NAGAMATSU[1], H. FUJIMOTO[1], M. ISHIBASHI[1], O. SHIROTA[2], S. SEKITA[2], M. SATAKE[2], J. RUIZ[3], F. A. FLORES[3] and S. YUENYONGSAWAD[4]
[1]*Chiba University, Graduate School of Pharmaceutical Sciences, Chiba, Japan;* [2]*National Institute of Health Sciences, Tokyo, Japan;* [3]*Herbario Etnobotanico Amazonico, Iquitos, Peru;* [4]*Prince of Songkla University, Songkla, Thailand*

1. INTRODUCTION

With a view to the efficient and sustainable use of natural resources, we have studied the pharmacological effects and active components of traditional medicines to gain scientific understanding and develop rational use and also to obtain lead-molecules for conventional medicines.

Chuchuhuasi[1] is an important herbal medicine in the Amazonian basin in Peru and other South American countries and has been used as a general tonic and for the treatment of rheumatism, anti-inflammatory analgesia, cancer, etc. The stem bark of some *Maytenus* species (Celastraceae) including the title plant is utilized as this herbal medicine.

From the clinical use of Chuchuhuasi, we focused on Prostaglandin (PG) inhibition and anti-oxidative effect of the herb. PGs, especially PG E_2, are mediators of inflammation and pain, and are considered to play an important role in inflammatory diseases such as rheumatoid arthritis and hyperalgesia and in cancer. Recently, PG receptor subtypes, EP1~4, FP, DP, IP and TP were identified, and they showed that PGs have diverse physiological actions.[2] Therefore, the inhibition of PG receptors such as EP receptor subtypes, together with the inhibition of COX-1 and COX-2 in the biosynthetic pathways, can be estimated in the action mechanism of the herbal efficacy. On the other hand, free radicals and oxidative stress are involved in many diseases including inflammation, rheumatism and cancer, and this fact prompted us to follow another activity-guide for this plant. Radical scavenging and anti-oxidative drugs are beneficial in anti-rheumatic therapeutics and in cancer prevention and therapy as well.[3]

This proceeding showed the activity-oriented separation of a Chuchuhuasi, *Maytenus amazonica* Mart. by dual assay guides of the inhibition of PG E_1 and/or E_2-induced contraction in guinea pig ileum and radical scavenging activity in the 2,2-diphenyl-1-picrylhydrazyl (DPPH) method, and the active components of the plant.

2. ISOLATION OF THE ACTIVE COMPONENTS BY DUAL BIOASSAY GUIDES

Chuchuhuasi was purchased at the market in Iquitos, Peru, and was identified as *Maytenus amazonica* Mart. The stem bark (1.34 kg) was extracted with methanol at room temperature to give 145 g of the extract. The extract indicated some major, pale yellow spots with a reddish-purple background on TLC by DPPH-spray reagent, and also caused inhibition of PG E_1 and PG E_2-induced contractions in guinea pig ileum at the concentration of 1×10^{-4} and 3×10^{-4} mg/ml, respectively. The separation was performed by both activities as assay-guides.

After the partition with ethyl acetate, *n*-butanol and water, PG inhibition was observed mostly in the ethyl acetate fraction, but the activity was weak in the *n*-butanol fraction. The DPPH positive spots were also observed in both fractions. The ethyl acetate fraction was further separated by Sephadex LH-20 column chromatography to concentrate the PG inhibitory activity to fr. 1-A. On the other hand, the major spots by DPPH-spray were observed in fr. 1-B. The separation of fr. 1-B was then achieved by repeated chromatography on silica gel and Sephadex LH-20, and the major components, MA-1 and MA-2, having DPPH-radical scavenging activity were obtained. The PG inhibitory fraction, fr. 1-A, was separated by silica gel and ODS column chromatography, and from the active fractions MA-4 ~ MA-6 were obtained, although the potency of the inhibition did not increase. The separation of the *n*-butanol fraction, which showed weak activity by both assay-guides, proceeded to give MA-3 as a DPPH-positive compound, together with MA-1, -2 and –4.

Bioactive Components of a Peruvian Herbal Medicine 87

3. STRUCTURES OF THE COMPOUNDS

The structures of the isolated compounds were estimated by spectroscopic method, and were finally identified by the comparison with the published data. These are the first isolations from *M. amazonica*.

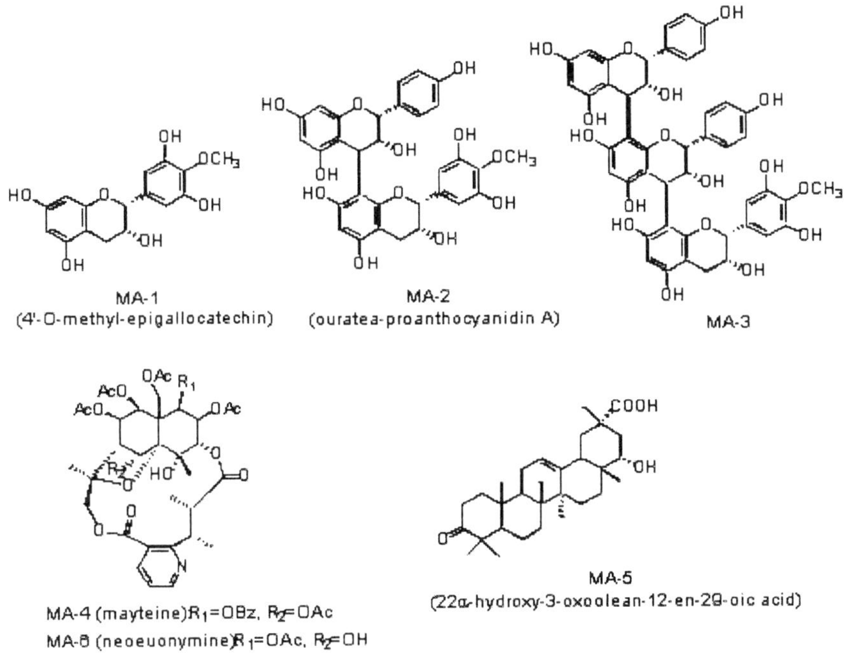

MA-1
(4'-O-methyl-epigallocatechin)

MA-2
(ouratea-proanthocyanidin A)

MA-3

MA-4 (mayteine) R_1=OBz, R_2=OAc
MA-6 (neoeuonymine) R_1=OAc, R_2=OH

MA-5
(22α-hydroxy-3-oxoolean-12-en-29-oic acid)

Figure 1. Structures of the isolated compounds

Among the DPPH-positive compounds, MA-1~-3, the former two were identified as 4'-O-methyl-epigallocatechin (ouratea-catechin) and ouratea-proanthocyanidin A, respectively. They have been already isolated from *Maytenus laevis*.[4] MA-3 has the molecular formula of $C_{46}H_{41}O_{12}$ by the high resolution FAB-MS spectrum, and the relative structure was estimated by 1D- and 2D-NMRs. The absolute structure was determined by the CD spectrum, which is superimposable on that of MA-2. The positive couplet at 200-220 nm indicated 4R-stereochemistry.[5] The structure shown in Fig. 1 has been already reported in *Heisteria pallida*.[6]

The compounds isolated from the PG inhibitory fraction, MA-4 ~ MA-6, were finally identified as sesquiterpene alkaloids, mayteine and neoeuonymine, and 22α-hydroxy-3-oxoolean-12-en-29-oic acid, respectively. Mayteine was isolated from other *Maytenus* sp. such as *M.*

guianensis[7], and neoeuonymine from *Euonymus sieboldianai*[8]. The oleanoic compound was also reported in *Tripterygium wilfordi*[9].

4. ACTIVITIES OF THE ISOLATED COMPOUNDS AND THE RELATIVES

The major DPPH positive compounds, MA-1 (4'-*O*-methyl-epigallocatechin) and MA-2 (ouratea-proanthocyanidin A), were evaluated for their ED_{50} value by the method of DPPH at 520 nm (butylated hydroxytoluene as a positive control, ED_{50} 17.7 ± 0.4 µg/ml). Both compounds gave ED_{50} 7.8 ± 0.5 µg/ml, and ED_{50} 11.4 ± 0.5 µg/ml, respectively. MA-3 was too labile for the value to be measured.

MA-4 (mayteine) at the concentration of 1×10^{-4} g/ml indicated strong PG inhibition on the contraction induced by both $PG\ E_1$ (3×10^{-7} M) and $PG\ E_2$ (1×10^{-7} M) in guinea pig ileum. MA-5 applied at 3×10^{-5} g/ml showed weak inhibition.

As many mayteine-related compounds have been isolated from Chuchuhuasi plants, some of them were tested for their PG inhibition. The compounds without a benzoyl at C-1 or a hydroxy group at C-4, such as euonymine and ebenifoline E-IV, did not show the inhibition at the concentration of 1×10^{-4} g/ml, while a wilforine-type compound such as euojaponine F showed the activity.

euonymine: R_1=OAc, R_2=OH
ebenifoline E-IV: R_1=OBz, R_2=H

euojaponine F

5. CONCLUSION

Chuchuhuasi (*Maytenus amazonica* Mart.) showed both PG inhibition and DPPH radical scavenging effects, which may be contribute to the efficacy of this herb, such as anti-inflammatory analgesia, anti-rheumatism and anti-cancer properties.

The radical scavenging effects are mostly caused by 4-O-methyl-epigallocatechin and the related compounds, while the PG inhibition in the Magnus method is ascribed to some sesquiterpene alkaloids such as mayteine, and possibly oleanoic acid derivatives. The COX-1 and COX-2 inhibition of flavan-3-ols was recently reported, although their activity seems to be less potent.[10]

These components in other *Maytenus* sp. similarly named Chuchuhuasi may have a role in their efficacy as well.

REFERENCES

1. Flores, F.A., 1999, Inventario Taxonomico de la Flora de la Amazonia Peruviana, Herbario Etnobotánico Amazónico, Iquitos-Peru, pp. 54; Duke, J.A., Vasquez, R., 1994, Amazonian Ethnobotanical Dictionary, CRC Press, London, pp. 114.
2. Abramovitz, M., Metters, K.M., 1998, Prostanoid Receptors, In *Annual Reports in Medicinal Chemistry-33* (A.M. Doherty, ed.), Academic Press, San Diego, pp. 223-32.
3. Bauerova, K., Bezek, A., 1999, *Gen. Physiol. Biophys.*, **18**, 15-20.
4. Gonzalez, J.G., Monache, G.D., Monache, F.D., Marini-Bettolò, G.B., 1982, *J. Ethnophrmacol.*, **5**, 73-7.
5. Barrett, M.W., Klyne, W., Scopes, P.M., Fletcher, A.C., Porter, L.J., Haslam, E., 1979, Plant Proanthocyanidins. Part 6. Chiroptical Studies. Part 95. Circular Dichroism of Procyanidins, *J. C. S. Perkin I*, 2375-7.
6. Dirsch, V., Neszmélyi, A., Wagner H., 1993, *Phytochemistry*, **34**, 291-3.
7. de Sousa, J.R., Pinheiro, J.A., Ribeiro, E.F., de Souza, E., Maia, J.G.S., 1986, *Phytochemistry*, **25**, 1776-.8.
8. Yamada, K., Sugiura, K., Shizuri, Y., Wada, H., Hirata, Y., 1977, *Tetrahedron*, **33**, 1725-8.
9. Kutney, J.P., Hewitt, G.M., Lee, G., Piotrowska, K., Roberts, M., Rettig, S.J., 1992, *Can. J. Chem.*, **70**, 1455-80.
10. Noreen, Y., Serrano, G., Perera, P., Bohlin, L., 1998, *Planta Med.*, **64**, 520-4.

Discovery of Natural Products from Indonesian Tropical Rainforest Plants: Chamodiversity of *Artocarpus* (Moraceae)

SJAMSUL ARIFIN ACHMAD, EUIS HOLISOTAN HAKIM, LUKMAN MAKMUR, DIDIN MUJAHIDIN, LIA DEWI JULIAWATY and YANA MAOLANA SYAH
Institute Teknologi Bandung, Department of Chemistry, Jalan Ganeca 10, Bandung 40132, Indonesia

1. INTRODUCTION

It is well known that Man has utilized the bioresources available in the environment for timber, food, clothes, fuel, and medicine. It is also well known that plants produce a bewildering array of natural chemicals many of which are used in modern health care. Consequently, many natural product chemicals produced by plants and microorganisms have been screened to identify compounds from which medicines for the treatment of malaria, cancer, AIDS, etc. might be developed.

Artocarpus is a group of tropical plants. It is one of the major genus of the plant family Moraceae, consisting of about 50 species, the greatest diversity of which is in Southeast Asia[1,2]. Many species of these plants have been used in many countries as folk medicine[3],.and in Indonesia some of the species have been used traditionally against inflammation, malarial fever, etc[4]. These plants also display a diverse range of secondary metabolites, including stilbenes, arylbenzofuranes, flavonoids, Diels-Alder type adducts, triterpenes and steroids. Many of these metabolites exhibited interesting and useful biological activities, including antitumor promoter, cytotoxic activities against human hepatoma PLC/PRF/5, human murine leukaemia P388 and KB cell-lines, inhibition of arachidonate 5-lipoxygenase and platelet agregation, and anti-inflammatory[5]. Thus, during the last couple of years, a systematic evaluation on a number of Indonesian species of *Artocarpus* had been carried out in our laboratory. The work involved selection of suitable plant materials followed by laboratory investigations for chemical isolation, characterization, structure elucidation and biological evaluation of the isolated compounds. We now report the discovery of these compounds including a number of novel compounds and their cytotoxicity against murine leukemia P-388 cell lines.

2. RESULTS AND DISCUSSION

Several species of *Artocarpus*, such as *A. champeden, A. lanceifolius, A. maingayi, A. altilis, A. rotunda, A. scortechenii, A. teysmannii,* and *A. bracteata*, collected from many parts of Indonesia, had been investigated. The appropriate tissues of each of the plants were extracted exhaustively by cold percolation with organic solvents, hexane and methanol, followed by liquid-liquid partitions with benzene, chloroform, acetone and finally with ethyl acetate. Chromatographic separations of the fractions thus obtained, using vacuum-liquid, column and centrifugal partition chromatography techniques, yielded pure compounds, the structures of which were determined based on UV, IR, ^{1}H and ^{13}C NMR, including 2D ^{1}H-^{1}H COSY, HMQC and HMBC, and MS spectroscopic analyses. While, cytotoxic assay against P-388 cell-lines was carried out using the cells supplied by the Japan Foundation for Cancer Research and maintained in RPMI-1640 medium supplemented with 5% fetal calf serum and kanamycin.

Artocarpus champeden **Spreng**. is a large evergreen tree, up to 20 m tall, common in lowland rainforest. Our investigation of this species yielded some new compounds, named cyclochampedol or artoindonesianin (1), artoindonesianins A (2) and B (3), and artoindonesian E (4), together with artocarpin (5), cycloartocarpin (6), chaplasin (7), heteroflavanone A (8), and artocarpanone (9)[6-9].

***Artocarpus lanceifolius* Roxb.** is a large evergreen tree, distributed in lowland rainforest of the islands of Sumatera, Bangka and Borneo. This species had also yielded three new novel flavonoids, named artoindonesianins G – I (10 – 12), together with five known compounds, namely artelastin (13), cycloartobiloxanthone (14), artonol-B (15), artelastochromene (16), and β-resorcylaldehyde (17)[10-12].

Artocarpus maingayii **King** is a medium-sized evergreen tree up to 40 m tall, commonly found in lowland forest of Sumatera. This species had also yielded one novel flavonoid, named artoindonesianin D (18) together with cudraflavone A (19), in addition to artocarpin (5), cycloartocarpin (6) and chaplasin (7)[13].

Artocarpus altilis **(Parkinson) Fosberg**. is a native of tropical Asia, and Indonesia is the centre of genetic diversity. It is up to 30 m tall and evergreen in the humid tropics. This species so far yielded a new stilbene derivative, named artoindonesianin F (20), together with artoindonesianin B

(3), cycloartobiloxanthone (14), artonol B (15), morusin (21), and artonin E (22)[14-15].

Artocarpus rotunda (**Hout**) **Panzer** is a large evergreen tree and common in lowland rainforest. This species yielded a new geranylated flavone derivative named artoindone-sianin L (23) together with an interesting quinonodihydroxanthone, artonin O (24), cycloartobiloxanthone (14), artonin M (25) and artonin E (22)[16].

Artocarpus scortechinii **King.** is a large evergreen tree, distributed in Sumatera and common in lowland rainforest. This species yielded artonin E (22) and norartocarpetin (26), together with two other flavonoids, namely artobiloxanthone (27) and 5`-hydroxycudraflavone A (28), in addition to a novel modified flavonoid derivative, a dihydroxanthone artonol A (29)[17].

Artocarpus teysmannii **Miq.** is a large evergreen tree up to 45 m tall, distributed in Sumatera, Borneo, Celebes and Molucus. This species yielded a novel modified flavonoid derivative, named artoindonesianin C (30), another prenylated flavonoid artonin J (31), cycloartobiloxanthone (14) and artonol B (15)[18-19].

Artocarpus bracteata **Hook.**, furthermore, yielded a new prenylated chalcone artoindonesianin J (32), in addition to kanzonol C (33) and carpachromene (34)[20,21].

3. BIOGENESIS OF FLAVONOIDS FROM *ARTOCARPUS*

The general features of the phenolic constituents isolated from several species of *Artocarpus* described above, revealed that this group of compounds may be differentiated into two classes of compounds, flavonoids and modified flavonoids. Biogenetically, it may be suggested that the first reaction step is the formation of flavone derivatives, such as norartocarpetin (26), followed by hydroxylation and prenylation of the flavone skeleton, thus form 3-prenylated flavone derivatives, such as artocarpin (5), morusin (21) and artonin E (22). This type of compounds serve as precursors for several different structural types, namely pyranoflavones, such as artoindonesianin (1) and cycloartocarpin (6), and oxepinoflavones represented by artoindonesianin B (3) and chaplasin (7)[22]. The first step toward the formation of the modified flavonoid class of compounds is the formation of the C-C bond between C-6` and the prenyl substituent at C-3, by oxidative cyclization reaction, to give dihydrobenzoxanthone molecules such as artobiloxanthone (27)[23]. These dihydrobenzoxanthone derivatives again serve as exclusive precursors for several other different structural types, namely furanodihydrobenzoxanthones, i.e. artoindonesianin A (2), quinonodihydroxanthones type of compound, i.e. artonin O (24), cyclopentenoxanthones, i.e. artoindonesia-nin C (30), dihydroxanthones, i..e. artonol A (29), and xanthonolides, such as artonol B (15)[24].

4. BIOLOGICAL ACTIVITIES OF FLAVONOIDS FROM *ARTOCARPUS*

The biological activities of some selected flavonoids so far isolated from the species of *Artocarpus* were evaluated against murine leukemia P-388 cell lines. The flavonoid type of compounds showed significantly potent cytotoxic activity, exemplified by artoindonesianin (1) (LC_{50} 0.2 µg/mL), artoindonesianin A (2) (21.0 µg/mL), artoindonesianin B (3) (3.9 µg/mL), artoindonesianin C (30) (6.2 µg/mL), artoindonesianin G (10) (0.7 µg/mL}, artoindonesianin H (11) (1.8 µg/mL), artoindone-sianin I (12) (1.8 µg/mL), artoindonesianin L (23) (0.6 µg/mL), artocarpin (5) (1.8 µg/mL), and artonin E (22) (0.06 µg/mL). While, the modified flavonoid class of compounds such as artonol B (15) showed weaker cytotoxic activity, LC_{50} >100 µg/mL. Thus, it may be concluded that the presence of an isoprenyl moiety attached to C-3 position together with free hydroxyl groups in ring B of the flavone skeleton are important factors for the cytotoxic activity. Some of these flavonoid derivatives also inhibited amino acid transport in *Bombyx mori* midgut[25].

5. CONCLUSIONS

A detailed study of chemical constituents of some *Artocarpus* species have led to a conclusion that *Artocarpus* produce two different major classes of phenolic compounds, namely flavonoids and modified flavonoids. Various types of compounds belong to the flavonoid group are characterized by interesting and unique features. Chemically, the molecular structures contain a C-γ,γ-dimethylallyl substitruent at position C-3 and hydroxyl groups at positions C-2`,4` or C-2`,4`,5` of ring B of the flavone skeleton. It is noteworthy that this unique structural features of the flavonoid group correlate very significantly with the potent cytotoxicity of these compounds against murine leukemia tumor P-388 cell lines.

On the basis of the above results, there is little doubt that many more chemicals of these two classes of compounds and other related structural variations will be disclosed when more *Artocarpus* species are examined. Further works are in progress.

ACKNOWLEDGEMENTS

This work was supported, in part, by a grant from the Graduate Team Research Grant Batch IV, 1999/2001, University Research for Graduate Education (URGE) Project, Ministry of National Education, Republic of Indonesia.

REFERENCES

1. Venkataraman, K., 1972, *Phytochemistry*, **11**, 1571-86.
2. Verheij, E.W.M., Coronel, R.E., (eds.), 1992, *Plant Resources of South-east Asia No.2, Edible Fruits and Nuts*, PROSEA, Bogor, Indonesia, pp. 79-95.
3. Perry, L.M., 1980, Medicinal Plants of East and Southeast Asia, MIT Press, Cambridge, pp. 269-71.
4. Heyne, K., 1987, The Useful Indonesian Plants, Sarana Wana Jaya Foundation, Jakarta, pp. 668-83.
5. Nomura, T., Hano, Y., Aida, M., 1998, *Heterocycles*, **47**(2), 1179-1205.
6. Achmad, S.A., Hakim, E.H., Juliawaty, L.D., Makmur, L., Aimi, N., Ghisalberti, E.L., 1996, *J. Nat. Prod.*, **59**(9), 878-9.
7. Hakim, E.H., Asnizar, Kurniadewi, Ghofar, T.A., Achmad, S.A., Aimi, N., Makmur, L., Mujahidin, D., Takayama, H., Tamin, R., 1999, *Proc. ITB,* **31**(2), 57-62.
8. Hakim, E.H., Fahriyati, A., Kau, M.S., Achmad, S.A., Makmur, L., Ghisalberti, E.L., Nomura, T., 1999, *J. Nat. Prod.*, **62**(4), 613-5.
9. Hakim, E.H., Aripin A., Achmad, S.A., Aimi, N., Kitajima, M., Makmur, L., Mujahidin, D., Syah, Y.M., Takayama, H., 2001, *Proc. ITB* , 2001 (in press).

10. Hakim, E.H., Achmad, S.A., Makmur, L., Syah, Y.M., 2000, *Proceedings of the 8th International Symposium on Natural Products Chemistry*, Karachi.
11. Mujahidin, D., Achmad, S.A., Syah, Y.M., Aimi, N., Hakim, E.H., Kitajima, M., Makmur, L., Takayam, H., Tamin, R., 2000, *Proc. ITB*, **32**(2), 41-6.
12. Syah, Y.M., Achmad, S.A., Ghisalberti, E.L., Hakim, E.H., Makmur, L., Mujahidin, D., 2001, *Fitoterapia*, 2001 (in press).
13. Hakim, E.H., Afrida, Eliza, Achmad, S.A., Aimi, N., Kitajima, M., Makmur, L., Mujahidin, D., Syah, Y.M., Takayama, H., 2000, *Proc. ITB,* **32**(1), 13-9
14. Adimurti, V., Makmur, L., Achmad, S.A., Aimi, N., Hakim, E.H., Kitajima, M., Mujahidin, D., Syah, Y.M., Takayama, H. , 2001, *Proc. ITB*, (in press).
15. Erwin, Hakim, E.H., Achmad, S.A., Aimi, N., Kitajima, M., Makmur, L., Syah, Y.M., Takayama, H., 2000, *Bull. Soc. Nat. Prod. Chem (Indonesia)*, **1**(1), 20-7.
16. Suhartati T., Achmad, S.A., Aimi, N., Hakim, E.H, Kitajima, M., Takayama, H., Takeya, K., 2001, *Fitoterapia*, (in press).
17. Ferlinahayati, Hakim, E.H., Achmad, S.A., Aimi, N., Kitajima, M., Makmur, L., 1999, *Proceedings National Seminar on Chemistry of Natural Products '99*, Jakarta, pp. 162-7.
18. Makmur, L., Syamsurizal, Tukiran, Syamsu, Y., Achmad, S.A., Aimi, N., Hakim, E.H., Kitajima, M., Mujahidin, D., Takayama, H., 1999, *Proc. ITB,* **31**(2), 63-8.
19. Makmur, L., Syamsurizal, Tukiran, Achmad, S.A., Aimi, N., Hakim, E.H., Kitajima, M., Takayama, H., 2000, *J. Nat. Prod.*, **63**(2), 243-4.
20. Ersam, T., Achmad, S.A., Ghisalberti, E.L., Hakim, E.H., Tamin, R., 1999, *J. Mat. Sains*, **4**(2), 172-7.
21. Ersam, T., Achmad, S.A., Ghisalberti, E.L., Hakim, E.H., Makmur, L., Syah, Y.M., 2001, *J. Chem. Res.*, (in press).
22. Aida, M., Yamagami, Y., Hano, Y., Nomura, T., 1996, *Heterocycles*, **43**(12), 2561-5.
23. Hano, Y., Aida, M., Shiina, M., Nomura, T., 1989, *Heterocycles*, **29**(8), 1447-53.
24. Aida, M., Yamaguchi, N., Hano, Y., Nomura, T., 1997, *Heterocycles*, **45**(1), 163-75.
25. Parenti, P., Pizzigoni, A., Hanozet, G., Hakim, E.H., Makmur, L., Achmad, S.A., Giordana, B., 1998, *Biochem. Biophys. Res. Comm.*, **244**, 445-8.

Seminal Findings on a Novel Enzyme: Mechanism of Biochemical Action of 4-Methylcoumarins, Constituents of Medicinal and Edible Plants

VIRINDER S. PARMAR[1,4], HANUMANTHARAO G. RAJ[2], ASHOK K. PRASAD[1], SUBHASH C. JAIN[1], CARL E. OLSEN[3] and ARTHUR C. WATTERSON[4]

[1]*University of Delhi, Department of Chemistry, Delhi-110 007, India;* [2]*University of Delhi, V.P. Chest Institute, Department of Biochemistry, Delhi-110 007, India;* [3]*Royal Veterinary and Agricultural University, Department of Chemistry, DK-1871 Frederiksberg C, Copenhagen, Denmark;* [4]*University of Massachusetts, Institute for Nano Science Engineering and Technology, Department of Chemistry, One University Avenue, Lowell, MA 01854, USA*

1. INTRODUCTION

Coumarin is a naturally occurring compound widely present in plants and microorganisms, it finds extensive use in perfumery industry. Although a large number of coumarin derivatives have been reported to have diverse biological activities[1,2], the main focus appears to be on the anticoagulant action of some of the coumarin analogues[3]. The extensive work carried out by Lake and coworkers[4] has highlighted the toxicity of coumarin compounds due to the double bond between C-3 and C-4 in them. These studies have established the formation of a 3,4-epoxide, the products arising out of the scission of the epoxide moiety and their interactions with cellular macromolecules resulting in the toxic action of the parent coumarins[4]. 4-Methylcoumarins which occur in the edible plant fenugreek (*Trigonella foenumgraecum*) and other edible and medicinal plants[5-12] show a wide range of biological activities and are relatively nontoxic even at higher doses[13-17]. For the past several years, we have concentrated on the mode of biological action of substituted 4-methylcoumarins with emphasis on the behavior of their hydroxy and acetoxy substituted analogues.

2. A NOVEL TRANSACETYLAE IN THE LIVER MICROSOMES

Our discovery of the novel enzyme, acetoxycoumarin: protein transacetylase originated form the observation that the acetoxy 4-methylcoumarins offered highest protection against the liver microsome catalysed aflatoxin B_1 (AFB_1)-DNA binding, followed by their hydroxy and methoxy derivatives[18]. These findings led to the preliminary inference that the acetoxy groups may be involved in the acetylation of lysine residues in the cytochrome P-450 active site similar to the action of chloramphenicol[19]. We then proceeded to examine the nature of the inhibition of the P-450-linked mixed function oxidases (MFO) including the epoxidation of AFB_1 by 7,8-diacetoxy-4-methylcoumarin (DAMC) which was used as a model acetoxy compound. DAMC when preincubated with liver microsomes, followed by the incubation with AFB_1 and DNA along with cofactor NADPH was found to produce the time dependant inhibition of AFB_1 binding to DNA; AFB_1-DNA binding was inhibited up to 90% by 90µM concentration of DAMC at 45 min of pre-incubation and neared total inhibition at 60 min[20]. Similarly, preincubation of rat liver microsomes with DAMC led to inactivation of CYP1A and CYP2B as demonstrated by inhibition of dealkylation of ethoxyresorufin and pentoxyresorufin[20]. These observations clearly indicated that DAMC caused microsome-catalysed suicidal inactivation of P-450-linked MFO. It was interesting to note that this drastic inhibition of MFO was nearly abolished by thiol blocking agents, such as *para* hydroxymercuribenzoate and iodoacetamide. Thus the involvement of a novel enzyme (containing an active thiol group) termed "transacetylase", more specifically "DAMC: protein transacetylase (TAase)" that can possibly transfer the acetyl groups from acetylated xenobiotics, such as DAMC to apoprotein of P-450 resulting in its mechanism-based inactivation was postulated. The inhibitory action of DAMC as described above was independent of the presence of NADPH in the preincubation mixture. These findings clearly ruled out the role of oxidative metabolism of DAMC in effecting the mechanism-based inhibition of P-450 catalysed AFB_1 epoxidation (measured as AFB_1-DNA binding). Mechanism-based inhibition of P-450 linked MFO by definition necessitates at least one catalytic P-450 cycle during or subsequent to oxygen transfer step when the drug is activated by oxidation to its inhibitory species[21]. Since DAMC without undergoing oxidative metabolism caused mechanism based inhibition, we described such an inhibition as pseudo-mechanism based inhibition as opposed to the classical inhibitors of P-450 which needed oxidative metabolism to elicit metabolite complexation with the enzyme leading to suicide inactivation[19].

An effort was made to examine whether any other protein in the P-450 cycle is modified by DAMC catalysed by microsomal TAase in a manner similar to P-450 linked MFO. Accordingly, liver microsomes were preincubated with DAMC, followed by the addition of cytochrome C and NADPH in order to assay NADPH cytochrome C reductase, another important enzyme participating in the microsomal electron transfer[22]. The catalytic activity of reductase was dramatically enhanced by 600% by 25 µM concentration of DAMC after 10 min of preincubation. Also, it was observed that DAMC, contrary to its inhibitory action on hepatic P-450 caused kinetically discernible hyperbolic activation of liver microsomal NADPH cytochrome C reductase[22]. The profound activation of reductase described above was totally abolished by PHMB or iodoacetamide as in the case of P-450-linked MFO. Subsequently we have found that the reductase activity is significantly activated at even 2µM concentration of DAMC. It is pertinent to record that Nadler and Strobel[23] observed activation of reductase catalytic activity when specific reductase lysine residues were modified using acetic anhydride. Such non-enzymatic activation of reductase is known, while our results demonstrated for the first time the possible enzymatic acetylation of the reductase. These observations strengthened the role of microsomal transacetylase in the possible transfer of acetyl groups of DAMC to receptor proteins, such as P-450 and NADPH cytochrome C reductase resulting in the dramatic alteration of their catalytic activities. An effort was also made to examine whether any cytosolic enzyme protein could be a target for transacetylase catalysed action of DAMC. Cytosolic glutathione S-transferase (GST) was indeed found to undergo mechanism-based inhibition due to DAMC.[24] An elegant assay procedure was developed for the rat liver microsomal DAMC: protein transacetylase (TAase) based on the irreversible inhibition of GST when the latter was preincubated with DAMC and microsomes, the extent of inhibition of GST under the said conditions represented quantitatively the catalytic activity of the transacetylase[24]. Using this assay procedure, rat liver TAase was characterized; the enzyme exhibited hyperbolic kinetics, Km and V_{max} values were obtained by varying DAMC and GST concentrations, thereby establishing the fact that the reaction of TAase with DAMC and GST is bimolecular[24].

Liver microsome catalysed inhibition of GST by DAMC was abolished by the thiol blocking agents PHMB and iodoacetamide as in the case with P-450-linked MFO and the reductase. One of the products of the transacetylase reaction, *i.e.* 7,8-dihydroxy-4-methylcoumarin (DHMC) was identified[24] and the effort to demonstrate transacetylase catalysed protein acetylation *perse* is being persued by analysis of the modified protein by MALDI-TOF mass spectrometry.

3. QUANTITATIVE STRUCTURE ACTIVITY RELATIONSHIP(QSAR) STUDIES WITH TAASE

We have investigated the specificity for TAase with respect to the number and position of the acetoxy groups on the benzenoid as well as the lactone rings of the coumarin system governing the efficient transfer of acetyl group(s) to proteins(s). In order to address this issue, coumarins bearing one acetoxy group separately at C-3 and C-4 positions and also 4-methylcoumarins bearing one acetoxy group separately at C-7, C-6 and C-5 positions were synthesized and their specificities to rat liver microsomal transacetylase were examined[25]. Also the activity of the diacetoxycoumarin, 7,8-diacetoxy-4-methylcoumarin (DAMC) was compared with those of the monoacetoxy coumarins listed above in the assessment of their specificity to TAase. Among the monoacetoxycoumarins, 3-acetoxycoumarin (3-AC) when used as a substrate for TAase showed negligible activity, while 7-acetoxy-4-methylcoumarin (7-AMC) exhibited maximum TAase activity, followed by 6-acetoxycoumarin, 5-acetoxycoumarin and 4-acetoxycoumarin. DAMC, the diacetoxycoumarin showed almost double the TAase activity as compared to that of 7-acetoxy-4-methylcoumarin[25]. These results conclusively proved that a high degree of acetyl group transfer capability is conferred when the acetoxy group on the benzenoid ring of the coumarin system is in closer proximity to the oxygen heteroatom[25]. Our attention was also focused on the importance of the pyran carbonyl group of acetoxycoumarins in determining the specificity for the TAase. For this purpose, we synthesized 7-acetoxybenzopyran and compared its activity with that of 7-acetoxy-4-methylcoumarin. The absolute requirement of the carbonyl group on the pyran ring of coumarin was exemplified by the observation that 7-acetoxybenzopyran was not a substrate for the TAase (unpublished data). The optimized structures of various acetoxycoumarins were analyzed with a view to offer explanation to the positional specificity of the acetoxy group(s) to TAase[25]. We observed that in DAMC, the C-7 acetoxy group helps the C-8 acetoxy group to orient itself most favorably towards the oxygen heteroatom of the coumarin moiety. The relative orientation of acetoxy group towards the oxygen heteroatom is maximum in the case of 7-AMC, while it is least in the case of 3-AC, this phenomenon presumably controls the TAase mediated transfer of acetyl group to proteins[25].

4. TAASE MEDIATED BIOLOGICAL EFFECTS OF ACETOXYCOUMARINS

The aforementioned property of liver microsomal transacetylase in effectively inhibiting P-450 linked MFO may prove useful in modulating the clastogenic/ mutagenic effects of chemicals. Accordingly, an effort was made to investigate the activity of various acetoxy 4-methylcoumarins to inhibit the genotoxic changes due to aflatoxin B_1 (AFB_1) by pretreating the rats with DAMC, followed by the administration of ^3H AFB_1 which resulted in a significant reduction in AFB_1 binding to liver DNA[26]. Several other acetoxy 4-methylcoumarins, such as 7-AMC, 5-acetamido-6-acetoxy-4-methylcoumarin and DHMC were tested for their efficacies to prevent AFB_1-induced micronuclei formation in lung and bone marrow cells. DAMC was found to be most effective, while the effects of monoacetoxycoumarins were approximately half that of DAMC; DHMC, the deacetylated product of DAMC was hardly effective in protection against AFB_1-induced genotoxicity[26]. Similar results were also obtained on the inhibitory action of acetoxycoumarins on benzene-induced genotoxicity[27]. These investigations highlighted that microsomal TAase mediated the biological action of acetoxycoumarins.

5. ISOLATION AND PURIFICATION OF DAMC: PROTEIN TRANSACETYLASE

The aforementioned observations strongly furnished the evidence of the existence of TAase in the microsomes of several tissues of rat, such as liver, lung and kidney. Several domestic animal species viz. buffalo, sheep, goat and dog were screened for hepatic TAase activity. Buffalo liver was found to contain TAase activity comparable to that of rat liver (unpublished data). Since TAase is a protein of unknown identity, isolation and purification of the enzyme in large amounts will be feasible from the livers of large animals and warrants the elucidation of physicochemical characterization of the enzyme protein. Hence, we set out to isolate and purify TAase from buffalo liver. For this purpose, microsomes from buffalo liver, obtained from the local slaughter house were prepared and solubilised using 1M phosphate buffer according to the published procedure[28]. The solubilized supernatant was subjected to chromatography on DEAE Sepharose, followed by gel filtration using Sephacryl HR-2000. The enzyme activity was detected in a sharp peak which was concentrated and resolved on SDS-PAGE, resulting in the appearance of one major band (stained with Coomassie brilliant blue) having molecular weight in the range of 63KDa. The product of buffalo liver

6. ANTIOXIDANT AND RADICAL SCAVENGING PROPERTIES OF ACETOXYCOUMARINS

As a part of our engagement in the study of chemistry and biological effects of 4-methylcoumarins, we have studied the mechanism of antioxidant action of substituted 4-methylcoumarins. A large number of 4-methylcoumarins bearing different functionalities were examined for their effects on NADPH-catalysed liver microsomal lipid peroxidation with a view to establish structure-activity relationship[29]. The dihydroxy and diacetoxy 4-methylcoumarins displayed remarkable ability to inhibit membrane lipid peroxidation[29,30]. DAMC and DHMC were found to possess extraordinary ability to terminate radical chain reactions and inhibit propagation of lipid peroxidation along with excellent radical scavenging potency[29]. The antioxidant action of DHMC is evident from the fact that the *ortho* dihydroxy system is able to form a highly stable radical through delocalization of the unpaired electron due to extensive conjugation, thus enabling such compounds to acquire the radical scavenging property. But it was not clear as to how a diacetoxy coumarin (e.g. DAMC) can efficiently scavenge the radical like DPPH and reactive oxygen radicals. Apart from the conversion of DAMC to DHMC by the action of a microsomal de/transacetylase as demonstrated by us[31], we postulated the conversion of DAMC to DHMC in the presence of free radical initiators, such as DPPH, oxygen radical, etc. through the formation of the reactive ketene species[29]. In order to explain the antioxidant action of diacetoxy 4-methylcoumarin (DAMC), pulse radiolysis studies were undertaken where DAMC and DHMC were separately reacted with the system generating azide radicals and the resultant transient spectra were recorded[32]. The transient spectra of both DAMC and DHMC displayed a peak at 410 nm characteristic of phenoxyl radical. The rate constants for the formation of phenoxyl radical from DHMC and DAMC were 34×10^8 $m^{-1}s^{-1}$ and 6.2×10^8 $m^{-1}s^{-1}$, respectively[32]. We proposed for the first time that the free radical mediated oxidation of DAMC may initially produce a radical cation that may lose an acetyl carbocation to produce a phenoxyl radical which is justified by the lower rate constant for the formation of the phenoxyl radical from DAMC. We concluded from the results that the mechanism of the antioxidant action of DAMC followed the pathway similar to that of DHMC involving the formation of an extensively resonance stabilized coumarin-phenoxyl radical[29].

6.1 Conclusions and future directions

Our studies have demonstrated that acetoxy 4-methylcoumarins are biologically active compounds whose actions are partly mediated through a unique membrane bound enzyme acetoxycoumarin: protein transacetylase (TAase). TAase action tantamounts to effect acetyl CoA-independent protein acetylation. Further studies on acetoxycoumarins warrant the following:

Elucidation of TAase protein characteristics: search for its inhibitors, aminoacid sequence, secondary and tertiary structures.

TAase mediated modulation of biosynthetic P-450 involved in the synthesis of steroids, prostaglandins, etc. by acetoxycoumarins which may disclose new targets.

Examine other acetoxy polyphenols for biological effects in the light of acetoxycoumarins.

Occurrence of TAase in microorganisms with a view to develop potential antimicrobial agents.

ACKNOWLEDGEMENTS

The financial support from DANIDA and INSET to our work in these areas is gratefully acknowledged. We are thankful to a large number of co-workers who are mentioned in the references for their committed and dedicated hardwork.

REFERENCES

1. Egan, D., Kennedy, R.O., Moran, E., Cox, D., Prosser, E., Thormens, R., 1990, *Drug Metab. Rev.,* **22**, 503.
2. Murray, R.D.H., Medez,J., Brown, S.A., 1982, *The Natural Coumarins*, John Wiley and Sons, New York.
3. Takata, J., Kasube, Y., Handa, M., Iwasak, H.S., 1999, *Biol. Pharm. Bull.,* **22**, 172.
4. Lake, B.G.,1999, *Fd. Chem. Toxicol.* **37**, 423.
5. Bhardwaj, D.K.,i, R.,Seshadri, T.R., Singh, R., 1977, *Indian J.Chem.,* **15B**, 94.
6. Khurana, S.K., Krishnamoorthy, V., Parmar, V.S., Sanduja, R., Chawla, H.L., *Phytochemistry,* **21**, 2145.
7. Parmar, V.S., Jha, H.N., Sanduja,S.K., Sanduja, R. , *Z. Naturforsch,,* **37B**, 521.
8. Parmar, V.S.,Singh, S., Rathore. J.S., 1984, *J. Chem. Res. S,* 378 .
9. Parmar, V.S., Rathore, J.S., Singh, S., Jain, A.K., Gupta. S.R., 1985, *Phytochemistry* **24**, 871.

10. Chawla, H.M., Chibber, S.S., 1977, *Indian J. Chem.*, **15B**, 492.
11. Gonzalez, A.G., Fraga, B.M., Hernandez, M.G., Luis., J.G., *Phytochemistry* **11**, 2115.
12. Venturella, P., Bellino, A., Piozzi, F., 1974, *Tetrahedron Lett.* **12**, 979.
13. Lake, B.G., Garg, T.J.B., Evans, J.G., Lewis, D.F.V., Bleamand, A., Hue, H.L., 1980, *Toxicol. Appl. Pharmacol.*, **9**, 11.
14. Takeda, S., Aburada, M., 1981, *J. Pharmacobio-Dyn.*, **4**, 724.
15. Tyagi, A., Dixit, V.P., Joshi, B.C., 1980, *Naturwissenschaften*, **67**, 104.
16. Deana, A.A., Stocker, G.E., Schultz, E.M., Smith, R.L., Crageo, E.J., Russo, H.F., Watson, L.S., 1983, *J. Med. Chem.*, **26**, 580.
17. Yang, C.H., Chiang, C., Liu, K., Peng, S., Wang, R., 1981, *Chem. Abstr.*, **95**, 161758.
18. Raj, H.G., Gupta, S., Biswas, G., Singh, S., Singh, A., Jha, A., Bisht, K.S., Sharma, S.K., Jain, S.C., Parmar V.S., 1996, *Bioorg. Med. Chem.*, **4**, 2225.
19. Halpert, J.R., 1981, *Biochem. Pharmacol.*, **30**, 875.
20. Raj, H.G., Parmar, V.S., Jain, S.C., Goel, S., Singh, A., Gupta, K., Rohil, V., Tyagi, Y.K., Jha, H.N., Olsen, C.E., Wengel, J., 1998, *Bioorg. Med. Chem.*, **6**, 1895.
21. Williams, D.A., 1995, In *Principles of Medicinal Chemistry* (Foye, W.O., Lemkeand, T.L., Williams, D.A., eds.), BI Waverly Pvt. Ltd., pp 83-140.
22. Raj, H.G., Parmar, V.S., Jain, S.C., Goel, S., Singh, A., Tyagi, Y.K., Jha, H.N., Olsen, C.E., Wengel, J., 1999, *Bioorg. Med. Chem.*, **7**, 369.
23. Nadler, S.G., Strobel, H. W., 1988, *Arch. Biochem. Biophys.*, **261**, 418.
24. Raj, H.G., Parmar, V.S., Jain, S. C., Kohli, E., Ahmad, N., Goel, S., Tyagi, Y. K., Sharma, S.K., Wengel, J., Olsen, C.E., 2000, *Bioorg. Med. Chem.*, **8**, 1707.
25. Raj, H.G., Kohli, E., Goswami, R., Goel, S., Rastogi, R.. C., Jain, S.C., Wengel, J., Olsen,C.E., Parmar, V.S., 2001, *Bioorg. Med. Chem.*, **9**, 1085.
26. Raj, H.G., Kohli, E., Rohil, V., Dwarakanath, B.S., Parmar, V.S., Malik, S., Adhikari, J.S., Tyagi, Y.K., Goel, S., Gupta, K., Bose, M., Olsen C.E., 2001, *Mutat. Res.*, **494**, 31.
27. Raj, H.G., Malik, S., Parmar, V.S., Kohli, E., Tyagi, Y.K., Rohil, V., Dwarakanath, B.S., Adhikari, J.S., Bose, M., Jain, S.C., Olsen, C.E., 2001, *Teratogen. Carcinogen. Mutagen.*, **21**, 181.
28. Dey, A.C., Rahal, S., Rimsay, R.L., Senciall, I.R., 1981, *Anal. Biochem.*, **110**, 373.
29. Raj, H.G., Parmar, V.S., Jain, S.J., Goel,S., Poonam, Himanshu, Malhotra, S., Singh, A., Olsen, C.E., Wengel, J., 1998, *Bioorg. Med. Chem.*, **6**, 833.
30. Raj, H.G., Sharma, R.K., Garg, B.S., Parmar, V.S., Jain, S. C., Goel, S., Tyagi, Y.K., Singh, A., Olsen, C.E., Wengel, J., 1998, *Bioorg. Med. Chem.*, **6**, 2205.
31. Raj, H.G., Parmar, V. S., Jain, S.C., Goel, S., Tyagi, Y.K., Sharma, S.K., Olsen, C.E., Wengel, J., 2000, *Bioorg. Med. Chem.*, **8**, 233.
32. Raj, H.G., Parmar, V.S., Jain, S.C., Priyadarsini, K., Mittal, J.P., Goel, S., Das, S.K., Sharma, S.K., Olsen, C.E., Wengel, J., 1999, *Bioorg. Med. Chem.*, **7**, 2091.

Medicinal Plants – A Source of Potential Chemicals of Diverse Structures and Biological Activity

BINA S. SIDDIQUI, FARHANA AFSHAN, MUNAWWER RASHEED, NADEEM KARDAR, SABIRA BEGUM and SHAHEEN FAIZI
H.E.J. Research Institute of Chemistry, University of Karachi, Karachi 75270, Pakistan

1. INTRODUCTION

Biodiversity is defined as the total diversity and variability of living thing and of the system of which they are a part. This covers the total range of variation in and variability among systems and organisms, at the bioregional, landscape, ecosystem and habitat levels, at the various organism levels down to species, population and genes. The variability that we observe among individuals (phenotype) results from the interaction of genetic differences (genotype) with their surrounding environments. Genetic diversity can be assayed directly by surveying the actual genetic material (i.e. the genotype). The genetic diversity occurs in the form of nucleotide variation within the genome sometimes causing a change in a given protein. Hence we observe chemical diversity through biosynthetic processes controlled by specific enzymes (proteins).

Further, the circulatory movements of essential life elements derived from atmosphere, ocean, soil and bedrocks in our ecosystem lead to biogeochemical cycles. The most important of these elements involved in biogeochemical cycles are carbon, hydrogen oxygen, nitrogen (from air) and potassium, calcium, magnesium, phosphorous, sulphur, iron, copper, zinc etc. (from soil and bedrock). These elements in various combinations constitute the building blocks of various organic compounds. The balanced recycling of these elements in the ecosystems is important for the maintenance of the equilibrium state of the biotic community and an ecosystem[1].

1.1 Biodiversity in Plants

The concept of biodiversity was known to the taxonomists of pre-Linnaean times, it was only after the Rio convention that biodiversity in plants received much attention. Biodiversity in plant kingdom offers a potential renewable natural resource for utilization in the discovery of bioactive molecule. Study and conservation of biodiversity in medicinal plant species, is of great relevance as far as their sustainable utilization is concerned.

1.2 Medicinal Plants in South-Asian Region

The south-Asian region, spreading over ten countries and bearing the highest mountain chain and one of the largest deserts on the earth is characterized by a wide spectrum of topographic, edaphic, climatic and other physical features. The number of medicinal plants in south Asian region is said to be between 3500-4000 species of higher plants. The number of plants with confirmed therapeutic value ranges around 400 only. Out of these, the plants yielding raw materials for the production of drugs and other health and cosmetic products, hardly exceed 150-160 in number.

While the concept of medicine from plants is not new, the enormous, unexplored chemical diversity available within plant kingdom holds tremendous promise for additional weapons in the continuing struggle against diseases.

1.3 Search for New Leads

The search for new leads involves an interdisciplinary effort and contribution of researchers working in different fields including ethnobotany, ethnomedicine, chemistry, pharmacognosy, pharmacology, biochemistry, modern medicine and pharmacy etc.

Considering the reputed medicinal and other biological significance a number of plant are currently studied in our laboratory. In the present talk, some new leads with potential biological significance will be illustrated. Variations observed in chemical structures in respect to different parts of the plant and in respect to the plants growing in different ecological zones are also described.

1.4 Constituents of *Azadirachta indica* A. Juss.

A large number of compounds have been isolated from various parts of this tree by us, many of which displayed important pest controlling properties[2-4]. More recent studies[5] led to the isolation of three new compounds from the leaves namely, desfurano-desacetylnimbin-17-one (**I**), meliatetraone (**II**), and melianol (**III**). It is noteworthy that **II** is the first octanortriterpenoid bearing the cleaved ring C of nimbin[6] while **III** has the previously unknown seven carbon side chain (mono-nor) in the nimbin skeleton. Hence **I** and **III** represent interesting biogenetic relationship with nimbin which has a four carbon furan ring at C-17 (i.e. a tetranortriterpenoid). The biosynthesis of **I** may be rationalized to proceed from **III**, a postulated biosynthetic precursor of deacetyl nimbin (or nimbin). Compound **II** on the other hand represents a tetranortriterpenoid lacking C-21, which is unprecedented.

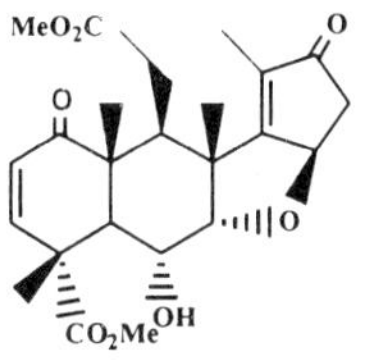

Desfurano-desacetyl nimbin-17-one (I)

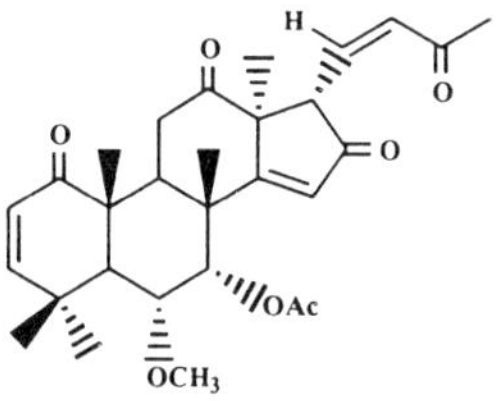

Meliatetraone (II)

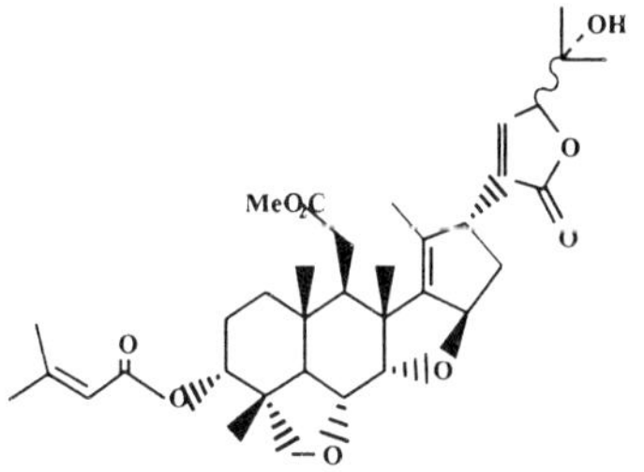

Melianol (III)

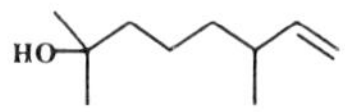

Dihydromyrecenol (2,6-dimethyl-7-octen-2-ol) (IV)

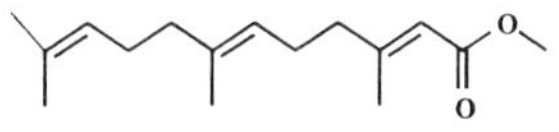

Methyl (2E, 6E)-farnesoate (V)

Figure 1. Monoterpenes and Sesquiterpenes of *Azadirachta indica*.

From the fruit coatings, five new terpenoidal constituents have been isolated and their structures elucidated. These include one monoterpene (**IV**), one sesquiterpene(**V**) and three triterpenes(**VI-VIII**)[7]. It may be noted that earlier to these studies, mono and sesqui-terpenes are very rarely reported from this source.

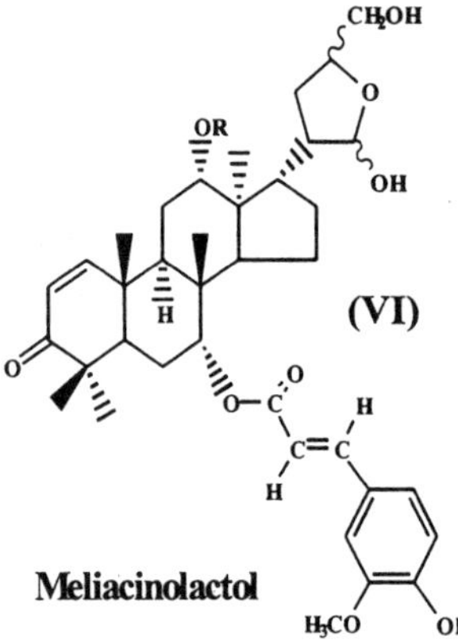

(VI)

Meliacinolactol

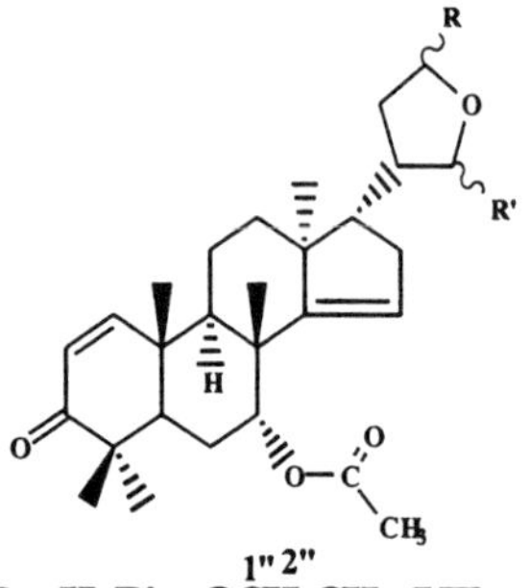

Limocin C R = H, R' = $OC^{1''}H_2C^{2''}H_3$ **(VII)**

Limocin D R = $OC^{1''}H_2C^{2''}H_3$, R' = H **(VIII)**

Figure 2. Triterpenes from *Azadirachta indica.*

1.5 Constituents of *Lawsonia alba* Lam.

Two new constituents representing interesting biosynthetic pathways have been obtained from the aerial parts of this plant. These are lawsonilin (**IX**) and lawsonicin (**X**). **IX** is derived from a flavonoid skeletal rearrangement.**X** represnts naphthaquinone adduct[8].

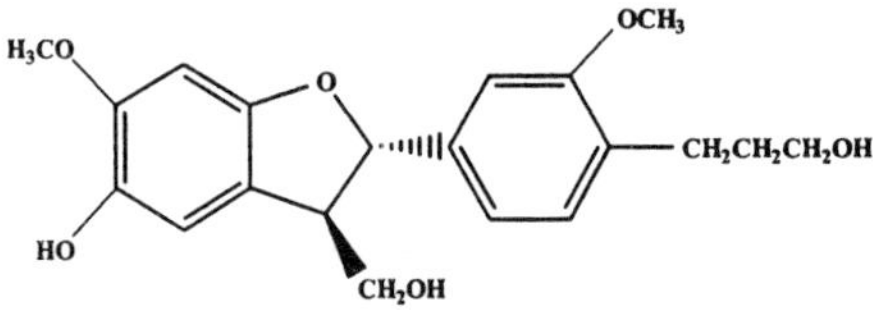

Lawsonilin (IX)

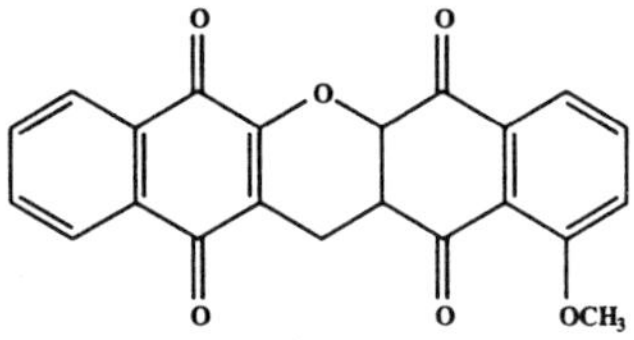

Lawsonicin (X)

Figure 3. Constituents of *Lawsonia alba.*

1.6 Constituents of *Piper nigrum* Linn

A plant of potential medicinal and pesticidal significance has recently been explored for its constituents possessing pesticidal activity.Several new compounds were obtained and their structures elucidated including compounds (**XI-XX**)[9]. These were tested against *Aedes agyptii* larvae (4th instar). Compounds **XI, XIII,** and **XV** are synthetic nitro derivatives prepared from piperine. It is interesting to note that nitration of piperine also resulted in the degradation product (**XI**) which was the most active compound in the series tested[10].

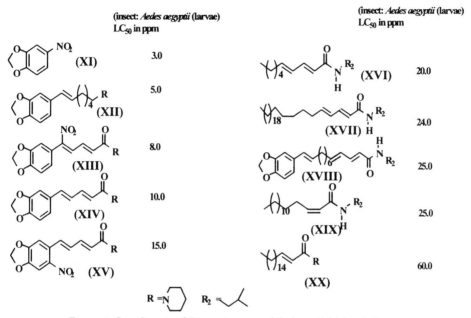

Figure 4. Constituents of *Piper nigrum* and their pesticidal activity

2. CHEMICAL DIVERSITY

Biogeochemical cycles and genetic diversity play an important role in the variability of chemical metabolites and their classes within the same plant species. This is very well manifested by the example of *Azadirachta indica* one of the most studied plants in our group as far as the chemistry is concerned. It contributes a stunning diversity of skeletal types of tri- and di-terpenoids from its different parts each decorated with a variety of substituents and skeletal rearrangements resulting during their biosynthesis, presumably due to genetic variations.

Chart I displays various skeletal modifications observed within the triterpenoids[11,12] and Chart II those of diterpenoids[13]. Table 1 illustrates the variations with respect to the parts studied and presents an interesting picture. Table 2 shows the difference in the fruit seeds and coatings (pericarp).

114

CHART-I **TRITERPENOIDAL SKELETA FROM**
AZADIRACHTA INDICA

Tirucallanes — C_8 Side chain

Apotirucallanes — C_8 Side chain

Tetranortriterpenoids — C_4 Side chain

Hexanortriterpenoids — C_2 Side chain

Octanortriterpenoids

Nononortriterpenoids

CHART-I -CONT... RING-C SECO TETRANORTRIPENOIDS

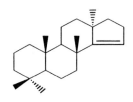

RING D SECOTETRANORTRITERPENOIDS

CHART-II DITERPENOIDAL SKELETA FROM A.INDICA

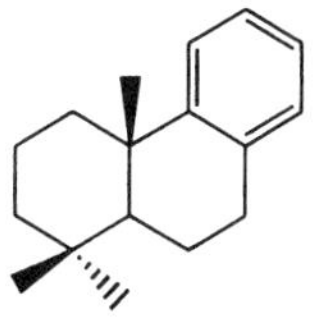

Podocarpanes

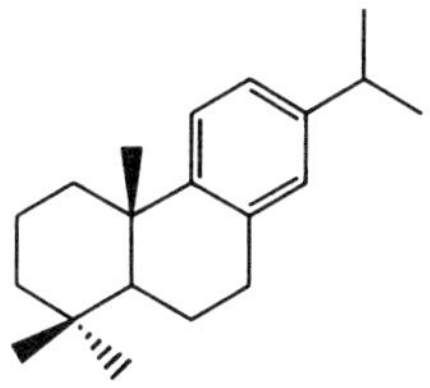

Abtietanes

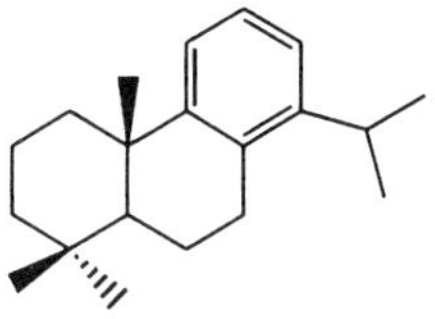

Totarol

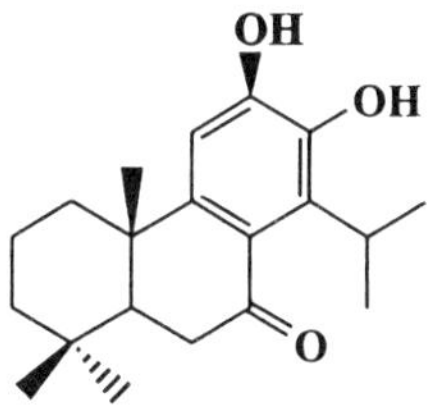

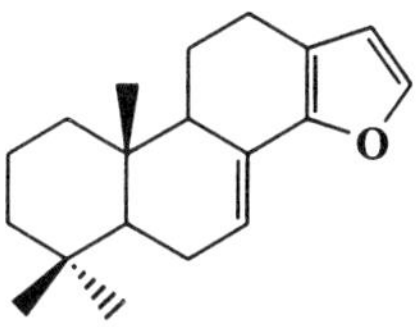

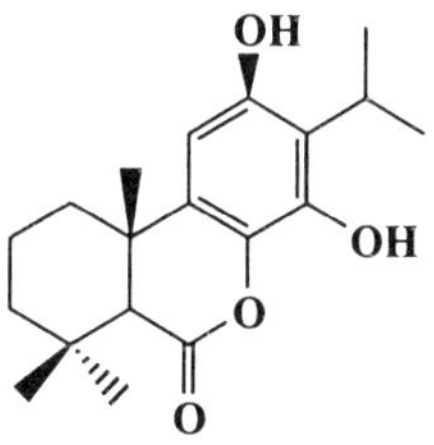

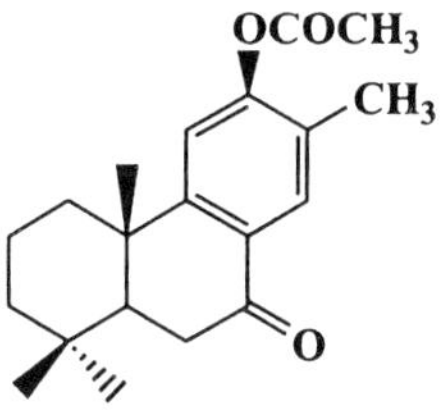

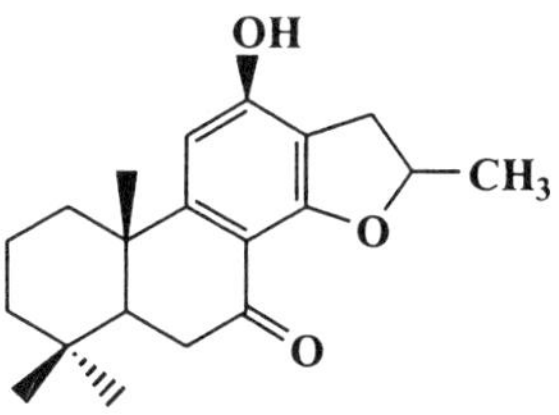

An important observation is the absence of azadirachtin, the most active pest controlling principle of the tree in the coatings.

As expected, effect of ecotype on the formation of plant metabolites is also very obvious (Table 3) from one of the studies undertaken on the evaluation of azadirachtin and salannin contents, two important bioactive constituents of *Azadirachta indica* growing in different ecological zones.

Table 1. Variations observed in the constituents of different parts of *Azadirachta indica*

	Leaves	Fruits	Coating	Twigs	Stem bark	Root bark	Oil
Triterpenoids							
Tirucallanes (Euphanes)	+	+	+	-	-	-	-
Apo-tirucallanes	-	+	+	-	-	-	-
Tetranortriterpenoids							
(All the rings intact)	+	+	+	+	-	-	+
(Ring C seco)	+	-	-	+	+	(1)+	-
(Ring D seco)	-	+	+	-	-	+	-
Hexanortriterpenoids	-	+	-	-	-	+(1)	-
Octanortriterpenoids	-	-	+	-	-	-	-
Diterpenoids							
Abietanes	-	-	-	-	+	+	-
Podocarpanes	-	-	-	-	+	+	-
Miscellaneous	-	-	-	-	+	+	-
Sesquiterpenes	-	-	+	-	-	-	-
Monoterpenes	-	-	+	-	-	-	-

Table 2. Major constituents of dried fruit seeds (FFS) and dried fruit coating (FFC)

SDS	SDC
Azadirachtin	-
Azadirone	Azadirone
-	7-Deacetyl-7α-Benzoylazadiradione
-	7-Deacetyl -7α Benzoylgedunin
-	7-Decetyl-7α-Benzoylepoxtazadiradione
3-Deacetyl-11-Desoxyazadirachtin	-
2', 3' – Dehydrosalannol	-
Deacetylnimbin	-
Epoxyazadiradione	Epoxy Azadiradione
Gedunin	Gedunin
Nimbin	-
Salannin	-

Table 3. Azadirachtin and Salannin content in Neem fruits of different regions of Sind Province (Calculated on dry weight basis of total fruit).

Sample No.	Azadirachtin %*	Salannin %*
XX	0.10	0.07
XXI	0.11	0.34
XXIV	0.11	0.32
II SB/DC	0.24	0.20
III	0.24	0.21
II	0.33	0.18
VI	0.46	0.22
XI	0.43	0.14
XI (NC) 55983	0.45	0.12
XII	0.43	0.12
XXII	0.48	0.31
XXIII	0.41	0.20
IV	0.54	0.27
VIII	0.56	0.10
XXV	0.83	0.42
V	1.07	0.42
XXVI	1.16	0.85
I	1.39	0.54
XXVIII	3.03	0.29

REFERENCES

1. Proceedings abstracts on regional training programme on *Biodiversity Systematic Evaluation and Monitoring with Emphasis on Medicinal Plants* held at National Botanical Research Institute, Lucknow, India, September 3-13, 2001 (sponsored by UNESCO).
2. Siddiqui, B.S., Afshan, F., Ghiasuddin, Faizi, S., Naqvi, S.N.H., Tariq, R.M., 1999, *J.Chem.Soc.Perkin Trans I,* **16**, 2367.
3. Siddiqui, B.S., Afshan, F., Ghiasuddin, Faizi, S., Naqvi, S.N.H., Tariq, R.M., 2000, *Phytochemistry,* **53**, 371.
4. Siddiqui, B.S., Rasheed, M., Ghiasuddin, Faizi, S., Naqvi, S.N.H., Tariq, R.M., 2000, *Tetrahedron,* **56**, 3547.
5. Siddiqui, B.S., Afshan, F., Faizi, S., 2001, *Tetrahedron*, **58**, 10281.
6. Harris, M., Henderson, R., McCrindle, R., Overtone, K.H., Turner, D.W., 1968, *Tetrahedron*, **24**, 1517.
7. Siddiqui, B.S., Rasheed, M., 2001, *Helv. Chim. Acta,* **84**, 1962.
8. Siddiqui, B.S., Kardar, N., Constituents of *Lawsonia alba* (Submitted).
9. Siddiqui, B.S., Gulzar, T., Begum, S., 2002, *Heterocycles,* (in press).
10. Siddiqui, B.S., Gulzar,T., Begum, S., *Nitration Studies on Piperine* (Submitted).
11. Siddiqui, S., Siddiqui, B.S., Faizi, S., Mahmood, T., 1988, *J. Nat. Prod.,* **51**, 30.
12. Siddiqui, S., Siddiqui, B.S., Ghiasuddin, Faizi, S., 1992, *Phytochemistry,* **31**, 4275.
13. Ara, I., Siddiqui, B.S., Faizi, S., Siddiqui, S., 1989, *J.Chem.Soc.Perkin Trans I,* 343.

Biodiversity in Turkish Folk Medicine

ERDEM YEŞİLADA
Gazi University, Faculty of Pharmacy, Department of Pharmacognosy, 06330 Ankara, Turkey

1. INTRODUCTION

Nature is a talented scientist capable to design and synthesize unique molecules that would never be achieved by man or even computers. Evaluation of these molecules to discover new leading medicines for human health is the supreme target of scientists in pharmacognosy. Certainly, it is impossible to screen or investigate thousands of natural sources for their possible biological effects.

As related to the recent progress in chemical and biological research techniques, especially after 1980's, field surveys have gained importance. In order to document more reliable information, new techniques were put forth and organized as a new science of "Ethnobotany", which provides data to facilitate drug discovery researches.

1.1 Is there a rich traditional medicine in Turkey?

Turkey is a country of rich flora. The number of higher plant species is over 10 000 and the ratio of endemism is also high, about 30%, which means about 3 500 of these plants grow only in Turkey. As a comparison, the number that reported for the continental Europe is about 11 000.

Furthermore, due to her geographical location, she becomes a bridge not only between the continents of Europe and Asia, also the east and west cultures. In her thousands years of history, dozens of civilizations reigned the peninsula from Hittites to Ionians, Greeks to Romans, Persians to Arabs. Finally a thousand year ago Turks migrated from Central Asia. Each civilization brought its own culture and was influenced by the local ones. During the Ottoman period, the borders of the empire extended to middle Europe, northern Africa, Middle-East and Caucasia. For 600 years Ottomans were influenced by the local cultures of these lands. After the collapse of the empire contracted to motherland, Anatolia, with its influenced culture. As a result of these historical events a great deal of cultural diversity exists in Anatolia.

Combination of these two factors, namely, *cultural diversity* and *rich flora*, are the main indicators for the existence of a rich traditional medicine in Anatolia.

1.2 Documentation of Turkish traditional medicine

Unfortunately, the number of scientific studies on Turkish traditional medicine is very poor. The book published by Prof. Baytop[1] is the only available scientific book on this subject. In this book utilization of plants were given in an encyclopedic format without distinguishing any regional difference. The author compiled the known information from the available references about the important medicinal plants including his personal notes obtained during his random surveys in Anatolia. There are also some papers published in the journals on this subject, but, unfortunately, only a few could be regarded as scientific and were localized only in a small fraction of the country and almost all were written in Turkish.

An extensive project was conducted in the years 1986, 1991, 1992, 1994 and 1995 in 55 provinces and 160 villages throughout Anatolia by our research team in order to document the traditional medicine in Turkey. Results were published in eleven consecutive papers in international and reputable journals[2-12]. Recently, observations obtained during these field studies are evaluated by using some techniques in semi-quantitative ethnobotany and published as a chapter in a serial book of eight volumes *(Recent Progress in Medicinal Plants)* in the volume entitled *"Ethnomedicine and Pharmacognosy"*[13].

2. "DATA BANK OF TURKISH FOLK MEDICINE" (TUHIB)

For the evaluation of the large number of information compiled during our field surveys mentioned above, we assembled a database system, which consist of 2 512 data. This data bank is then extended to present form, i.e. TUHIB, by entering the available knowledge assured from 24 scientific papers, which were published by various other Turkish scientists[14-37]. The name TUHIB is created from the Turkish name of data bank as "**TÜ**rk **H**alk **İ**laçları **B**ilgi **B**ankası", which means "Data Bank of Turkish Folk Medicine". The information involved in the Data Bank are selected according to the following criteria:

A. *Source of the information:* 1) The address of Herbarium and Herbarium number of the specimen should be described appropriately. 2) The author name of the specimen should be specified. 3) The Journal, which the information was published, should only accept papers after the peer review of referees.

B. *Type of information* : 1) Wild / cultivated plants used as remedy. 2) Wild / cultivated edible plants. 3) Plants used as animal food. 4) Animal-originated remedies. 5) Human-originated remedies. 6) Mineral-originated remedies. 7) Poisonous wild plants. 7) Ethnobotanical utilization of wild plants. 8) Imported Herbal drugs and spices supplied by Herb dealers are excluded unless they are employed as a component of a folk remedy.

As shown in the selection criteria, edible wild and economic plants as well as poisonous plants are also included in the database system. The main purpose for this is to accumulate information for the toxicity assessment of the folk remedies. If a folk remedy is recorded as a food in any of the districts, this may offer clues for the toxicity assessment of the remedy and may be evaluated as a safety proof in further scientific studies.

In each data form the following information are recorded: Herbarium number; Latin name and author names; Plant family; Informant's name and address (if available); Date of information; Locality of information (Province, Town, Village); Local name(s); Folkloric utilizations (parts used, preparation of remedy, classification of effect).

3. EVALUATION OF THE DATA IN TUHIB

The information in TUHIB is compiled from totally 69 provinces in Turkey. There are 4 757 data forms, 7 963 information, 2 444 vernacular names of 1 024 plant species from 112 plant families.

Plant-originated drugs used as remedy in Turkish traditional medicine are obtained from 107 plant families. Lamiaceae, Rosaceae and Asteraceae are the most frequently employed plant families. The following list of plant families is arranged according to the number of remedies obtained from each family in our database system (Table 1).

As shown in Table 1, in addition to plant species several animal species are used as folk remedy such as; bee, scorpion, snake, boar, snail, cow, sheep, trout, mole, fox and etc. As human-originated remedies, urine or spittle are used to treat some dermatological problems.

Table 1. Distribution of data according to origins (i.e. plant family, animal, mineral and human) and medicinal and non-medicinal utilizations

Family name	Total use	Non-medicinal uses	Medicinal uses
Lamiaceae	954	127	827
Rosaceae	914	97	817
Asteraceae	608	90	518
Plantaginaceae	375	4	371
Urticaceae	332	6	326
Pinaceae	275	11	264
Malvaceae	246	19	227
Liliaceae	230	43	187
Cucurbitaceae	226	5	221
Apiaceae	180	87	93
Solanaceae	174	7	167
Caprifoliaceae	164	5	159
Ranunculaceae	162	15	147
Fabaceae	163	50	113
Cupressaceae	155	11	144
Poaceae	132	19	113
Anacardiaceae	123	27	96
Hypericaceae	118	2	116
Polygonaceae	107	37	70
Araceae	100	5	95
Plant-originated	7380	874	6506
Animal-originated	129	4	125
Inorganic material	9		9
Human-originated	4		4
TOTAL	7522	878	6644

In order to evaluate the documented information, the data were first classified into two main categories according to medicinal (88.8 %) and non-medicinal (11.2 %) utilizations. Through the evaluation of 6 644 records in TUHIB, the rank order list of most common 20 ailments, which folk remedies are employed to treat are found as follows.

Table 2. Ailments treated with folk remedies

1. Hemorrhoids	451	11. Various gastric complaints	141
2. Rheumatism	376	12. Pain (to stop)	126
3. Abscess (for maturation)	374	13. Eczema	112
4. Stomach ache/peptic ulcer	310	14. Diarrhoea	109
5. Abdominal pain	288	15. Bronchitis	108
6. Wounds (for healing)	249	16. Worm in tooth/ eyes/	91
7. Cold, flu	192	17. Jaundice	76
8. Renal calculus	191	18. Cuts (as hemostatic)	70
9. Diabetes	164	19. Dermatological affections	66
10. Cough	144	20. Inflammation	63

Biodiversity in Turkish Folk Medicine 123

Remedies are further combined under 17 categories of indigenous uses based on the part of the human body affected by an illness or problem were distinguished. It is obvious that to set absolute borders for such a classification is very difficult. Most common types of disturbances are distinguished under following headlines.

4. MEDICINAL USES

4.1 Gastro-Intestinal disorders

The most frequently used traditional remedies are employed to palliate various gastro-intestinal ailments; *i.e.* 1451 (22.9 %) remedies classified under this heading are; *i.e.* ptosis, gastric descent, acute and chronic stomach complaints (141), peptic ulcers (98), stomach ache (212), flatulence (13), indigestion (33), hernia (1), against vomiting (11), as emetic (7), abdominal pain (288), food intoxications (7), spasms (2), colic (1), various intestinal problems (10), diarrhea (109), constipation (42), diseases of the rectum [i.e. hemorrhoids and fistulas (451)], various hepatic and liver complaints (10) and gallbladder problems [i.e. as choleretic (2), for biliary problems (10)] as well as those defined as digestive. Plants used to treat these types of ailments are listed below according to their order of utilization frequency.

Table 3. Gastro-Intestinal disorders

Plant name	Part	CF
Rosa sp. [*canina, montana*]	Fr, RB, R, L	168
Teucrium sp. [*chamaedrys, polium*]	Hb	70
Mentha sp. [*longifolia, piperita, pulegium, spicata*]	Hb	58
Malva sp. [*neglecta, nicaensis, sylvestris*]	Hb, L	51
Anthemis sp. [*austriaca, cotula, cretica, pseudocotula, tinctoria, wildemanniana, Bellis perennis, Matricaria recutita*	Fl, Hb	50
Origanum sp. [*dubium, majorana, minutiflorum, onites, vulgare*]	Hb	50
Achillea sp. [*biebersteinii, millefolium, schischkinii, setacea, wilhemsii, lycaonica*]	Fl, Hb, Fr	49
Thymus sp. [*atticus, longicaulis, praecox, pseudopulegioides, zygioides*]	Hb, VO	47
Hypericum sp. [*atomarium, lydium, olympicum, orientale, perforatum, scabrum, triquetifolium*]	Fl/Hb	42
Rubus sp. [*canescens, discolor, hirtus, sanctus*]	R, YS, Fr, L	40
Juniperus sp. [*communis, drupacea, excelsa, foetidissima, oxycedrus, sabina.*]	Fr, L, Tr, R	41
Salvia sp.[*aethiopsis, crypthanta, dichroantha, fruticosa, hypargeia, russelii, tomentosa, triloba, verticulata*]	Hb, VO	39

Urtica sp. [*dioica, membranacea, pilulifera, urens*]	L, Hb, R, S	39
Arum sp. [*balansanum, dentruncatum, elongatum, euxinum, hirta, italicum, orientale*]	R, Fr, L	37
Pinus sp. [*brutia, nigra, sylvestris*]	Re, Tr, Fr, Fm, Br, YS	31
Ecbalium elaterium.	Fr	27
Equisetum sp. [*arvense, palustre, talmateia*]	Hb	23

Abbreviations in Tables: Br:bark, Bn:branch, B:bulb, CF:citation frequency, Hb:herb, Fm:floem, Fl:flower, Fr:fruit, L:leaf, LX:latex, Re:resine, R:root, RB:rootbark, S:seed, Sk:stalk, St:stylus, Tr:tar, Tu:tuber, VO:volatile oil, YS:young shoot

4.2 Dermatological problems

1149 remedies (18.2 %) are used to treat this kind of problem. Traditional remedies used to treat infectious or inflammatory skin problems *i.e.* abscesses (to promote suppuration) (374), erysipelas (13), scrofula (1), allergic ailments (14), several dermatological problems including vesicles and rashes or to remove slivers/nails from the skin (66), to treat furuncle (2), eczema (112), pimples (4), as wound healing (249), for inflammatory wounds (42), gangrenous wounds (3), wound worms (2), sprains (10), bruises (17), cysts (1), burns or sunburns (45); to treat hangnail (1), against perspiration in foot (1), those employed as homeostatic (70), antiseptic (33) or as siccative for wounds/burns (17); as well as those to treat psoriasis (3), vitiligo (2), some hair problems [dandruff, allopecia (14), ringworm (2), oily hair (9)] and to remove wart (40) and callus (1) are evaluated under this title.

Table 4. Dermatological problems

Plant name	Part	CF
Plantago sp. [*major, lanceolata*]	L	282
Allium sp. [*cepa, sativum, ampeloprassum*]	B	56
Urtica sp. [*dioica, urens*]	Hb, R	44
Pinus sp. [*nigra, brutia, slyvestris*]	Re, Tr, YS	44
Malva sp. [*neglecta, sylvestris, nicaensis*]	L, Hb, R	42
Rubus sp. [*caesius, canascens, discolor, hirtus, sanctus, tereticaulis*]	L, R	40
Rosa sp. [*canina, montana*]	Fr, R, L	35
Hypericum sp. [*lydium, olympicum, perforatum, scabrum, tetrapterum*]	Fl/Hb	32

4.3 Lung and upper respiratory track ailments

601 remedies (9.5 %) used to treat various lung and upper respiratory track disorders; *i.e.* bronchitis (108), sinusitis (25), tonsillitis (3), sore throat (26), cough (144), lung disorders (3), colds and catarrh (192), shortness of

breath (36), asthma (42), hoarseness (3), snorring (1), nose bleeding (3) and those used as expectorant (15) are combined under this title.

Table 5. Lung and upper respiratory track ailments

Plant name	Part	CF
Ecbalium elaterium	Fr	55
Rosa sp. [canina, montana]	Fr, R, L	46
Pinus sp. [brutia, nigra, sylvestris]	Bn, Re, Tr, R, Fm, Fr, YS	39
Juniperus sp. [communis, drupacea, excelsa, foetidissima, oxycedrus, sabina]	Fr, YS, Tr	37
Tilia sp. [argentea, rubra]	Fl	33
Cydonia oblonga	L	27
Thymus sp. [longicaulis, praecox, pseudopulegioides, sibthorpii, transcaucasicus]	Hb	27
Salvia sp. [fruticosa, tomentosa, verticillata]	Hb	25

4.4 Skeleto-Muscular problems

466 Remedies (7.4 %) that are used to treat some skeleto-muscular disorders; *i.e.* rheumatoid arthritis and rheumatism (376), inflammation and edema (39), lumbago (11), osteoporosis (1), calcinosis (1), decalcificant for bones (9), to fix fractures (17), to treat paralysis or used as muscle relaxant (2), against muscular pain (5) are combined under this title.

Table 6. Skeleto-Muscular problems

Plant name	Part	CF
Urtica sp.[dioica, haussknechtii, pilulifera, urens]	Hb, L	102
Sambucus sp. [ebulus, nigra]	Hb, L, R	55
Ranunculus sp.[arvensis, caucasicus, cappadocicus, kochii muricatus, neopolitanus, repens, sericeus]	Hb, Fl	49
Helleborus orientalis	R	29
Juglans regia	L	14
Cistus laurifolius	L	13

4.5 Urological complaints

425 plant remedies (6.7 %) obtained from 92 plant genus to treat; calculi in the kidney and urethra (191), kidney inflammation (11), various kidney problems (12), dysuria (14), nocturia (7), cystitis (3), prostatitis (13) and those used as diuretic (46) are classified under this group.

Table 7. Urological complaints

Plant name	Part	CF
Equisetum sp. [*arvense, hyemale, palustre, talmateia*]	Hb	58
Zea mays	St	36
Helichrysum sp. [*arenarium, armenium, graveolens, orientale, pamphylicum, plicatum, sanguinea, stoechas*]	Fl	25
Ecbalium elaterium	R, Fr	16
Juniperus sp. [*oxycedrus, drupacea*]	YS, L, Fr, VO	15
Viburnum sp. [*opulus, tinus*]	Fr	10
Ceterach officinarum	Hb	10

4.6 Parasitic and microbial infections

360 plant remedies (5.7 %) obtained from 101 plant genus are employed for the treatment of various types of infectious diseases such as sinusitis (25), hepatitis (jaundice) (76), erysipelas (13), dermatophyte and lichen infections (22), fungal infections (11), malaria (16), mumps (5), dysentery, pneumonia (9), tuberculosis (22), leishmaniasis (1), gonorhea (5), herpes (2), aphtha (13), scabies (25), rabies (1), gonorrhoea, parasitic diseases (5) including those used as anthelmintic (66) are combined under this group.

Table 8. Parasitic and microbial infections

Plant name	Part	CF
Ecbalium elaterium	Fr	51
Pinus sp.[*brutia, nigra, slyvestris*]	Fm, YS, Re, Tr	20
Picea orientalis	Re, Fr	19
Cicer arientinum	S	13
Plantago major	L	10

4.7 Central Nervous-System related diseases

278 Remedies (4.4 %) obtained from 92 plant genus that are used to treat some CNS-related disorders; *i.e.* to reduce fever (46), sunstroke (47), pain or headache (126), neuralgia (1), dizziness (1), for epilepsy (3), to induce sleep or as sedative (28), or any condition, which might be related with the symptoms of depression (2), anaesthetics (1), to treat alcoholism (2), to treat paralysis (7) and tremor (1) and remedies attributed to possess central effects (13) are classified under this title.

Table 9. Central Nervous System-related disorders

Plant name	Part	CF
Solanum tuberosum	T	28
Teucrium sp. [*chamaedrys, polium*]	L	22
Urtica sp. [*dioica, urens*]	L	21
Ecbalium elaterium	Fr	16
Sambucus ebulus	L	15
Centaurea sp.[*calcitrapa, iberica, jacea, solstitialis* L.]	L, Fl	11
Juglans regia	L	7

4.8 Endocrine and metabolic diseases

176 remedies (2.8 %) obtained from 77 plant genus are classified under this title, such as; as hypoglycemic or to treat diabetes (164), as lactagoque (4), to treat goitre (9) and obesity (3).

Table 10. Endocrine and metabolic diseases

Plant name	Part	CF
Rosa canina	Fr	17
Rubus sp.[*canescen, hirtus, sanctus*]	R, L	14
Teucrium sp.[*chamaedrys, polium*]	Fl/Hb	13
Laurocerasus officinalis	S	11
Pyrus sp.[*amygdaliformis, elaeagnifolia*]	Fr	7
Myrtus communis	L, VO	7
Sorbus domestica	L	7

4.9 Immune system and neoplastic diseases

Since immunological problems could not be defined by lay people, 174 remedies (2.7 %) obtained from 70 plant genus, which are described as panacea (38), as tonic (46) or as prophylactic (3) or to treat cancer (66) and internal diseases (21) were combined under this title.

Table 11. Immune system and neoplastic diseases

Plant name	Part	CF
Urtica sp. [*dioica, pilulifera, urens*]	Hb, R	50
Rosa canina	Fr, R	13
Crataegus sp.[*microphylla, orientalis, pontica, pseudoheterophylla*]	Fr, R	11
Malva sp. [*neglecta, nicaensis*]	Hb, R	10
Pinus sp. [*nigra, slyvestris*]	Fr, Fm	9
Hypericum perforatum	Fl/Hb	8

4.10 Genital system problems

120 Remedies (1.9 %) obtained from 60 plant genus are used to treat gynaecological disorders such as; to regulate menstruation, amenorhoe and

dysmenorrhoea (12), various unspecified woman disorders (18), uterine inflammation (2), infertility (41), to induce abortion (27), for birth control (1) or as aphrodisiac (18) or to suppress libido (1) are grouped under this title.

Table 12. Genital system problems

Plant name	Part	CF
Solanum melongena	Sk	9
Malva sp.	R	8
Rosa canina	R	6
Rubus sp. [ideus, sanctus]	R, L	6
Arum incomptum	R	4

4.11 Folk illnesses

Ailments, which the symptoms could not be defined with the pathological conditions of orthodox medicine are classified under this title such as; vascular stifness (14), worm in tooth (65), eyes (24) or ears (2). There are 123 remedies (2.0 %) from 9 plant genus.

Table 13. Folk illnesses

Plant name	Part	CF
Hyoscyamus sp. [H. aureus, H. niger, reticulatus]	S	69
Datura stramonium L.		
Ulmus glabra.	RB, R	7
Scrophularia sp. [libanotica, striata, umbrosa]	Hb	3

Among the many records of folkloric utilization small worms with black head, which induce severe pain in eyes, ears or tooth is worth mentioning here as an outstanding folk illness. According to Townsend (1934), the belief of fictitious small worms that cause a cavity in tooth or eye diseases originated from Anatolia. The seeds of some Solanaceous plants *i.e.* *Hyoscyamus niger, H. reticulatus, H. aureus* or *Datura stramonium* are used for the treatment of such cases. Applicants put a pot of boiling water on a dying fire, then cover the pot and the fire with a big blanket. Seeds of the plant are spread on the fire and a patient with his head under the blanket inhales the vapor and smoke of the seeds through his mouth to relieve toothache, or exposes his eyes to heal itching or pain in eyes, or ears to stop ear-pain. It is said that during the treatment small worms with black head and white body come out of the patient's mouth, ears or eyes into the water pot. Of course such utilization seems completely nonsense, especially the informants claimed that the toothache or the eyesight of the patient was completely recovered after this application. It should be emphasized that the use of different plants from the same family for the same purpose is still present and applied in Anatolia. During our field studies, we had also chance to explain the etiology of this condition. The source of the worms is a fly, which is attracted by the sheep or cows. Especially during shepherding when the shepherd fall into sleep to spend time, or the peasants sleep in or near stalls, the fly laid egg either in his/her mouth inside a tooth cave (which is very common in rural parts), ears or eyes. These eggs give terrible pain to the patient and thus they apply this treatment to get rid of these worms.

Another interesting and widespread "folk illness" which is recorded in the vicinity of Ankara is *"damar tutukluğu"*. Literally means "vessel stiffness". They described the condition as having symptoms of pain and spasticity in extremities. The folk remedy prepared by boiling the stem/root barks or roots of *Ulmus glabra* in milk are used to soften and dissolve the obstruction in the blood vessels and eventually to stop pain.

"Kütnü" or *"tatlı yara"* are unspecified dermatological problems and may also be accepted as folk illnesses. *Colutea cilicica* and *Scrophularia* species are used in the treatment of these symptoms.

"To remove water from spleen" is another folk illness, which is mainly described with pain and feeling of swelling in abdomen and latex of *Goundelia tournefortii* is chewed for the treatment of this symptom.

4.12 Cardio-Vascular complaints

104 remedies (1.64 %) obtained from 43 plant genus are used to treat cardiovascular complaints; *i.e.* to treat cardiac problems (30), vascular problems (7), arteriosclerosis (20) or used as hypotensive (35), as vasodilatator (6), as blood diluent (1) or cardiotonic (3), or to lower cholesterol (2).

Table 14. Cardio-Vascular complaints

Plant name	Part	CF
Melissa oficinalis	Hb	15
Allium sativum	B	14
Crataegus sp. [*monogyna, orientalis, pontica, pseudoheterophylla*]	Fr	11
Rosa canina	Fr	8
Sorbus sp. [*torminalis, umbellata*]	Fr, L	8
Cirsium hypoleucum	L	6

4.13 Animal bite-repellent-anti repellent

Plants, which are employed to keep some insects or animals; *i.e.* mosquito-, snake-, flea-, moth-, etc., away from house or body (repellent) (9) or used as insecticide (2) or to attract especially bees for apiculture (antirepellent) (9), or those to treat scorpion bite (31), snake-bite (23), insect-bite (9) and animal-bite (3) are grouped under this category. In addition, a drop of the latex of *Euphorbia* species is dropped inside water in order to clean the spring water from larvae and insects. 86 records (1.4 %) from 32 plant genus are available.

Table 15. Animal Bite-Repellent-Antirepellent

Plant name	Part	CF
Allium sp. [*cepa, sativum, scorodoprassum*]	B	5
Euphorbia sp. [*macroclada, microshaera*]	LX	5
Laurus nobilis	YS, L	4
Achillea sp. [*biebersteinii, vermicularis*]	Fl/Hb	4
Cedrus libani	Tr	4
Pinus sp. [*brutia, nigra*]	Tr	4

4.14 Oral hygiene or oral diseases

Several plant remedies (47 records; 0.74 %) from 29 plant genus are used to palliate oral complaints, *i.e.* toothache (21), oral wounds (8), oral/tooth ailments (7), oral hygiene (6) or tooth care (5). It is noteworthy that plants used to treat toothache induced by worms are not classified here but under folk illnesses.

Table 16. Oral hygiene or oral diseases

Plant name	Part	CF
Plantago major.	S	9
Morus nigra.	Fr	5
Helleborus orientalis	R	4
Origanum sp. [*onites, vulgare, minutiflorum*]	Hb	4
Gundelia tournefortii	LX	4

4.15 Ophthalmologic and Otologic problems

Under this title totally 53 remedies (0.85 %) are recorded. 31 remedies obtained from 21 plant genus are employed to treat several ophtalmological complaints *i.e.* unspecified eye diseases (14), blepharitis (6), sore eyes (5), or blotshot (4) and sty (2) in eyes. While 23 remedies obtained from 15 plant genus are used to treat ear disturbances, *i.e.* otalgia and otitis (22).

Table 17. Ophthalmologic and Otologic problems

Plant name	Part	CF
Solanum tuberosum	T	8
Sambucus sp. [*ebulus, nigra*]	Fl	4
Helichrysum sp. [*arenarium, graveolens*]	Fl	7
Teucrium polium	Hb	3

4.16 Animal diseases

46 records (0.73 %) are used to treat various animal diseases. Among these type of cases inflammatory wounds or udder, edema in legs, against pain after castration etc. are classified.

4.17 Haematopoietic and lenfoid system

36 remedies (0.57 %) to treat anemia (11), increase blood (3), blood builder (2), thallisemia (1), spleen ailments (3) and against malaria (11) are categorized under this title.

Table 18. Haematopoietic and lenfoid system

Plant name	Part	CF
Rubus sp. [*canescens, discolor*]	Fr	7
Rosa canina	Fr	3

5. ETHNOBOTANICAL USES

5.1 Food

Among the wild plants in the database a total of 580 records (9.2 %) are reported to be used as food, either for human (432 records) or for animal (57). Additionally 91 records are employed as tea. It should be noticed that teas are also believed to good for health.

5.2 Other Types of uses

Other than above mentioned data, there are 129 records thos are described to be used for various purposes such as ethnobotanical (90), as dye (20), for agricultural/animal breeding (2). Meanwhile 9 wild plants are reported to be poisonous.

6. CONCLUSION AND REMARKS

The frequency and preference order of utilization of the plant remedies show variation depending upon several factors. The rank order list of plants that are employed most frequently in traditional medicine is listed in Table 2. Among the plants used in Turkish folk medicine, *Plantago* species (*P. major* or *P. lanceolata*) has a prominent place. The most common way of utilization of this remedy is the application of a fresh leaf on inflammatory wound to induce suppuration of abscess. This application is practiced all through Anatolia may be due to the widespread presence and abundance of this plant in nature. However, the other plants listed in Table also widespread in Anatolia, but their utilization purposes may vary according to location.

Table 19. Rank-order list of plant species according to their citation frequencies for the treatment of any ailment

Genus name	Total use	Main species
Plantago sp.	371	*major* (220), *lanceolata* (66)
Rosa sp.	354	*canina* (282), *montana*
Urtica sp.	327	*dioica* ( ), *urens* ( )
Sideritis sp.	194	
Malva sp.	188	*slyvestris* (18), *neglecta* (65), *nicaensis* (33)
Ecbalium elaterium	176	
Rubus sp.	175	
Pinus sp.	173	*nigra* (63), *sylvestris* (14), *brutia* (43)
Teucrium sp.	158	*polium* ( ), *chamaedrys* (17)
Sambucus sp.	147	*ebulus* ( ), *nigra* ( )
Juniperus sp.	146	*oxycedrus, sabina, drupacea, foetidissima*
Salvia sp.	145	*Triloba*
Allium sp.	140	*sativum* (42), *cepa* (65)
Thymus sp.	121	
Hypericum sp.	118	
Origanum sp.	97	
Hordeum vulgare	87	
Achillea sp.	86	*biebersteinii* (21)
Equisetum sp.	83	*talmateia* (76)
Ranunculus sp.	80	
Arum sp.	76	
Hyoscyamus sp.	76	
Pistacia sp.	77	
Quercus sp.	72	
Euphorbia sp.	66	
Anthemis sp.	65	
Helichrysum sp.	64	
Juglans regia	62	
Crataegus sp.	59	
Rumex sp.	56	
Solanum sp.	55	
Helleborus orientalis	54	
Verbascum sp.	54	
Zea mays	54	
Cydonia oblonga	50	

One of the most important outcomes brought up by TUHIB is the number of plant species employed in Turkish folk medicine. Due to the lack of information, the number of plant species employed as folk remedy in Turkish folk medicine was estimated about 500. However, through evaluation of the data accumulated in TUHIB, the number of wild / cultivated plant species employed as folk medicine in Turkey is concluded as 1.011. It should be taken into consideration that this number is obtained only from the available scientific studies have been published so far and entered in TUHIB. No doubt, does not reflect the exact number that will never be achieved, since information is lost rapidly due to the modernization of the society. Especially, with the influence of recent development of transportation and communication media, easier access to modern medicine, migration from villages to cities and several other factors, the knowledge of Turkish traditional medicine is faced to be lost. Actually the young generation is reluctant to transmit this knowledge and eventually this unregistered information is lost with the death of informants. Moreover due to the widen settlement areas in countryside, obtaining the wild plant species have become gradually difficult and this situation cause people to become reluctant to use folk remedies.

Even under these circumstances, scientists have still managed to collect plenty of knowledge by detailed small-scale studies i.e. in a town or province. It should be emphasized that this kind of extensive and scheduled field studies should be performed throughout Turkey within near future.

No doubt, reliability of the information extracted from the database is extremely important. It is reasonable to note that reliability is dependent upon the increasing number of citation. Actually, this is more dependent to the method employed for the collection of information. Since various scientists employ different methods, at least for the evaluation of the results from TUHIB, this parameter is worthwhile. Succeeding experimental studies have revealed that for the information collected during our field surveys, this threshold of reliability may be decreased to lower citation values, even once. For example, during our field surveys in south Anatolia, we recorded that *Spartium junceum* and *Cistus laurifolius,* were used for the treatment of stomachache in only one location. However, following in vivo studies showed that both to possess very potent antiulcerogenic activity and the active components were then chemically defined through bioassay-guided fractionation[38,39].

REFERENCES

1. Baytop, T., 1999, Türkiye'de Bitkiler ile Tedavi, Geçmişte ve Bugün (Therapy with medicinal plants in Turkey, Past and Present), 2nd. ed., Nobel Tıp Kitapevi, İstanbul, p.480.
2. Fujita, T., Sezik, E., Tabata, M, Yeşilada, E., Honda, G., Takeda, Y., Tanaka, T., Takaishi, Y., 1995, *Economic Botany*, **49**, 406-22.
3. Honda, G., Yeşilada, E., Tabata, M., Sezik, E., Fujita, T., Takeda, Y., Takaishi, Y.,Tanaka, T., 1996, *Journal of Ethnopharmacology*, **53**, 75-87.
4. Sezik, E., Tabata, M., Yeşilada, E., Honda, G., Goto, K., Ikeshiro, Y., 1991, *Journal of Ethnopharmacology*, **35**, 191-6.
5. Sezik, E., Zor, M, Yeşilada, E., 1992, *International Journal of Pharmacognosy*, **30**, 233-6.
6. Sezik, E., Yeşilada, E., Tabata, M, Honda, G., Takaishi, Y., Fujita, T., Tanaka, T., Takeda, Y., 1997, *Economic Botany*, **51**, 195-211.
7. Sezik, E., Yeşilada, E., Honda, G., Takaishi, Y., Takeda, Y., Tanaka, T., 2001, *Journal of Ethnopharmacology*, **75**, 95-111.
8. Tabata, M., Sezik, E., Honda, G., Yeşilada, E., Fukui, H., Goto, K., Ikeshiro, Y., 1991, *International Journal of Pharmacognosy*, **32**, 3-12.
9. Yeşilada, E., Honda, G., Sezik, E., Tabata, M., Goto, K., Ikeshiro, Y., 1993, *Journal of Ethnopharmacology*, **39**, 31-8.
10. Yeşilada, E., Honda, G., Sezik, E., Tabata, M., Fujita, T., Tanaka, T., Takeda, Y., Takaishi, Y., 1995, *Journal of Ethnopharmacology*, **46**, 133-52.
11. Yeşilada, E., Sezik, E., 1998, *Zeitschrift für Phytotherapie*, **19**, 132-8.
12. Yeşilada, E., Sezik, E., Honda, G., Takaishi, Y., Takeda, Y., Tanaka, T., 1999, *Journal of Ethnopharmacology*, **64**, 195-210.
13. Yeşilada, E., Sezik E., 2002, A Survey on the Traditional Medicine in Turkey: Semi-quantitative Evaluation of the results. In *Recent Progress in Medicinal Plants*. Vol.VII. "*Ethnomedicine and Pharmacognosy*" (Singh, V.K., Govil, J.N., eds.), Gurdip Singh, Research Periodicals & Book Publishing House, New Delhi.
14. Akalın, E., Alpınar, K., 1994, *Ege Üniversitesi Eczacılık Fakültesi Dergisi*, **2**, 1-11.
15. Alpınar, K., 1979, *Bitki*, **6**, 243-9.
16. Alpınar, K., 1987, Batı Türkiye'nin *Arum* L. türlerinin yöresel ad ve kullanılışları (Local uses and names for *Arum* L. species of West Turkey). *Proceedings of the VI. Symposium on Plant Originated Crude Drugs*, May 1986, (Şener, B. ed.), Gazi Univ. Yayınları, Ankara, pp. 237-96.
17. Çubukçu, B., Özhatay, N., 1987, Anadolu halk ilaçları hakkında bir araştırma (A study on Anatolian folk medicines), *III. Milletlerarası Türk Folklor Kongresi Bildirleri*, **4**, 103-15.
18. Çubukçu, B., Atay, M., Sarıyar, G., Özhatay, N., 1994, *Geleneksel ve Folklorik Droglar Dergisi*, **1**, 1-58.
19. Ertuğ, F. 2000, *Economic Botany*, **52**, 155-82.
20. İlçim, A., Varol, Ö., 1996, *OT Sistematik Botanik Dergisi*, **3**, 69-74.
21. Melikoğlu, G., Çubukçu, B, 1989, Giresun ili halk ilaçları (Folk drugs in Giresun). *Proceedings of the VIII. Symposium on Plant Originated Crude Drugs*, (Çubukçu, B., Sarıyar, G., Mat, A. eds.), İstanbul Univ. Yayınları, İstanbul, pp. 251-5.
22. Özçelik, H., 1987, *DOĞA Turkish Journal of Botany*, **11**, 316-21.
23. Özçelik, H., Ay, G., Öztürk, M., 1990, Doğu ve Güneydoğu Anadolu'nun ekonomik yönden önemli bazı bitkileri (Some economically important species of east and southeast Anatolia), *X. Ulusal Biyoloji Kongresi, 18-20 Temmuz 1990*, Erzurum, pp.1-10.
24. Tonbul, S., Altan, Y., 1989, *Fırat Üniversitesi Dergisi*, **3**, 267-78.
25. Townsend, B.R. 1944, *Bulletin of the History of Medicine*, **15**, 37-8.

26. Tuzlacı, E., 1977, *Acta Biologica* **27**, 9-12.
27. Tuzlacı, E., 1985, *Mar. Üniv. Ecz. Der.*, **1**, 101-6.
28. Tuzlacı, E., Erol, M.K., 1999, *Fitoterapia*, **70**, 593-610.
29. Tuzlacı, E., Tolon, E., 2000, *Fitoterapia*, **71**, 673-85.
30. Tuzlacı, E., Aymaz P.E., 2001, *Fitoterapia*, **72**, 323-43.
31. Tümen, G., Sekendiz, O.A., 1993, Balıkesir ve merkez köylerinde halk ilacı olarak kullanılan bitkiler (The plants used as folk medicine in the city of Balıkesir and its vicinity). *Proceedings of the VIII. Symposium on Plant Originated Crude Drugs*, Istanbul, May 1989, (Çubukçu, B., Sarıyar, G., Mat, A. eds.), İstanbul Univ. Yayınları, Istanbul, pp. 347-53.
32. Vural, M., Karavelioğulları, F.A., Polat, H., 1997, *OT Sistematik Botanik Dergisi*, **4**, 117-24.
33. Yazıcıoğlu, A., Tuzlacı, E., 1996, *Fitoterapia*, **67**, 307-18.
34. Yıldırımlı, Ş., 1987, Bolkar Dağları'nın yerel bitki adları ve tıbbi bitkileri. *Proceedings of the VI. Symposium on Plant Originated Crude Drugs*, May 1986, (Şener, B., ed.), Gazi Univ. Yayınları, Ankara, pp. 279-85.
35. Yıldırımlı, Ş., 1985, *Doğa Bilim Dergisi A2*, **9**, 593-7.
36. Yıldırımlı, Ş. 1994, *OT Sistematik Botanik Dergisi*, **1**, 7-12.
37. Yıldırımlı, Ş., 1994, *OT Sistematik Botanik Dergisi*, **1**, 43-6.
38. Yeşilada, E. and Gürbüz, İ., Ergun, E., 1997, *Journal of Ethnopharmacology*, **55**, 201-11
39. Yeşilada, E., Takaishi, Y., Fujita, T., Sezik, E., 2000, *Journal of Ethnopharmacology*, **70**, 219-26.

Biodiversity of Phenylethanoid Glycosides

İHSAN ÇALIŞ
Hacettepe University, Faculty of Pharmacy, Department of Pharmacognosy, 06100 Ankara, Turkey

1. INTRODUCTION

Phenylethanoid glycosides are a group of natural products widely distributed in the plant kingdom, most of which have been isolated from medicinal plants, and their patterns of distrubition have been suggested to be valuable taxonomic markers. The most widely studied families are Scrophulariaceae, Oleaceae, Plantaginaceae, Lamiaceae and Orobanchaceae[1,2]. Structurally, they are characterized as caffeic acid and hydroxyphenylethyl moieties attached to a β-glucopyranose unit by ester and glycosidic linkages, respectively. Allose, arabinose, apiose, galactose, lyxose, rhamnose and xylose may also be attached to the glucose residue. Although in most cases glucose is the core of the molecule, it is replaced by allose in some compounds such as magnolosides A, B and C[3]. Since most of them contain caffeic acid as acyl moiety, they are also termed as caffeic acid glycoside esters, phenylpropanoid glycosides or caffeoyl phenylethanoid glycosides. The most prevalent compound is acteoside (**1** = verbascoside, kusaginin, orabanchin), 3,4-dihydroxyphenethyl-[α-L-rhamnopyranosyl-(1→3)]-4'-O-caffeoyl-β-D-glucopyranoside. This compound was first isolated from *Verbascum sinuatum* and named as verbascoside. However,

[1] Acteoside (= Verbascoside)

the complete structure of acteoside was first reported at 1968[4].

Our studies on this type of compounds date back to a note reported by Harborne[5], in which caffeic acid ester distribution in higher plants had been discussed. Besides the well-known quinic acid esters chlorogenic acid and rosmarinic acid, compounds with similar properties to orobanchin have been suggested as familial chemical markers in some families and even at the generic level in Lamiaceae. Along these lines, we have studied the plants from Lamiaceae (*Galeopsis*[6], *Stachys*[7,8], *Phlomis*[9-19], *Leonurus*[20], *Scutellaria*[15,21,22], *Marrubium*[23], *Sideritis*[24], *Teucrium*[25]), Scrophulariaceae (*Scrophularia*[26,27], *Pedicularis*[28-31], *Lagotis*[32], *Rhynchocorys*[33], *Veronica*[34], *Digitalis*[35-37], *Euphrasia*[38], *Verbascum*[39]), Globulariaceae (*Globularia*[40,41]) and Oleaceae (*Fraxinus*[42]). Most of these plants are known as herbal drugs in traditional medicine and some of them are used as herbal teas in Turkey. Phytochemical investigations performed on the selected plants resulted in the isolation of over fifty different phenylethanoid glycosides of which twenty-seven had not been mentioned in the literature before. In addition, some of the compounds were screened for their antimicrobial[43], cytotoxic and cytostatic[15,44], antioxidant[45] and antiinflammatory[19] activities.

2. RESULTS AND DISCUSSION

The first study was performed on *Galeopsis pubescens* (Lamiaceae). From the fractions containing phenolic compounds, martynoside [6] and isomartynoside [7] were isolated[6]. At that time, approximately ten phenylethanoid glycosides were known. This prompted us to perform systematical studies on the plants of Scrophulariaceae and Lamiaceae. The preliminary studies were carried on the plants of Lamiaceae at the generic level, especially on the subfamilies Ajugoideae, Scutellaroideae and Nepetoideae (= Stachyoideae).

Phenylethanoid glycosides isolated in our laboratory can be classified into four groups according to their sugar moieties: i) monosaccharides, ii) disaccharides, iii) trisaccharides, and iv) tetrasaccharides.

i) **Monosaccharides**.- Calceolarioside A [2] from *Globularia orientalis*[41] and fuhsioside [3] from *Veronica fuhsii* were isolated.[34]

[2] Calceolarioside A

[3] Fuhsioside

ii) **Disaccharides**.- This group can be divided into several subgroups according to their oligosaccharidic moieties:
a) Rhamnose-glucose containing glycosides (acteoside-type) [4-9].

Compounds	R_1	R_2	R_3
[4] Isoacteoside	H	trans-caffeoyl	H
[5] cis-acteoside	H	cis-caffeoyl	H
[6] Martynoside	trans-feruloyl	H	Me
[7] Isomartynoside	H	trans-feruloyl	Me
[8] Darendoside B	H	H	Me
[9] Leucosceptoside A	trans-feruloyl	H	H

b) Glucose-glucose containing glycosides [10-13].
This type of glycosides contain glucose as a second sugar unit on the core sugar. The glycosidation site is either the C-3(OH) of the core sugar as in plantamajoside [10] or the C-6(OH) group as in lugrandoside [13].

[10] Plantamajoside

Compounds	R_1	R_2
[11] Ferruginoside A	trans-caffeoyl	H
[12] Ferruginoside B	H	H
[13] Lugrandoside	H	trans-caffeoyl

c) Rhamnose-glucose containing glycosides (β-hydroxyacteoside type) [14].

[14] β-hydroxy-acteoside

d) Rhamnose-glucose containing glycosides (crenatoside-type) [15, 16].

Compounds	R_1	R_2
[15] Crenatoside (= Oraposide)	*trans*-caffeoyl	H
[16] Isocrenatoside	H	*trans*-caffeoyl

Crenatoside [15] was first reported from *Orobanche crenata* (Orobanchaceae)[46]. One year later, the same compound has been reported under the name of oraposide from *O. rapum-genistae*[47].

e) Apiose-glucose containing glycosides [17, 18].

[17a] **Darendoside A**: R = H
[17] **Hattushoside**: R = syringyl
[18] **Fimbrilloside**: R = vanilloyl

iii) **Trisaccharides**.- Acteoside [1], a common phenylethanoid diglycoside constructs the basic structure of this group. The third sugar attached can be apiose, arabinose, xylose, lyxose, rhamnose or glucose. The substitution takes place either on the C-6 [19-29] (Group A) or C-2 [30, 31] (Group B) positions of the core sugar, glucose. Alternatively, C-2 [32-40] (Group C) or C-4 [41, 42] (Group D) positions of the rhamnose moiety can be substituted. Maxoside [43] is the only trisaccharidic phenylethanoid glycoside consisting of three β-D-glucose units (Group E).

Biodiversity of Phenylethanoid Glycosides

Group A [**19-29**].

Compounds	R_1	R_2	R_3
[19] **Angoroside A**	*trans*-caffeoyl	H	α-L-arabinopyranosyl
[20] **Angoroside B**	*trans*-caffeoyl	Me	α-L-arabinopyranosyl
[21] **Angoroside C**	*trans*-feruloyl	Me	α-L-arabinopyranosyl
[22] Forsythoside B	*trans*-caffeoyl	H	β-D-apiofuranosyl
[23] **Alyssonoside**	*trans*-feruloyl	H	β-D-apiofuranosyl
[24] Leucosceptoside B	*trans*-feruloyl	Me	β-D-apiofuranosyl
[25] Arenarioside	*trans*-caffeoyl	H	β-D-xylopyranosyl
[26] Poluimoside	*trans*-feruloyl	H	α-L-rhamnopyranosyl
[27] **Ferruginoside C**	*trans*-feruloyl	Me	α-L-rhamnopyranosyl
[28] Echinacoside	*trans*-caffeoyl	H	β-D-glucopyranosyl
[29] **Wiedemannioside C**	*trans*-feruloyl	H	β-D-glucopyranosyl

Group B [30, 31].

[30] Ehrenoside: $R_1 = R_2 = H$
[31] **Lagotoside**: $R_1 = R_2 = CH_3$

Group C [**32-40**].

Compounds	R_1	R_2	R_3
[32] **Lavandulifolioside**	*trans*-caffeoyl	H	α-L-arabinopyranosyl
[33] **Leonoside A**	*trans*-feruloyl	H	α-L-arabinopyranosyl
[34] **Leonoside B**	*trans*-feruloyl	Me	α-L-arabinopyranosyl
[35] Teucrioside	*trans*-caffeoyl	H	β-D-lyxopyranosyl
[36] **Phlinoside A**	*trans*-caffeoyl	H	β-D-glucopyranosyl
[37] **Phlinoside B**	*trans*-caffeoyl	H	β-D-xylopyranosyl
[38] **Phlinoside C**	*trans*-caffeoyl	H	α-L-rhamnopyranosyl
[39] **Phlinoside D**	*trans*-feruloyl	H	β-D-xylopyranosyl
[40] **Phlinoside E**	*trans*-feruloyl	H	α-L-rhamnopyranosyl

Group D [41, 42].

[41] **Trichosanthoside A**: R = H
[42] **Rossicaside A**: R = CH$_2$OH

Group E [43].

[43] Maxoside

iv) **Tetrasaccharides**.- Trichosanthoside B [44], is only the third example of a tetrasaccharidic phenylethanoid glycoside which was isolated from *Globularia trichosantha* (Globulariaceae).[40] The two previous examples are magnolioside C[3] and ballotetroside[48], which were isolated from *Magnolia obovata* (Magnoliaceae) and *Ballota nigra* (Lamiaceae), respectively.

[44] **Trichosanthoside B**

v) **Acetylated Phenylethanoid Glycosides [45–48]**.- The phenylethanoid glycosides with several acetyl groups on the saccharidic portion were only obtained from *Verbascum wiedemannianum* in our studies[39].

Compounds	R_1	R_2	R_3	R_4	R_5	R_6
[45] **Wiedemannioside A**	Me	Me	H	H	H	Ac
[46] **Wiedemannioside B**	Me	Me	H	Ac	Ac	Ac
[47] **Wiedemannioside D**	H	Me	Ac	H	H	α-L-rhamnopyranosyl
[48] **Wiedemannioside E**	H	Me	Ac	Ac	H	α-L-rhamnopyranosyl

The phenylethanoid glycosides [1-48] isolated in our laboratory and their sources are listed in Table 1.

3. BIOLOGICAL ACTIVITIES

Biological activities of phenylethanoid glycosides prompted several pharmacological studies and they have been claimed as the active components of some medicinal remedies used in traditional medicine. Especially, the main compound acteoside [1] has been screened for its different biological activities[1,2]. Some of the phenylethanoid glycosides isolated in our laboratory were also screened for their biological activities as given below.

The cytotoxic and cytostatic activities of some phenylethanoid glycosides such as acteoside [1], martynoside [6], leucosceptoside A [9], angorosides A, B, C [19–21], forsythoside B [22], poluimoside [26], teucrioside [35] and phlinoside B [37] were investigated by the dye exclusion method using 3-[4,5-dimethylthiazol-2-yl]-2,5-diphenyltetrazolium bromide (MTT). Caffeic acid containing phenylethanoid glycosides were found to exhibit activity against several cancer cells, like dRLh-84 cell (rat hepatoma), Heta cell (human epithelial carcinoma), S-180 (sarcoma) and P-388/01 cell (mouse lymphoid neoplasma)[15,44]. However, they did not effect the growth and viability of primary-cultured rat hepatocytes. Studies on the structure-activity relationship indicated that *ortho*-dihydroxy aromatic systems of phenylethanoid glycosides are necessary for their cytotoxic and cytostatic activities.

Table 1. Phenylethanoid Glycosides reported between 1984-2001.

Plant name	Phenylethanoid glycosides	References
LAMIACEAE		
Galeopsis pubescens	6, 7	6
Stachys lavandulifolia	1, 32	7
S. macrantha	1, 6, 9, 32	8
Phlomis linearis	1, 36, 37, 38, 39, 40	9, 10
P. grandiflora var. fimbrilligera	17, 18, 22, 23, 24	11
P. pungens var. pungens	17, 22, 23, 24	12
P. pungens var. hirta	22, 23, 24	13
P. bourgei	1, 22, 24	14
P. armeniaca	1, 9, 22, 37	15
P. tuberosa	1, 22	16
P. siehana	1, 5, 6, 9, 14, 22	17
P. longifolia var. longifolia	1, 9, 22	18
P. lycia	22, 23, 24	19
Leonurus glaucescens	1, 32, 33, 34	20
Scutellaria albida subsp. colchica	1, 6, 9	21
S. orientalis subsp. pinnatifida	1, 6, 8, 9, 17a	22
S. salvifolia	1, 6, 9, 35, 38	15
Marrubium alysson	1, 6, 9, 23, 24	23
Sideritis lycia	1, 6, 32,	24
Teucrium polium and T. chamaedrys*	26, 35*	25
SCROPHULARIACEAE		
Scrophularia scopolii	1, 19, 20, 21	26, 27
Pedicularis nordmanniana	1, 6, 9, 22	28
P. condensate	1, 28	29
P. pontica	28	30
P. comosa var. acmodonta	1, 6, 9, 22, 28	31
Lagotis stolonifera	1, 10, 30, 31	32
Rhynchocorys stricta	1, 6, 9	33
Veronica fuhsii	3, 10	34
Digitalis ferruginea	11, 12, 13, 27	35, 36
D. cariensis	13, 43	37
Euphrasia pectinata	1, 9	38
Verbascum wiedemannianum	1, 6, 24, 28, 29, 45 - 48	39
GLOBULARIACEAE		
Globularia trichosantha	1, 15, 16, 25, 41, 42, 44	40
G. orientalis	1, 2, 9	41
G. davisiana	1, 4, 9	unpublished
G. cordifolia	1, 6, 9, 42	unpublished
OLEACEAE		
Fraxinus angustifolia	1	42

Acteoside [1] induced cell death in promyelocytic leukemia HL-60 cells with an IC$_{50}$ value of 26.7 µM. Analysis of extracted DNA on agarose gel electrophoresis revealed that acteoside induced the internucleosomal breakdown of chromatin DNA characteristic of apoptosis (programable cell death). Apoptosis-specific DNA fragmentation was clearly detectable 4h after treatment with acteoside and was independent of the cell cycle phase. These data indicated that acteoside induces apoptosis in HL-60 cells[49].

Acteoside [1] was also found to induce interleukin (IL)-1, IL-6, and tumor necrosis factor-α (TNF-α) in macrophage-like cell line J774.A1 at 1-100 ng/ml. In addition, when the stimulatory action of acteoside was studied using the bovine glomerular endothelial cell line GEN-T, it was shown that acteoside stimulated IL-6 production. These activities were not abolished by treatment with Polymixin B, which inactivates lipopolysaccharide (LPS), indicating that the action was not a contamination of LPS[50].

Table 2. Antimicrobial Activity of some Phenylethanoid Glycosides.

Microorganisms	MIC (mg/ml)				(IU/ml)
	Angorosides			Acteoside	Penicilin
	A [19]	B [20]	C [21]	[1]	
Gram positive					
Staphylococcus aureus ATCC 25923	1.56	3.13	-	1.56	0.39
Staphylococcus aureus	1.56	3.13	-	3.13	200
Streptococcus faecalis MN 10541	6.25	12.5	-	12.5	1.56
Streptococcus faecalis	6.25	12.5	-	12.5	1.56

Some phenylethanoid glycosides, such as acteoside [1], angorosides A, B and C [19-21] were tested for their antimicrobial activities against some gram positive and gram negative bacteria and one yeast. The results indicated that the phenylethanoid glycosides having caffeoyl and 3,4-dihydroxyphenethyl moieties showed higher antimicrobial activity against gram positive microorganisms, while angoroside C, which has feruloyl and 3-hydroxy,4-methoxyphenetyl moieties showed no activity in the tested doses (Table 2). On the other hand, angoroside B differed from angoroside A and acteoside chemically only in the type of aglycone moiety exluding a methyl group etherified on C-4 hydroxyl group had a weaker activity. These results indicated that the antimicrobial activities of phenylethanoid glycosides might be attributed to their phenolic moieties, which have *ortho*-dihydroxy functionalities[43].

Phenylethanoid glycosides are widely distributed in different plant families. Various plants used in traditional medicine contain significant amounts of these compounds. The strong radical scavenger activity of some phenylethanoid glycosides in *in-vitro* assays is already known. Because of the small number of compounds tested and the different test systems used hitherto, exact statements on the structure-activity relationships and structural requirements concerning their radical scavenger activity were not possible. Therefore, twentyone phenylethanoid glycosides and corresponding free acids (caffeic acid, ferulic acid, vanillic acid and syringic acid) were tested for their effects on oxygen radical production by human polymorphonuclear neutrophils (PMNs)[45]. The cells were stimulated with the chemoattractant formyl-methionyl-leucyl-phenylalanine (FMLP) and generated oxygen species were detected by luminal-augmented chemiluminescence measurements. These reactive oxygen species (ROS), e.g., superoxide anions, singlet oxygen or hydroxyl radicals, have been proposed to induce cellular damage which plays an important role in cancer, inflammatory and ageing processes. All phenylethanoid glycosides acylated with phenolic acids showed strong antioxidant activity whereas the deacyl derivatives were more than 30-fold less active. Therefore, the antioxidant activity is mainly related to the number of aromatic hydroxy and methoxy groups and the structure of the acyl moiety (C_6-C_1 or C_6-C_3). In contrast, modification of the sugar chain by replacement of hydroxy groups by methoxy groups in the acyl or the aglycone moieties is of minor importance. The position of the acyl moiety is without significance. Free acids were found to be less active compared to the phenylethanoid glycosides.

Phenylethanoid glycosides [**1, 6, 9, 32**] isolated from *Sideritis lycia* together with some flavonoid glycosides were screened for their anti-inflammatory activity using carrageenan-induced mouse paw edema. Although flavonoid glycosides showed higher activity than the phenylethanoid glycosides, the gastric ulceration effect of phenylethanoid glycosides was found to be less than flavonoid glycosides[24].

4. CONCLUSION

The above mentioned biological activities of phenylethanoid glycosides were studied in house or in other laboratories by collaborative researches. Their antifungal, antiviral, immunosuppresive, cardiovascular, antifeedant and analgesic effects, and the activities on CNS and platelet aggregation, and a protective effect on decrease of sex and learning behavior in mice are well summarized in the previous reviews[1,2,51]. It is also well know that they are generally enzyme and hormone inhibitors. Several enzymes involved in some pathological processes are inhibited by these glycosides, including 5-lipoxygenase, cyclic-AMP phosphodiesterase, aldose reductase and protein kinase C. Their activities were found to be mainly dependent on their antioxidant properties, their amphiphilic features and partial affinity for intracellular membrane systems. They are also less toxic and more soluble than flavonoids[52].

Their significant occurrence in the plant kingdom is also discussed for the chemosystematic importance. The presence of both the phenylethanoid and iridoid glycosides in several plant families is well known and of taxonomic and biogenetic importance. Phenylethanoid glycosides have been used as chemosystematic markers also in infrafamiliar and infragenic studies[1,2]. A recent study performed on the leaves of 365 specimen representing 355 species and varieties of 110 genera of Lamiaceae from the point of view of distribution and taxonomic implications of some phenolics showed that two chemical characters give strong support to the subfamily of Lamiaceae: Rosmarinic acid confined to Nepetoideae and phenylethanoid glycosides confined to Lamioideae[53]. β-hydroxyacteoside [**14**] was also solely confined to Lamioideae.

ACKNOWLEDGEMENTS

The results presented here were mainly carried out in our laboratory. I would like to thank to all academic staff participated in these studies, who are co-authors or authors listed in the references.

REFERENCES

1. Jimenez, C., Riguera, R.,1994, *Natural Product Reports,* **11**, 591-606.
2. Cometa, F., Tomassini, L., Nicoletti, M., Pieretti, S., 1993, *Fitoterapia,* **LXIV**, 195-217.
3. Hasegawa, T., Fukuyama, Y., Yamada, T., Nakagawa, K.,1988, *Chem. Pharm. Bull.,* **36**, 1245-8.
4. Birkofer, L., Kaiser, C., Thomas, U.,1968, *Z. Naturforsch.*, **23b**, 1051-8.
5. Harborne, J. B., 1966, *Z. Naturforsch.* Teil B, **21**, 604-5.
6. Çalış, İ., Lahloub, M.F., Rogenmoser, E., Sticher, O., 1984, *Phytochemistry,* **23**, 2313-5.
7. Başaran, A.A., Çalış, İ., Anklin, C., Nishibe, S., Sticher, O., 1988, *Helv. Chim Acta,* **71**, 1483-90.
8. Çalış, İ., Başaran, A.A., Saraçoğlu, ,İ., Sticher, O., 1992, *Phytochemistry,* **31**, 167-9.
9. Çalış, İ., Başaran, A.A., Saraçoğlu, İ., Sticher, O., Rüedi, P., 1990, *Phytochemistry,* **29**, 1253-7.
10. Çalış, İ., Başaran, A.A., Saraçoğlu, İ., Sticher, O., Rüedi, P.,1991, *Phytochemistry,* **30**, 3073-5.
11. Çalış, İ., Heilmann, J., Harput, U.S., Schuhly, W., Sticher, O., 1999, "Phenylethanoids and iridoid glycosides from *Phlomis grandiflora* var. *fimbrilligera* and their antioxidative activity". Joint Meeting of ASP, AFERP, GA and PSE. *2000 Years of Natural Products Research – Past, Present and Future*, July 26-30, 1999, Amsterdam.
12. Saraçoğlu, İ., Kojima, K., Harput, U.S., Ogihara, Y.,1998, *Chem. Pharm. Bull.,* **46**,726-7.
13. Harput, U.S., Saraçoğlu, I., Tulemis, F., Akay, C., Cevheroğlu, S., 1998, *Hacettepe University J. of Faculty of Pharmacy,* **18**, 1-7.
14. Harput, U.S., Saraçoğlu, I., Çalış, I., 1999, *Hacettepe University J. of Faculty of Pharmacy,* **19**, 1-11.

15. Saraçoğlu, İ., Inoue, M., Çalış, İ., Ogihara, Y., 1995, *Biol. Pharm. Bull.*, **18**, 1396-1400.
16. Ersöz, T., Ivancheva, S., Akbay, P., Sticher, O., Çalış, İ., 2001, *Z. Naturforsch.*, **56c**, 695-8.
17. Ersöz, T., Harput, U.S., Çalış, İ., Dönmez, A.A., 2001, *Turk. J. Chem.*, (in press).
18. Ersöz, T., Schuhly, W., Popov, S., Handjieva, N., Sticher, O., Çalış, İ., 2001, *Nat. Prod. Letters*, (in press).
19. Saraçoğlu, İ., Harput, U.S., Çalış, İ., Ogihara, Y. 2001, *Turk. J. Chem.*, (in press).
20. Çalış, İ., Ersöz, T., Taşdemir, D., Rüedi, P., 1992, *Phytochemistry*, **31**, 357-9.
21. Saraçoğlu, İ., Ersöz, T., Çalış, İ., 1992, *Hacettepe University J. of Faculty of Pharmacy*, **12**, 65-70.
22. Çalış, İ., Saraçoğlu, İ., Başaran, A.A., Sticher, O., 1993, *Phytochemistry*, **32**, 1621-3.
23. Çalış, İ., Hosny, M., Khalifa, M., Rüedi, P., 1992, *Phytochemistry*, **31**, 3624-6.
24. Akcos, Y., Ezer, N., Çalış, İ., Demirdamar, R., Tel, B.C., 1999, *Pharm. Biol.*, **37**, 118-22.
25. Bedir, E., Çalış, İ., 1997, *Hacettepe University J. of Faculty of Pharmacy*, **17**, 9-16.
26. Çalış, İ., Gross, G.-A., Sticher, O., 1987, *Phytochemistry*, **26**, 2057-61.
27. Çalış, İ., Gross, G.-A., Sticher, O., 1988, *Phytochemistry*, **27**, 1465-8.
28. Akdemir, Z., Çalış, İ., Junior, P., 1991, *Planta Med.*, **57**, 584-5.
29. Akdemir, Z., Çalış, İ., Junior, P., 1991, *Phytochemistry*, **30**, 2401-2.
30. Akdemir, Z., Çalış, İ., 1991, *DOĞA T. J. of Pharm.*, **1**, 67-75.
31. Akdemir, Z., Çalış, İ., 1992, *DOĞA T. J. of Pharm.*, **2**, 63-70.
32. Çalış, İ., Taşdemir, D., Wright, A.D., Sticher, O., 1991, *Helv. Chim. Acta*, **74**, 1273-7.
33. Çalış, İ., Saraçoğlu, İ., Kitagawa, S., Nishibe, S., 1988, *DOĞA T. J. of Med. and Pharm.*, **12**, 234-8.
34. Özipek, M., Saraçoğlu, İ., Kojima, K., Ogihara, Y., Çalış, İ., 1999, *Chem. Pharm. Bull.*, **47**, 561-2.
35. Çalış, İ., Taşdemir, D., Sticher, O., Nishibe, S., 1999, *Chem. Pharm. Bull.*, **47**, 1305-7.
36. Çalış, İ., Akbay, P., Kuruüzüm, A., Yalçın, F.N., Şahin, P., Pauli, G., 1999, *Pharmazie*, **54**, 926-30.
37. Kırmızıbekmez, H., Ersöz, T., Çalış, İ., 2000, Phenylethanoid Glycosides from *Digitalis cariensis*, In *Proceedings of the XIIIth Symposium on Plant Originated Crude Drugs*. September 20-22, 2000, İstanbul.
38. Ersöz, T., Berkman, M.Z., Taşdemir, D., Ireland, C.M., Çalış, İ., 2000, *J. Nat. Prod.*, **63**, 1449-50.
39. Abou Gazar, H., 2001, Ph.D. Thesis, Hacettepe University, Institute of Health Sciences, Ankara.
40. Çalış, İ., Kırmızıbekmez, H., Rüegger, H., Sticher, O., 1999, *J. Nat. Prod.*, **62**, 1165-8.
41. Kırmızıbekmez, H., Sticher, O., Taşdemir, D., Ireland, C.M., Çalış, İ., 2000, Antioxidative compounds from *Globularia orientalis*. In *Proceedings of the Int. Congress and 48th Annual Meeting of the Society for Medicinal Plant Research (GA). Natural Products Research in the New Millennium*. September 3-7, 2000, Zurich.
42. Hosny, M., 1992, Ph. D. Thesis, Hacettepe University, Institute of Health Sciences, Ankara.
43. Çalış, İ., Saraçoğlu, İ., Zor, M., Alaçam, R., 1988, *DOĞA T. J. of Med. and Pharm.*, **12**, 230-3.
44. Saraçoğlu, İ., Çalış, İ., Inoue, M., Ogihara, Y., 1997, *Fitoterapia*, **LXVIII**, 434-8.
45. Heilmann, J., Çalış, İ., Kırmızıbekmez, H., Schuhly, W., Harput, S., Sticher, O., 2000, *Planta Med.*, **66**, 746-8.
46. Afifi, M.S., Lahloub, M.F., El-Khayaat, S.A., Anklin, C.G., Ruegger, H., Sticher, O., 1993, *Planta Med.*, **59**, 293-390.

47. Andary, C., Wylde, R., Maury, L., Heitz, A., Dubourg, A., Nishibe, S., 1994, *Phytochemistry*, **37**, 855-7.
48. Seidel, V., Bailleul, F., Libot, F., Tillequin, F., 1997, *Phytochemistry*, **44**, 691-3.
49. Inoue, M., Sakuma, Z., Ogihara, Y., Saraçoğlu, İ., 1988, *Biol. Pharm. Bull.*, **21**, 81-3.
50. Inoue, M., Ueda, M., Ogihara, Y., Saraçoğlu, İ., 1988, *Biol. Pharm. Bull.*, **21**, 1394-5.
51. Andary, C., 1993, Caffeic acid glycoside esters and pharmacology. In *Polyphenolic Phenomena* (Scalbert, A., ed.), INRA Editions, Paris, pp. 237-45.
52. Nishibe, S., 1989, Structure Elucidation and Biological Activities of Phenylpropanoids, Coumarins and Lignans from Medicinal Plants. In *Studies in Natural Products Chemistry, Structure Elucidation* (Part B) (Rahman, A.-ur., ed.), Vol.5, Elsevier, Amsterdam, pp. 505-48.
53. Pedersen, J.A., 2000, *Biochem. Syst. Ecol.*, **28**, 229-53.

The Chemo- and Biodiversity of Endophytes

REN XIANG TAN and REN XIN ZOU
Institute of Functional Biomolecules, School of Life Sciences, Nanjing University, Nanjing 210093, China

1. INTRODUCTION

An endophyte is a bacterial or fungal microorganism, which spends the whole or part of its life cycle colonizing inter- and/or intra-cellularly inside the healthy tissues of the host plant, typically causing no apparent symptom of disease[1]. Owing to its chemo- and biodiversity, the endophyte plays a multiple physiological and ecological role in the process of endophyte-plant and endophyte-plant-herbivore interactions. It has been well ascertained that the colony of endophytes may enhance the hosts' growth by increasing the plant tolerance to abiotic and biotic stresses such as drought, salinity, heavy metals as well as attacks of or consumptions by microbial pathogens, nematodes, insects and mammal herbivores. A growing pile of evidences has indicated that the "host-helping" effects are ascribable to the production of the bioactive compounds by endophytes which are accepted as well from the bio-resource viewpoint as a reservoir of "special microorganisms" being a rich source of novel agrochemical and/or drug leads[2].

2. BIODIVERSITY OF THE ENDOPHYTES

2.1 Ubiquitous distribution

Endophytic bacteria and/or fungi have been found to be ubiquitous in almost all vascular plant species examined to date[2,3]. Moreover, marine algae[4] and ferns[5-7] were also reported to be colonized by certain endophytic fungi. A survey of the endophyte-colonized plants was given in Table 1, indicating that the economically important species were most frequently studied. It is estimated that every plant species harbors endophytic fungi and/or bacteria, and the amount of endophytes in nature may be over 500,000 species[8].

Table 1. List of reported endophyte-harboring plants[2]

Kingdom	Family	Plant name
Algae		Unknown
Pteridophytes	Hypolepidaceae	*Pteridium aquilinum*
		Christela dentata
Gymnosperms	Araucariaceae	*Wollemia nobilis*
	Cupressaceae	*Chamaecyparis thyoides*
	Pinaceae	*Abies alba, Abies balsamea, Abies* sp.
		Larix laricina, Larix sp.
		Picea abies, Picea mariana, Picea sitchensis, Picea sp.
		Pinus mugo ssp. *uncinata, P. resinosa, P. banksiana, P. thunbergii, P. densiflora, P. strobus*
	Taxaceae	*Taxus brevifolia, T. wallachiana, T. baccata, T. mairei, T. yunnanensis,*
		Torreya grandifolia
		Torreya taxifolia
	Taxodiaceae	*Sequoia sempervirens*
		Taxodium distichum
Angiosperms-Monocotyledonae	Gramineae	*Achnatherum inebrians*
		Agrostis alba, A. hiemalis
		Ammophila reviligulata
		Rachyelytrum erectum
		Brachypodium sylvaticum
		Bromus erectus
		Cymbopogon flexuosus
		Dactylis glomerata
		Danthonia spicata
		Echinopogum ovatus
		Elymus spp., *E.canadensis*
		Festuca arizonica, F. arundinacea, F. rubra
		Glyceria striata
		Hordeum bogdanii, H. brevisubulatum, H. comosum
		Hordeum bogdanii, H. brevisubulatum subsp. *violaceum*
		Lolium spp.,*L. multiflorum, L. pratense*
		Oryzae sativa
		Phleum pratense
		Poa ampla

		Saccharum officinarum
		Sorghum bicolor
		Stipa robusta
		Triticum aestivum
		Triticum dichasians, T. tripsacoides, T. columnare, T. cylindricum, T. monococcum, T. neglecta, T. recta, T. triunciale, T. turgidum, T. umbellulatum
		Zea luxurians, Z. mays, Zea sp.
	Orchidaceae	*Cypripedium parviflorum*
		Lepanthes sp.
		Rhizanthella gardneri
		R. slateri
	Palmaceae	*Licuala ramsayi*
		Licuala sp.
		Euterpe oleracea
		Sabal bermudana
		Livistona chinensis
		Trachycapus fortunei
	Rapateaceae	*Stegolepis guianensis*
	Zingiberaceae	*Amomum siamense*
Angiosperms-Dicotyledonae	Aceraceae	*Acer pseudoplata*nus
		Acer macrophyllum
	Anacardiaceae	*Mangifera*
	Berberidaceae	*Berberis oregana*
	Betulaceae	*Alnus* spp., *A. glutinosa, A. rubra, Betula nana, B. pendula, B. pubescens, B. pubescens* var. *tortuosa*
	Chenopodiaceae	*Beta vulgaris*
	Compositae	*Baccharis coridifolia*
		B. artemisioides
	Cruciferae	*Brassica napus*
	Ericaceae	*Enkianthus perulatus, Pieris japonica, Rhododendron indicum, R. macrosepalum, R. mucronulatum* var. *ciliatum, R. obtusum, R. pulchrum* var. *speciosum, R. reticulatum Cassiope mertensiana, C. tetragonal, Empetrum nigrum, Gaultheria humifusa, Kalmia polifolia, Loiseleuria procumbens, Menziesia ferruginea, Phyllodoce empetriformis, P. glanduliflora, Rhododendron albiflorum, Vaccinium membranaceum, V. myrtilloides, V. scoparium, V. uliginosum, V. vitis-idaea*
		Vaccinium angustifolium
		Vaccinium myrtillus
	Fabaceae	*Medicago sativa*
		Trifolium pratense

Fagaceae	*Castanea satira*
	Fagus crenata, F. sylvatica
	Quercus alba, Q. mailandica, Q. velutina, Q. cerris, Q. emoryi, Q. garryana, Q. petraea, Q. robur
Gentianiaceae	*Fragraea bodenii*
Haloragaceae	*Myriophyllum spicatum*
Lauraceae	*Cinnamomum zeylanicum*
Leguminosae	*Vigna radiata*
	Prosopis juliflora
Loranthaceae	*Tristerix aphyllus*
Malvaceae	*Gossypium hirsutum*
Myrtaceae	*Eucalyptus grandis, E. nitens, E. globulus*
	Leptospermum scoparium
Oleaceae	*Fraxinus excelsior*
Rhamnaceae	*Zizyphus nummularia*
Rosaceae	*Dryas octopetala*
	Malus pumila
	Prunus lusitanica
	Rubus parviflorus
	R. spectabilis
Rubiaceae	*Coffea arabica, C. arabica*
Rutaceae	*Citrus* spp., *C. jambhiri*
Salicaceae	*Salix fragilis, S. glauca*
Sapotaceae	*Manilkara bidentata*
Solanaceae	*Lycopersicon esculentum*
Vitaceae	*Vitis vinifera*

2.2 Diversity of endophyte species

Endophytes are important components of microbial biodiversity. Generally, several to decades of endophytic microbes can be isolated from a single host species[2]. For instance, up to 34 isolates of endophytic fungi were obtained from single Norway spruce needle[9]. In tropical plants, hundreds species could be isolated from a single plant[10]. Similar observation is possible with endophytic bacteria[11-13].

Commonly, at least one species among the isolated assemblage of endophytes showed host specificity, and often represented as a new taxon. In Table 2, some new endophytic fungal taxa were summarized for exemplification.

Table 2. First-time reported new taxa of endophytes[2]

Endophytes	Host plant
Acremonium chilense	*Dactylis glomerata*
Chalara angustata	*Quercus petraea, Q. robur*
Cryptosporiopsis radicicola	*Quercus robur*
Dactylaria endophytica	*Prunus lusitanica*
Discostroma tricellulare	*Rhododendron indicum* and other Ericaceous plants
Epichloë amarillans	*Agrostis hiemalis*
Epichloë brachyelytri, E. elymi E. glyceriae	*Brachyelytrum erectum, Elymus* sp., *Glyceria striata*
Leptomelanconium abietis	*Abies balsamea*
Letendraeopsis palmarum	*Euterpe oleracea*
Muscodor albus	*Cinnamomum zeylanicum*
Mycoleptodiscus sp.	*Chamaecyparis thyoides*
Myrothecium groenlandicum	*Betula nana*
Ophiognomonia cryptica	*Quercus emoryi*
Pestalotiopsis jesteri	*Fragraea bodenii*
Prosthemium asterosporum	*Betula pendula*
Phialocephala victorinii	*Cypripedium parviflorum*
Piriformospora indica	*Zizyphus nummularia*
	Prosopis juliflora
Scytalidium vaccinii	*Vaccinium angustifolium*
Seimatoantlerium nepalense	*Taxus wallachiana*
Stagonospora pteridiicola	*Pteridium aquilinum*
Taxomyces andreanae	*Taxus brevifolia*
Thanatephorus gardneri	*Rhizanthella gardneri, Rh. Slateri*

The endophyte population was significantly affected by environmental conditions under which the host is growing. The endophyte profile of the same plant may differ strikingly with ecological environments. Moreover, genotypic diversity has been found in single endophyte species originated from conifers, birch, grasses[1], and others[14].

3. CHEMO-DIVERSITY

3.1 Diversified chemical structures

Endophytes could produce structurally diversified secondary metabolites *in planta* and/or *in vitro* cultivation. More than 140 different compounds were documented up to date, including alkaloids (amines, amides, indole derivatives, and pyrrolizidines), steroids, terpenoids, isocoumarin derivatives, quinones, flavonoids, phenylpropanoids and lignans, peptides, phenol and phenolic acids, aliphatic and chlorinated compounds, and other metabolites[1]. The structural diversity is somewhat related to the diversity of bioactivities they possess.

3.2 Diversified functional metabolites

Most of the above metabolites possess diversified bio-functions *in vivo* and/or *in vitro*, some of them serving possibly as drug or agrochemical leads, and the other contributing to improve the hosts' adaptability and competitiveness in nature.

3.2.1 Antimicrobial metabolites

The majority of investigated endophytes showed potent antifungal and/or antibacterial activity, many of such bioactive metabolites have been characterized from cultures of endophytes[1,2].

Cryptocandin, a new peptide antibiotics produced by the *Tripterygeum wilfordii* endophyte *Cryptosporiopsis* cf. *quercina* was displayed intensive antifungal activity against human pathogenic fungi[1]. Cryptocin, a new alkaloid isolated from the same endophytic fungus was strong growth-inhibitor against several phytopathogenic fungi[1].

A novel pentaketide produced by *Fusarium* sp. CR377, an endophyte from the stem of *Selaginella pallescens*, was shown to be active against *Candida albicns*[1]. Guanacastepene isolated from the liquid culture of an unidentified endophyte was observed to be avtive against many Gram positive and negative bacteria and the pathogenic fungus *Candida albicans*[1].

Colletotric acid, a new organic acid purified from the culture of *Artemisia mongolica* endophyte *Colletotrichum gloeosporioides*, was antimicrobial against to several bacteria and phytopathogenic fungi[1].Two structural similar new compounds, cytonic acids A and B isolated from the solid culture of an oak endophyte *Cytonaema* sp. F32027 were inhibitor against the assemblage of human cytomegalovirus[1].

Most recently, two novel cyclohexenone epoxides, jesterone and hydroxy-jesterone, were characterized from cultures of *Pestalotiopsis jesteri*, a new endophytic fungus originated from the rainforest plant *Fragraea bodenii*[15].The bioassay showed that these compounds displayed selective antimycotic activity against phytopathogenic oomycetes[16]. Ambuic acid, a novel cyclohexenone purified from cultures of *Pestalotiopsis* spp. and *Monochaetia* sp. isolated as endophytes from several rainforests, was found to be antifungal against several fusarium species, *Diplodia natelensis* and *Cephalosporium gramineum*[17].

3.2.2 Herbivore-toxic metabolites

Four alkaloids produced by various grass endophytes *in vitro* and/or *in planta*, namely the pyrrolopyrazine alkaloid (peramine), ergot alkaloids (e.g. ergovaline), indole diterpenes (e.g. lolitrem B), and pyrrolizidines (e.g. loline), were well known to be toxic to insects and/or mammal herbivores[1].

Several flavonoids characterized from endophyte-infected *Poa ampla* were found to be toxic to mosquito larvae[1].

A series of insecticidal metabolites were isolated from the cultures of endophytes originated from various needle trees[2].

3.2.3 Plant hormones

Plant hormones such as IAA, ethylene, cytokinins were observed to be produced by a number of endophytic fungi and bacteria[2]. It is considered that these plant growth regulators contribute partially to the endophyte-infected host's fast growth.

3.2.4 Allelopathic chemicals

Several analogs of lolines, the insecticidal alkaloids now proved to be produced in defined minimal media by *Neotyphodium uncinatum*[18], were demonstrated to contribute to the allelopathic properties of host grasses[1]. Allelopathic effect was also observed with other endophyte-infected plants[2], however, the active metabolites corresponding to this effect were not totally elucidated up to date.

3.2.5 Antitumor and other medicinal potential agents

In 1993, Stierle et al. first-time reported that Taxol, an antitumor diterpene produced tracely by *Taxus* species, could be alternatively produced by *Taxomyces andreanae*, an endophyte of *Taxus brevifolia*. Since then, many endophytic fungi originated from various Taxaceae and taxonomically related plants were found to be able to produce Taxol in cultures (Table 3).

Sequoiatones A and B, isolated from the culture of an endophytic fungus *Aspergillus parasiticus* from *Sequoia sempervirens*, were shown to inhibit the growth of the NCI human tumor 60 cell-line[1]. Three new cytochalasins isolated from *Rhinocladiella* sp., an endophytic fungus of *T. wilfordii*, were active against a series of human tumor cell-lines[1].

Oreganic acid originated from an unidentified endophyte of *Berberis oregana* and chaetomellic acids A and B from endophytic fungus *Chaetomella acutisea* were discovered to be selectively inhibitors of ras farnesyl-protein transferase (FPTase). Three novel quinones produced by endophyte *Coniothyrium* sp. and a new alkaloid TAN-1813 from endophytic *Phoma* sp. were found to be inhibitors of ras-farnesyltransferase. All these compounds have been demonstrated to be potent antitumor agents[2].

Subglutinols A and B, two diterpenes produced by endophytic *Fusarium subglutinans* originated from *T. wilfordii*, were nontoxic immunosuppressive agents[1].

Table 3. List of endophytes producing taxol[2]

Endophytes	Host Plant
Taxomyces andreanae, *Pestalotiopsis microspora*, *Pestalotia* spp., *Alternaria* spp., *Fusarium* spp., *Pithomyces* spp., *Monochaetia* spp.	*Taxus brevifolia*
Pestalotiopsis microspora *Seimatoantlerium nepalense*	*Taxus wallachiana*
fungi and actinomycetes	*Taxus baccata*
Tubercularia sp.	*Taxus mairei*
An unidentified fungal endophyte	*Taxus yunnanensis*
Pestalotiopsis microspora	*Taxodium distichum*
Periconia sp.	*Torreya grandifolia*
Pestalotiopsis guepinii	*Wollemia nobilis*
Stegolerium kukenani	*Stegolepis guianensis*

4. COMMENTS

Endophytes, hidden microbes with a great bio- and chemo-diversity, is a slightly opened big reservoir where human being can extract drugs and other materials. Some will-be-hot areas are the recognition and productivity of host-gene-carrying endophytes which could be alternative producers of important phytochemicals with "supply stress", the characterization of biologically and chemically particular secondary metabolites which possess greater pharmaceutical and/or agricultural potentials, and investigations of biomacromolecules such as enzymes and polysaccharides which are of medical and/or industrial significance.

REFERENCES

1. Tan, R. X, Zou, W. X., 2001, *Nat Prod Rep.*, **18**, 448-59.
2. Zou, W. X., Tan, R. X., 2001, *Acta Bot Sinica*, **43**, 881-92.
3. Sturz, A. V., Christie, B. R., Nowak, J., 2000, *Critical Rev Plant Sci.*, **19**, 1-30.
4. Stanley, S. J., 1992, *Can. J. Bot.*, **70**, 2089-2096.
5. Petrini, O., Fisher, P. J., Petrini, L. E., 1992 *Sydowia*, **44**, 282-93.

6. Fisher, P. J., Punithalingam, E., 1993, *Mycol. Res.*, **97**, 661-4.
7. Gabel, A., Studt, R., Metz, S., 1996, *Mycologia*, **88**, 635-41.
8. Strobel, G. A., Long, D. M., 1998, *ASM News*, **64**, 263-8.
9. Muller, M. M., Valjakka, R., Suokko, A., Hantula, J., 2001, *Molecular Ecology*, **10**, 1801-10.
10. Anorld, A. E., Maynard, Z., Gilbert, G. S., Coley, P., D., Kursar, T. A., 2000, *Ecol Lett.*, **3**, 267-74.
11. Germida, J., J., Siciliano, S. D., Freitas, J. R., de, Seib, A., M., 1998, *FEMS Microbiol., Ecol.*, **26**, 43-50.
12. Sturz, A. V., Christie, B. R., Matheson, B. G., Nowak, J., 1997,*Biol Fertil Soils*, **25**, 13-9.
13. Hallmann, J., Quadt-Hallmann, A., Mahaffee, W. F., et al, W. M., 2001, *Mycotaxon*, **77**, 512.
16. Li, J. Y., Strobel, G. A., 2001, *Phytochemistry,* **57**, 261-5. ., 1997, *Can J Microbiol.*, **43**, 895-914.
14. Li, J. Y., Strobel, G., Sidhu, R., et al., 1996, *Microbiology*, **142**, 2223-6.
15. Strobel, G., Li, J. Y., Ford, E., Worapong, J., Baird, G. I., Hess
17. Li, J. Y., Harper, J. K., Grant, D. M., Tombe, B. O., Bashyal, B., Hess, W. M., Strobel, G. A., 2001, *Phytochemistry*, **56**, 463-8.
18. Blankenship, J. D., Spiering, M. J., Wilkinson, H. H., Fannin, F. F., Bush, L. P., Schardl, C. L., 2001, *Phytochemistry*, **58**, 395-401.

Molecular Diversity and Specificity of Arthropod Toxins

EUGENE V. GRISHIN, T.M. VOLKOVA, YU.V. KOROLKOVA and K.A. PLUZHNIKOV
Shemyakin & Ovchinnikov Institute of Bioorganic Chemistry RAS, Moscow, Russia

1. INTRODUCTION

Arthropod toxins constitute a vast class of natural toxins. Their chemical properties and biological actions are quite diverse. The most known toxins are neurotoxins selectively interacting with functionally important components of the nerve cell membrane, i.e. mostly with the neuronal receptors or ion channels that block or modulate their functional activity. Some toxins can operate as channel-forming molecules. Arthropod toxins target insects, crustaceans or vertebrates. Molecular diversity of various toxins from spider, scorpion and ant venoms were analysed. They can be chemically divided into two major groups: polyamine and polypeptide neurotoxins.

2. RESULTS AND DISCUSSION

Spider venoms contain many compounds other than proteins and polypeptides that are neuroactive. Some of such compounds are well known neurotransmitters. Only one new, non-proteinaceous class of neurotoxic compounds has been discovered in spider venoms, the acylpolyamines[1]. The argiopine was first polyamine toxin (Fig.1) which structure was determined[2]. All polyamine neurotoxins from *Argiope lobata* spider consist of an aromatic unit, amino acid linker and polyamine chain. These toxins are able to block the ion channels of all types of glutamate receptors, which are important in many neural functions.

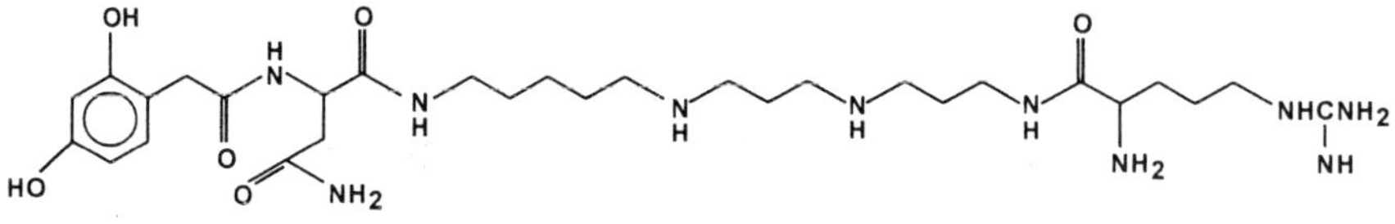

Figure 1. Chemical structure of argiopin (argiotoxin 636) –polyamine toxin from A.lobata spider venom

Scorpion venoms as the spiders ones are the reach source of active molecules differed from each other in the variety of specificities and selectivities. From the venoms of two scorpions *Orthrochirus scrobiculosus* and *Buthus eupeus* the toxins named OsK-1 and BeKm-1 correspondingly have been isolated[3,4]. Both toxins whose structures are presented on the Fig.2 display sequence homology to the other known scorpion toxins targeted toward potassium channels. The spatial structure of OsK-1 (Fig.3) resolved by NMR methods[5] contains the usual for all scorpion toxins a-b motif and are very similar to the three-dimensional structure of BeKm-1 toxin (unpublished data). But minor differences found in the toxin structures are important and result in their different biological functioning. The toxin OsK-1 was shown to block small-conductance Ca^{2+}-activated K^+-channels in neuroblastomaxglioma NG 108-15 hybrid cells (Kd 140 nM) which are insensitive to apamin [3], whereas BeKm-1 is the selective inhibitor of ERG types of K^+-channels (Kd 3.3 nM)[6]. Moreover it seems likely that this toxins use different surface of molecule to interact with receptors.

```
R P T D I K C S E S Y Q C F P V C K S R F G K T N G R C V N G F C D C F      BeKm-1
G V I I N V K C K I S R Q C L E P C K - K A G M R F G K C M N G K C H C T P K  OsK-1
```

Figure 2. Sequence alignment of OsK-1 and BeKm-1 toxins, homologous residues are shaded

The pore-forming activity is not typical for toxic components from Arachnida venoms. The family of high molecular mass neurotoxins – latrotoxins from the venom of the black widow spider *Latrodectus tredecimguttatus* is an exclusion. These toxins form cation-permeable membrane pores in target cells and provoke a massive and destructive transmitter release from presynaptic endings of various animals[7]. The α-latrotoxin (α-LTX) is mostly studied representative of the family and known as affecting the vertebrates[8]. The toxin initiates secretion of all known transmitters, evidently influencing the universal components of the secretion system. Although α-LTX has been used as a tool to study the mechanism of exocytosis for three decades its mode of action remain controversial.

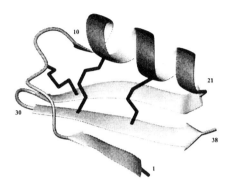

Figure 3. Schematic representation of the OsK-1 structure

The α-LTX molecule is shown to exist as a tetramer[9]. Its action is accompanied by the membrane depolarization and the Ca^{2+} influx into the presynaptic ending[10]. At this stage α-LTX is strongly bound to the membrane receptors - neurexin or latrophilin[11,12]. The black widow spider venom encourages also neurosecretion in different insect and crustacean preparations[13]. Its action on these preparations occurs due to the presence in the venom of insectotoxic proteins (latroinsectotoxins, LITs) and a latrocrustotoxin (α-LCT) affecting crustacean nerve endings[14]. Some of these toxins have a high affinity receptor on the presynaptic membrane and are capable of forming channels in lipid bilayers[15].

Cloning of the genes encoding latrotoxins revealed general principles of their molecular arrangement[16-18]. Alignment of the sequences of four latrotoxins shows that they contain conserved regions extending over the entire proteins. The latrotoxin molecule can be divided into four structural domains (Fig.4). The first domain made up of 14-38 amino acids is removed upon the protein maturation. The N-terminal domain consists of 450-480 amino acid residues. The central domain of latrotoxins is entirely formed by ankyrin repeats discovered also in the proteins, which compose cytoskeleton and involved in differentiation and transcription processes[19]. The forth latrotoxin C-terminal domain consisting of about 160 amino acid residues is removed upon the toxin maturation.

Figure 4. Schematic representation of latrotoxin structure. Cleavage of α-latrotoxin precursor by furin-like enzyme. The arrows mark the positions of post-translational cleavage

Genes encoding α-LTX, α-LIT and δ-LIT were sequenced through cDNA, and only one gene encoding α-LCT, at the genomic DNA level[20]. It turned out that the chromosomal gene of this toxin, despite its large size, is intronless. Then the overlapping fragments of the chromosomal DNA carrying genes for other toxins (α-LTX, α-LIT and δ-LIT) were PCR-amplified and cloned. Restriction analysis of the PCR products showed that all these genes are also intronless. This implies that the lack of introns is a common feature of black widow spider genes encoding high molecular mass neurotoxins[21].

Data available on the structure-functional properties of latrotoxins provide two models of their action[22]. The first stage is the toxin binding to the presynaptic receptor. Since ankyrin repeats according to many studies are involved in interprotein interaction, the reason arises for proposing that in case of latrotoxins they promote the formation of their tetramers. The next stage envisages the insertion of the fragments of several toxin molecules into the membrane with the formation of the cation channel, through which Ca^{2+} ions penetrate into the presynaptic ending. Of much importance here are two conservative hydrophobic regions found in the N-terminal domain of the latrotoxins. According to the other model the toxin binding to the receptor, which interacts with the neurosecretion proteins, activates the receptor and initiates secretion, if Ca^{2+} ions being absent. Noteworthy, the two models do not exclude each other; really the combine mechanisms can simultaneously take place.

Certain molecular latrotoxins properties are not typical of the majority of animal neurotoxins, but more inherent for channel-forming bacterial toxins. First of all the latrotoxins were synthesized as larger inactive precursors, post-translationally cleaved by furin-like protease in the N- and C-terminal parts (Fig.4). This affirmation was confirmed when functional expression of α-LTX in baculovirus system was accomplished[23]. After processing latrotoxins are oligomerized and form their active tetramer of molecular mass about 500 kDa. This tetramer is bound to its presynaptic receptor, then interacts with the presynaptic membrane forming rather a large pore; as a result neurotransmitter release is significantly increased.

Some Arthropod toxins can be considered as classical example of channel-forming molecules. So, a novel class of polypeptide toxins (ectatomins) has been discovered in the venom of tropical stinging ants *Ectatomma tuberculatum* and *E. qudridens*. The major toxic component of the *E. tuberculatum* venom was named ectatomin (Et-1). The molecular mass of ectatomin was estimated to be 7928 Da. The protein consists of two highly homologous polypeptide chains (47% identity), which are linked to each other by a disulfide bond with an internal disulfide bridge in each[24,25].

The affinity purified polyclonal antibodies against ectatomin allowed to identify and isolate a novel structural homologue of ectatomin (Et-2) in the *E. tuberculatum* venom and in the venom of closely related ant species *E. qudridens* (Eq-1 and Eq-2). The structure of all studied ectatomins was confirmed by cloning and sequencing of the corresponding cDNAs. It was shown that the main structural moiety of this toxin class (Et-1, Et-2, Eq-1, Eq-2) is two highly homologous polypeptide chains linked with a disulfide bond, each consisting of 34-40 amino acid residues with clearly presented clusters of positively charged lysine residues.

Et-1 spatial structure was determined by two-dimensional NMR spectroscopy techniques[26]. Ectatomin in aqueous solution forms a bundle of four α-helices. The spatial structure of the two ectatomin chains are similar, which accords with their rather high sequence homology. Each chain consists of two antiparallel α-helices connected by the hinge region of 4 residues. Two intrachain disulfide bridges connect the ends of the antiparallel α-helices and stabilize their relative positioning. The third disulfide bridge connects the two hinge regions. A substantial part of the molecular surface is positively charged, due to the large number of surface lysine side chains which are exposed to the solvent and mobile. Non-polar amino acid residues form a hydrophobic core of the protein. Both ectatomin chains have a pronounced amphiphilic structure. The prevalence of positively charged residues at the molecule surface ensures its interaction with negatively charged molecules (e.g., lipids).

The particular feature of the ectatomin structure suggest the possibility of its interaction with cell membranes. At the first stage, lipid molecules destroy the hydrophobic and electrostatic interactions between the A and B chains of the toxin and the chains separate. It is likely that after this stage a deployed Et-1 molecule sticks to the membrane surface, such that the hydrophobic surface of the amphiphilic helices are submerged into the lipid bilayer (Fig.5a). Then the two Et-1 molecules dimerize and penetrate into the lipid bilayer. There are many ways in which two Et-1 molecules can form a dimer in membrane bilayer. Two possibilities are presented in Figs.5b and

5c. Channel-forming properties of ectatomins may account for their toxic activity. On the other hand, high level of Et-1 toxic activity may have effects other than pore formation in plasma membrane. On the base the whole-cell perforated patch-clamp technique it was shown that ectatomin at concentrations of 0.01-10 nM inhibited cardiac L-type calcium current[27].

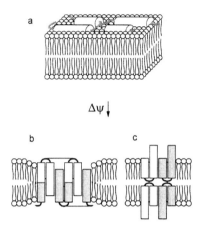

Figure 5. Scheme of ectatomin insertion into membranes suggested by the structure of Et-1 in aqueous solution
a - conformational rearrangement of the ectatomin molecule, so that hydrophobic surfaces of amphiphilic helices are submerged into the lipophilic part of the membrane;
b,c – two possible ways of ectatomin dimerization and channel formation in the presence of membrane potencial $\Delta\psi$

ACKNOWLEDGEMENTS

We thank our colleague A.S.Arseniev for providing the NMR data and for helpful discussion.

REFERENCES

1. McCormick, K.D., Meinwald J., 1993, *J. Chem. Ecology*, **19**, 2411-51.
2. Grishin, E.V., Volkova, T.M., Arseniev, A.S., Reshetova, O.S., Onoprienko, V.V., Magazanik, L.G., Antonov, S.M., Fedorova, I.M., 1986, *Bioorg Khim (Russia)*, **12**, 1121-24.
3. Grishin, E.V., Korolkova, Yu.V., Kozlov, S.A., Lipkin, A.V., Nosyreva, E.D., Pluzhnikov, K.A., Sukhanov, S.V., Volkova, T.M., 1996, *Pure & Appl. Chem.*, **68**, 2105-09.
4. Filippov, A.K., Kozlov, S.A., Pluzhnikov, K.A., Grishin, E.V., Brown, D.A., 1996, *FEBS Lett.*, **384**, 277-80.

5. Jaravin, V.A., Nolde, D.E., Reibarkh, M.J., Korolkova, Yu.V., Kozlov, S.A., Pluzhnikov, K.A., Grishin, E.V., Arseniev, A.S., 1997, *Biochemistry,* **36**,1223-32.
6. Korolkova, Yu.V., Kozlov, S.A., Lipkin, A.V., Pluzhnikov, K.A., Hadley, J.K., Filippov, A.K., Brown, D.A., Angelo, K., Strøbæk, D., Jespersen, T., Olesen, S.-P., Jensen, B.S., Grishin, E.V., 2001, *J. Biol. Chem.*, **276**, 9868-76.
7. Rosenthal, L., Meldolesi, J., 1989, *Pharmacology & Therapeutics,* **42**, 115-34.
8. Matteoli, M., Haimann, C., Torri-Tarelli, F., Polak, J.M., Ceccarelli, B., DeCamilli, P., *Proc. Natl. Acad. Sci. USA,* **85**, 7366-70.
9. Lunev, A.V., Demin, V.V., Zaitsev, O.I., Spadar, S.I., Grishin, E.V., 1991, *Bioorg. Khim. (Russia),* **17**, 1021-26.
10. Hurlbut, W.P., Iezzi, N., Fesce, R., Ceccarelli, B., *J. Physiol.*, **425**, 501-26.
11. Ushkaryov, Yu.A., Petrenko, A.G., Geppert, M., Sudhof, T.C., 1992, *Science,* **257**, 50-6.
12. Davletov, B.A., Shamotienko, O.G., Lelianova, V.G., Grishin, E.V., Ushkaryov, Yu.A., 1996, *J. Biol. Chem.*, **271**, 23239-45.
13. Fritz, L.C., Tzen, M.C., Mauro, A., 1980, *Nature,* **283**, 486-7.
14. Grishin, E., 1994, *Pure & Appl. Chem.*, **66**, 783-90.
15. Shatursky, O.Y., Pashkov, V.N., Bulgakov, O.V., Grishin, E.V., 1995, *Biochim. Biophys. Acta,* **1233**, 14-20.
16. Kiyatkin, N.I., Dulubova, I.E., Chekhovskaya I.A., Grishin, E.V., 1990, *FEBS Lett.,* **270**, 127-31.
17. Kiyatkin, N., Dulubova, I., Grishin, E., 1993, *Eur. J. Biochem.*, **213**, 121-7.
18. Dulubova, I.E., Krasnoperov, V.G., Khvotchev, M.V., Pluzhnikov, K.A., Volkova, T.M., Grishin, E.V., Vais, H., Bell, D.R., 1996, *J. Biol. Chem.*, **271**, 7535-43.
19. Michaely, P., Bennett, V., 1992, *Trends Cell Biol.*, **2**, 127-9.
20. Danilevich, V.N., Lukyanov, S.A., Grishin, E.V., 1999, *Russ J. Bioorg. Chem.*, **25**, 477-86.
21. Danilevich, V.N., Grishin, E.V., 2000, *Russ J. Bioorg. Chem.*, **26**, 838-43.
22. Grishin, E., 1998, *Toxicon,* **36**, 1693-701.
23. Volynski, K.E., Nosyreva, E.D., Ushkaryov, Yu.A., Grishin, E.V., 1999, *FEBS Lett.*, **442**, 25-8.
24. Arseniev, A.S., Pluzhnikov, K.A., Nolde, D.E., Sobol, A.G., Torgov, M.Yu., Sukhanov, S.V., Grishin, E.V., 1994, *FEBS Lett.,* **347**, 112-6.
25. Pluznikov, K.A., Nolde, D.E., Tertishnikova, S.M., Sukhanov, S.V., Sobol, A.G., Torgov, M.Yu., Filippov, A.K., Arseniev, A.S., Grishin, E.V., 1994, *Bioorg. Khim. (Russia),* **20**, 857-71.
26. Nolde D.E., Sobol A.G., Pluzhnikov K.A., Grishin E.V., Arseniev A.S., 1995, *J. Bio. NMR,* **219**, 1-13
27. Pluzhnikov K., Nosyreva E., Shevchenko L., Kokoz Yu., Shmalz D., Hucho F., Grishin E., 1999, *Eur. J. Biochem.*, **262**, 501-6.

Chemical Diversity of Coral Reef Organisms

TATSUO HIGA, MICHAEL C. ROY, JUNICHI TANAKA and IKUKO I. OHTANI
Department of Chemistry, Biology, and Marine Science, University of the Ryukyus, Nishihara, Okinawa 903-0213, Japan

1. INTRODUCTION

Coral reefs in the tropical and subtropical waters are habitats of diverse species of organisms, especially sessile invertebrates. These organisms have yielded numerous compounds having diverse chemical structures and biological activities. Some of them have been developed into anticancer agents which are now under clinical trials and some others as indispensable tools in biochemical research.

In this paper we present a brief overview of our research on coral reef organisms with an emphasis on the chemical diversity of their metabolites. The organisms used in our research encompass algae, mollusks, acorn worms, corals, sponges, and ascidians. We have found diverse classes of compounds ranging from simple aromatics to complex molecules of terpenoids, alkaloids, polyketides, and cyclic peptides. Many of these compounds showed significant biological activities including cytotoxic, antimicrobial, antiviral, and antimalarial activity.

2. ALGAE

The majority of algal secondary metabolites are terpenoids: sesquiterpenes from the red algal genus *Luarencia* and diterpenes from the brown algal family Dictyotaceae. Less abundant are monoterpenes from red algae and triterpenes from *Laurencia*. Other types of algal compounds include C_{15} acetogenins and halogenated aromatics.

The red alga *Portieria* (syn. *Chondrococcus*, *Desmia*) *hornemanni* is widely distributed in the Pacific and has been studied by several research groups. In all cases reported metabolites are halogenated monoterpenes. When we examined the alga collected in Okinawa, we obtained over a dozen cyclic and acyclic monoterpenes including cyclohexadienones **1-4**. Compound **1** was quite labile and spontaneously transformed into 4,5-dimethylbenzofuran on standing at room temperature[1]. Acetylation gave a stable monoacetate which showed in vitro antiviral activity.

From the brown alga *Dictyota spinulosa* we obtained 4-hydroxydictyodial (**5**) and the known dictyodial (**6**)[2] as feeding deterrent constituents. The diterpene **5** showed higher antifeeding potency than **6** in a test against the fish *Tilapia mossambica*. Recently a new diterpene, 3-bromobarekoxide (**7**) was obtained along with a number of sesquiterpenes from *Laurencia luzonensis*[3]. The absolute configuration of **7** was determined by X-ray. Chemical correlation of **7** with barekoxide (**8**), a known sponge metabolite, allowed determination of the absolute configuration and correction of the reported structure of the latter[4].

An earlier report on *Laurencia venusta* described the isolation of C_{15} acetogenins from a sample collected in Hokkaido[5]. From the same species collected in Okinawa we obtained a triterpene, venustatriol (**9**), together with two related triterpenes based on the squalene skeleton and some dozen known sesquiterpenes. Venustatriol exhibited significant antiviral activity[6].

Bromoindoles are the only metabolites which we and others have isolated from the alga *Laurencia brongniartii*[7]. We have so far isolated 19 indoles (e.g., **10**) including nine unpublished compounds. The alga is quite different from other *Laurencia* species, since it does not contain halogenated sesquiterpenes, C_{15} acetogenins, nor other metabolites characteristic to this genus. It may not be a *Laurencia* species and require taxonomic revision in the future. Some of these compounds showed antimicrobial activity and ichthyotoxicity. The sulfoxide **10** was optically active. It could be N-methylated with diazomethane and gave an unexpected product when treated with acetic anhydride and pyridine[7].

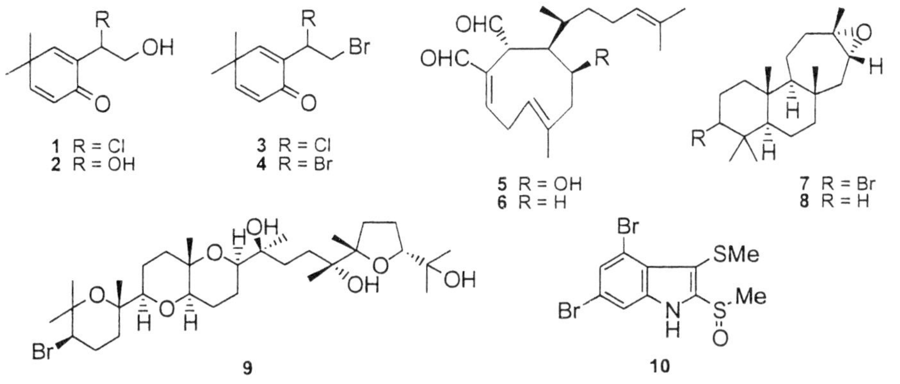

Figure 1.

3. MOLLUSKS

Among various mollusks, sea hares and nudibranchs belonging to the subclass *Opisthobranchia* have been main targets of chemical investigations. Sea hares usually contain toxic compounds sequestered from algae, while carnivorous nudibranchs possess compounds originated from invertebrate animals such as sponges. Therefore, their constituents vary depending on available diets.

Two classes of sesquiterpenes, chamigrane and cuparane, have been isolated from two different collections of the sea hare *Aplysia dactylomela*. One collection gave four halogenated chamigrenes (e.g.,**11**)[8] and another collection five cuparene-derived sesquiterpenes (e.g.,**12**)[9]. Obviously both classes are the compounds of *Laurencia* origin. Cyclolaurenol (**12**) and cupalaurenol were the first examples in which the positions of bromine and hydroxyl substituents are opposite from those usually found in related compounds. Compound **11** showed antiviral activity and **12** antifungal activity.

Sesquiterpenes containing an unusual functional group N=CCl$_2$ were first reported by Faulkner from a sponge in 1977[10]. About ten related compounds, carbonimidic dichlorides, have been described from three sponges prior to our recent isolation of the relatives, reticulidins (A: **13**) from the nudibranch *Reticulidia fungia*[11]. We subsequently found the reticulidins and new carbonimidic dichlorides from the sponge *Stylotella aurantium*[12]. These compounds exhibited significant cytotoxicity.

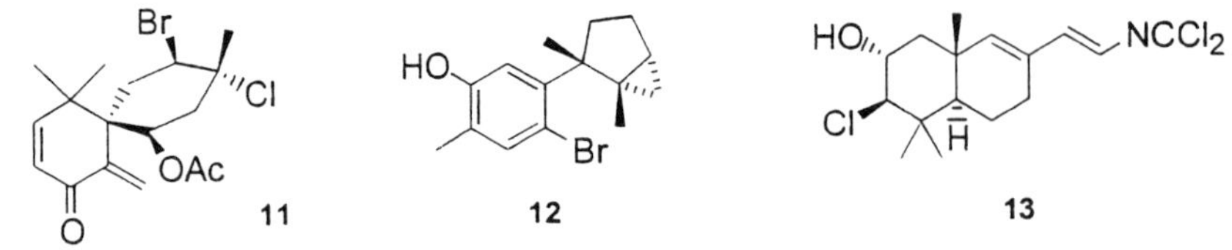

Figure 2.

4. ACORN WORMS

Acorn worms are fragile animals belonging to the order Enteropneusta (phylum Hemichordata) and living in sandy or muddy flats. We have so far examined six species of acorn worms, four from Japan and two from Hawaii. All of them contained halogenated phenols and/or indoles. A new species from Hawaii contained brominated cyclohexenes in addition to phenols and indoles. Simple halogenated phenoles and indoles were responsible for their peculiar odor described as "iodoform-like odor."

Both the Hawaiian *Ptychodera flava laysanica* and Okinawan *Ptychodera flava* contained a number of halogenated phenols (e.g., **14-16**) and indoles[13]. Their compositions were similar. Their odoriferous constituents were 3-chloroindole and 3-bromoindole. A species of *Glossobalanus* contained phenols with less number of bromine: that is, free hydroquinone, bromohydroquinone and 2,6-dibromohydroquinone. It also contained 4,6-dibromoindoles having no substitution at the C-3 position[14]. Its odor was due to 3-bromoindole. On the other hand 2,6-Dibromophenol was responsible for the odor of two species of *Balanoglossus* . Many of these metabolites showed antimicrobial activity.

In 1983 a new species of acorn worm was discovered from underwater caves of Maui, Hawaii. Chemical investigation on a specimen gave rise to the isolation of five new bromocyclohexenes including cyclohexenone **17**[15] . It also gave simple bromophenols and 3,4,6-tribromoindole. Biogenesis of the cyclohexenes has been proposed via epoxidation of 2,6-dibromophenol which was one of the constituents. The cyclohexenes exhibited potent cytotoxicity.

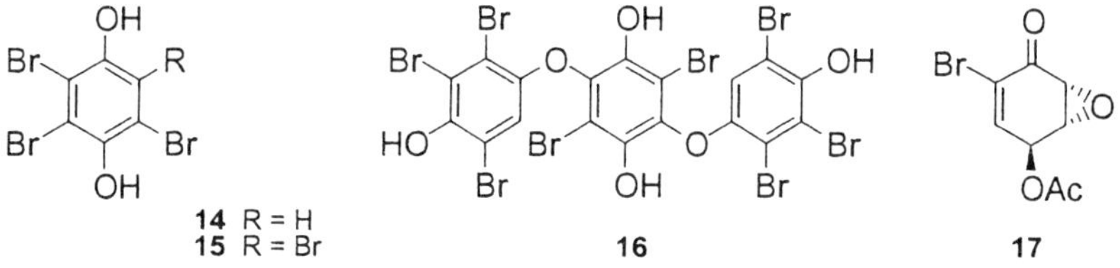

Figure 3.

5. CORALS

A large number of metabolites have been reported from alcyonaceans (soft corals) and gorgonians belonging to the subclass *Octocorallia*. They are mainly diterpenes of various structural classes.

A species of the gorgonian genus *Acalycigorgia* gave blue pigments, guaiazulene, linderazulene, and 2,3-dihydrolinderazulene which exhibited cytotoxicity[16]. A related cytotoxic sesquiterpene, echinofuran (**18**), has been isolated from the gorgonian *Echinogorgia praelonga*[17]. Four other species of *Acalycigorgia* gave a norditerpene, ginamallene (**19**), having a terminal allene functionality. It showed significant antifungal activity[18].

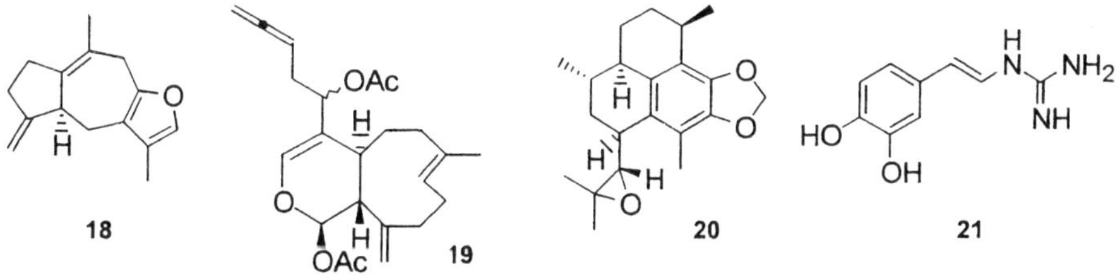

Figure 4.

The blue coral *Heliopora coerulea* yielded several diterpenes (e.g., helioporin A: **20**) related to pseudopterosins known from a Caribbean gorgonian. Some of the helioporins exhibited antiviral activity[1]. A stony coral belonging to the subclass Hexacorallia gave an antiviral phenol, tubastrine (**21**)[20]. We have also examined several species of reef-building hexacorals, some of which contained polyacetylenes (e.g., **22**) having antimicrobial activity[21].

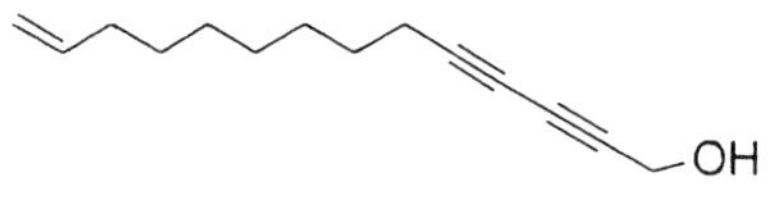

Figure 5.

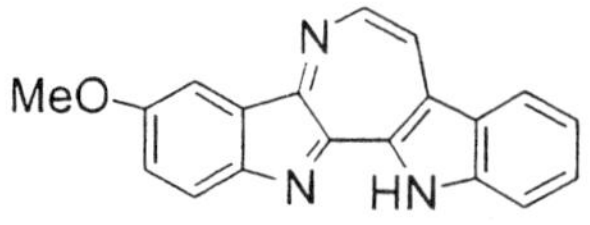

6. ASCIDIANS

Ascidians are a source of a variety of interesting molecules. We recently found a new class of bisindole pigments, iheyamins (e.g., **23**), from a colonial ascidian, *Polycitorella* sp.[22] The core of the iheyamins is new heteroaromatic system which opens up possibility to explore new heterocyclic chemistry. Iheyamins showed moderate cytotoxicity.

7. SPONGES

Sponges are a major source of diverse marine natural products. Many of them have unique structures with interesting biological activities. We have studied a number of species collected from Okinawa and other Western Pacific regions. Due to the space limitation discussion here is limited to some representative structures of several classes.

Terpenoids are abundant metabolites of sponges. We have reported sesquiterpenes[12, 23], diterpenes[24, 25], and sesterterpenes[26, 27]. Umabanol (**24**) is a unique tetracarbocyclic diterpene isolated from the sponge *Epipolasis kushimotoensis*[24]. The structure of umabanol is related to the tricyclic verrucosanes known from terrestrial plants and also from a marine sponge. Norsesterterpenes, mycaperoxides (e.g., **25**), isolated from a Thai sponge, *Mycale* sp., exhibited significant cytotoxicity and in vitro antiviral activity in an initial screening[27]. Further evaluation revealed in vivo cytostatic activity of **25** against human xenograft model lung carcinoma in mice.

A variety of macrolides have been described from sponges. We have discovered several unique macrolides. Zampanolide (**26**) was isolated as a minor constituent of a rare sponge tentatively identified as *Fasciospongia rimosa*[28]. It showed potent cytotoxicity against several tumor cell lines. From *Polyfibrospongia* sp. was found a unique macrolide, miyakolide (**27**), having structural features related to bryostatins, an anticancer agent under clinical trials[29]. However, miyakolide showed only marginal in vivo antitumor activity against P388 mouse leukemia. Misakinokide A from

Theonella sp. is a dimeric *bis*-lactone exhibiting potent cytotoxicity[30]. It shows a unique mode of action on actin cytoskeleton[31].

We have also reported a number of alkaloids or nitrogenous compounds including cyclic peptides from sponges. Of these compounds manzamines are the best known complex alkaloids. Manzamine A (**28**) has recently been shown to have potent in vivo antimalarial activity against the rodent parasite *Plasmodium berghei*[32].

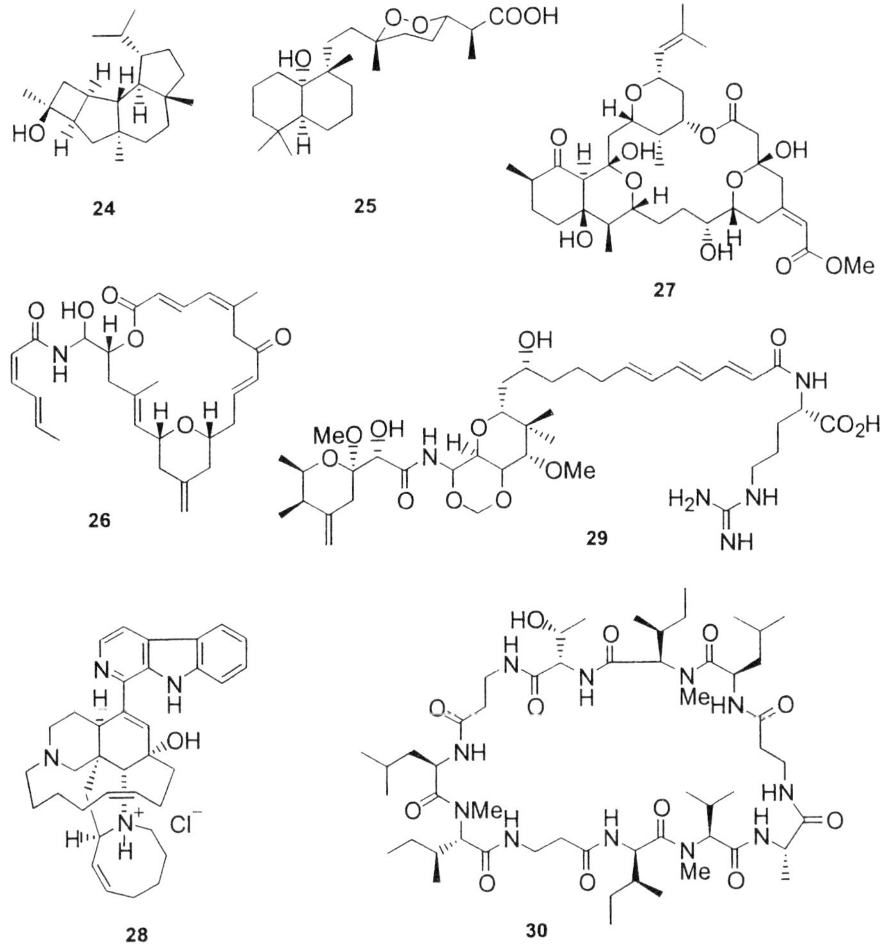

Figure 6.

Other nitrogenous compounds from Okinawan sponges include antiviral onnamide A (**29**)[33], antiviral hennoxazoles[34], echinoclathrines[35] having weak immunosuppressive activity and the cytotoxic cyclic peptide cupolamide A[36]. The latter was also isolated from an Indonesian collection of the same species *Theonella cupola*. Another Indonesian sponge, *Theonella swinhoei*, gave several new cyclic peptides and depsipeptides including barangamide A (**30**)[37].

8. CONCLUSION

Coral reefs harbor a great diversity of species. As shown by the above examples we have discovered a variety of natural products from various coral reef organisms. Although all of these compounds roughly fall under the categories of conventional structural classification, most of them are unique to marine organisms and have no terrestrial counterparts. In the coral reefs chemical diversity is as great as species diversity. Many compounds have biological activity and are potentially important as leads for new medicines and as biochemical research tools.

More than a dozen of our compounds have been targets of total synthesis in the laboratories around the world. Manzamine A, for example, has been challenged by at least ten groups of the world class synthetic chemists, two of which have so far completed the total syntheses.

REFERENCES

1. Higa, T., 1985, *Tetrahedron Lett.*, **26**, 2335-6.
2. Tanaka, J., Higa, T., 1984, *Chem. Lett.*, 231-2.
3. Kuniyoshi, M., Marma, M.S., Higa, T., Bernardinelli, G., Jefford, C.W., 2001, *J. Nat. Prod.*, **64**, 696-700.
4. Rudi, A., Kashman, Y., 1992, *J. Nat. Prod.*, **55**, 1408-14.
5. Suzuki, M., Kurosawa, E., 1980, *Chem. Lett.*, 1177-80.
6. Sakemi, S., Higa, T., Jefford, C.W., Bernardinelli, G., 1986, *Tetrahedron Lett.*, **27**, 4287-90.
7. Tanaka, J., Higa, T., Bernardinelli, G., Jefford, C.W., 1989, *Tetrahedron*, **45**, 7301-10.
8. Sakai, R., Higa, T., Jefford, C.W., Bernardinelli, G., 1986, *Helv. Chim. Acta*, **69**, 91-105.
9. Ichiba, T., Higa, T., 1986, *J. Org. Chem.*, **51**, 3364-6.
10. Wratten, S.J., Faulkner, D.J., 1977, *J. Am. Chem. Soc.*, **99**, 7367-8.
11. Tanaka, J., Higa, T., 1999, *J. Nat. Prod.*, **62**, 1339-40.
12. Musman, M., Tanaka, J., Higa, T., 2001, *J. Nat. Prod.*, **64**, 111-3.
13. Higa, T., Fujiyama, T. Scheuer, P.J., 1980, *Comp. Biochem. Physiol.*, **65B**, 525-30.
14. Higa, T., Ichiba, T., Okuda, R.K., 1985, *Experientia*, **41**, 1487-8.
15. Higa, T., Okuda, R.K., Severns, R.M., Scheuer, P.J., He, C.-H., Changfu, X., Clardy, J., 1987, *Tetrahedron*, **43**, 1063-70.
16. Sakemi, S., Higa, T., 1987, *Experientia*, **43**, 624-5.
17. Tanaka, J., Miki, H., Higa, T., 1992, *J. Nat. Prod.*, **55**, 1522-4.
18. Hokama, S., Tanaka, J., Higa, T., Fusetani, N., Asano, M., Matsunaga, S., Hashimoto, K., 1988, *Chem. Lett.*, 855-6.
19. Tanaka, J., Ogawa, N., Liang, J., Higa, T., Gravalos, D. G., 1993, *Tetrahedron*, **49**, 811-22.

20. Sakai, R., Higa, T., 1987, *Chem. Lett.*, 127-8.
21. Higa, T., Tanaka, J., Kohagura, T., Wauke, T., 1990, *Chem. Lett.*, 145-8.
22. Sasaki, T., Ohtani, I.I., Tanaka, J., Higa, T., 1999, *Tetrahedron Lett.*, **40**, 303-6.
23. Pham, A.T., Ichiba, T., Yoshida, W.Y., Scheuer, P.J., Uchida, T., Tanaka, J., Higa, T., 1991, *Tetrahedron Lett.*, **32**, 4843-6.
24. Tanaka, J., Nurrachmi, I., Higa, T., 1997, *Chem. Lett.*, 489-90.
25. Sharma, H.A., Tanaka, J., Higa, T., Lithgow, A., Bernardinelli, G., Jefford, C.W., 1992, *Tetrahedron Lett.*, **33**, 1593-6.
26. Musman, M., Ohtani, I.I., Nagaoka, D., Tanaka, J., Higa, T., 2001, *J. Nat. Prod.*, **64**, 350-2.
27. Tanaka, J., Higa, T., Suwanborirux, K., Kokpol, U., Bernardinelli, G., Jefford, C.W., 1993, *J. Org. Chem.*, **58**, 2999-3002.
28. Tanaka, J., Higa, T., 1996, *Tetrahedron Lett.*, **37**, 5535-8.
29. Higa, T., Tanaka, J., Komesu, M., Gravalos, D.G., Fernandez Puentes, J.L. Bernardinelli, G., Jefford, C.W., 1992, *J. Am. Chem. Soc.*, **114**, 7587-8.
30. Tanaka, J., Higa, T., Kobayashi, M., Kitagawa, I., 1990, *Chem. Pharm. Bull.*, **38**, 2967-70.
31. Terry, D.R., Spector, I., Higa, T., Bubb, M.R., 1997, *J. Biol. Chem.*, **272**, 7841-5.
32. Ang, K.K.H., Holmes, M.J., Higa, T., Hamann, M.T., Kara,U.A.K., 2000, *Agents Chemother.*, **44**, 1645-9.
33. Sakemi, S., Ichiba, T., Kohmoto, S., Saucy, G., Higa, T., 1988, *J. Am. Chem. Soc.*, **110**, 4851-3.
34. Ichiba, T., Yoshida, W.Y., Scheuer, P.J., Higa, T., Gravalos, D.G., 1991, *J. Am. Chem. Soc.*, **113**, 3173-4.
35. Kitamura, A., Tanaka, J., Ohtani, I.I., Higa, T., 1999, *Tetrahedron*, **55**, 2487-92.
36. Bonnington, L.S., Tanaka, J., Higa, T., Kimura, J., Yoshimura, Y., Nakao, Y., Yoshida, W.Y., Scheuer, P.J., 1997, *J. Org. Chem.*, **62**, 7765-7.
37. Roy, M.C., Ohtani, I.I., Ichiba, T., Tanaka, J., Satari, R, Higa, T., 2000, *Tetrahedron*, **56**, 9079-92.

Chemical Signals from Sponges and their Allelopathic Effects on Other Marine Animals

MARY J. GARSON
The University of Queensland, Department of Chemistry, Brisbane QLD 4072, Australia

1. INTRODUCTION

In their natural environments, sponges are subject to intense competition for space and for resources such as nutrients. Many of these organisms are soft-bodied, yet inhabit areas of intense predation pressure. Indeed, on coral reefs, sponges are the second most abundant biomass after corals. The ecological success of this group of colourful marine animals may be enhanced by use of a chemical defense strategy. A multitude of structurally-complex natural products representing all the major biosynthetic classes (terpene, alkaloid, polyketide etc) have been isolated from marine sponges[1]. Natural chemical signals (allelochemicals) from sponges are likely responsible for inducing or inhibiting the settlement of larvae of coral reef animals. While some of these chemicals are universally toxic to animal larvae, many others may differentially affect the physiology and development of specific types of larvae. One specialised group of marine animals, the nudibranchs, feed on marine sponges and may utilise toxic sponge chemicals as part of their own defensive strategy. In this paper, I discuss recent examples from our laboratory which illustrate the role played by sponge allelochemicals in underwater chemical "warfare"

2. A SPONGE ALLELOCHEMICAL INDUCES ASCIDIAN SETTLEMENT BUT INHIBITS METAMORPHOSIS

The alkaloid fraction of the tropical sponge *Haliclona* sp. is strongly cytotoxic and antifungal, is toxic to crustaceans and fish, kills neighboring coral tissue and deters feeding on the sponge tissue by reef fish[1].

The complex 3-alkylpiperidine alkaloid haliclonacyclamine A (**1**) was identified as the major bioactive metabolite of the sponge[2]. This chemical also has a very specific biological effect on *Herdmania curvata*, a solitary ascidian that inhabits the same reef crest and slope locations as *Haliclona* sp. at our study site at Heron Island. Competent larvae of *H. curvata*, when exposed to haliclonacyclamine A (HA) at 5, 10 or 25 µg, were induced to settle and initiate metamorphosis in greater numbers compared to filtered seawater (FSW) or KCl-elevated FSW controls[3].

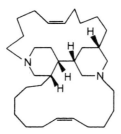

(**1**) Haliclonacyclamine A

At the three concentrations tested, HA consistently induced a higher rate of metamorphosis. However the HA-treated larvae became developmentally arrested about 4 - 6 h after the initiation of metamorphosis, just after tail resorption is completed, and undergo necrosis as evidenced by cellular lysis within 24 h. In contrast, the KCl-treated larvae had developed the typical morphological characteristics of juvenile ascidians. By incorporation of

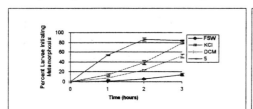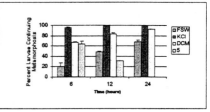

Figure 1. (a) Induction of *Herdmania curvata* larval settlement; (b) Subsequent inhibition of *Herdmania curvata* metamorphosis; in both (a) and (b) haliclonacyclamine A is at 5 µg/ml; FSW = filtered sea water and KCl is 50 µm-elevated FSW.

radiolabelled nucleotides, it was established that there is significant synthesis of mRNAs and proteins immediately after larval tail resorption. In contrast to the known transcriptional inhibitor, actinomycin D, HA does not affect overall transcriptional rates in *H. curvata* larvae and must therefore disrupt a different developmental process[3].

3. SPONGE-NUDIBRANCH INTERACTIONS

3.1 The Chemistry and Sponge Diet of *Asteronotus cespitosus* [1]

Shireen Fahey works on the systematics, molecular biology, and chemistry of the mollusc *Halgerda* spp., a genus that, to us is a chemical **unicorn** because these animals are infrequently found, at least on the East coast of Australia where we routinely dive. In her studies, the genus *Asteronotus* has been selected as a "sister group" for taxonomy, however the chemistry of this group of cryptic, well-camouflaged mollusc had not been previously studied. Extracts of the dorid nudibranch *Asteronotus cespitosus* from two geographically separate regions of Australia, and from the Philippines, were compared using thin-layer, high performance liquid and gas chromatography and ^{1}H NMR analysis (Figure 2)[4]. The major component detected in digestive tissue of specimens from the Great Barrier Reef in Northeastern Australia was 2-(2', 6'-dibromophenoxy) – 4,6-dibromophenol (**2**), with minor amounts of 2-(3', 5'-dibromo-2'-phenoxy) – 2,6-dibromoanisole (**3**). In a specimen collected from North-western Australia, only 2-(3', 5'-dibromo-2'-phenoxy) – 2,6-dibromoanisole was found, while a specimen from the Philippines contained 2-(2'-bromophenoxy-3,4,5,6-tetrabromophenol (**4**) together with a novel chlorinated pyrrolidone (**5**). In addition, the sesquiterpenes dehydroherbadysidolide (**6**) and spirodysin (**7**) were detected in the digestive organs and mantle tissue of the nudibranchs from the Great Barrier Reef and from the Philippines, whereas these chemicals were not found in the specimen from Northwestern Australia. All of the chemicals (**2**)-(**4**), (**6**)-(**7**) have previously been isolated from the tropical marine sponge *Dysidea herbacea*[5], as have chemicals closely related to (**5**)[6]. The bromophenol metabolites were strongly antibacterial and cytotoxic. When assayed against larvae of the solitary ascidian *Herdmania curvata*, bromophenols (**2**) and (**3**) inhibited larval development more strongly than the sesquiterpenes.

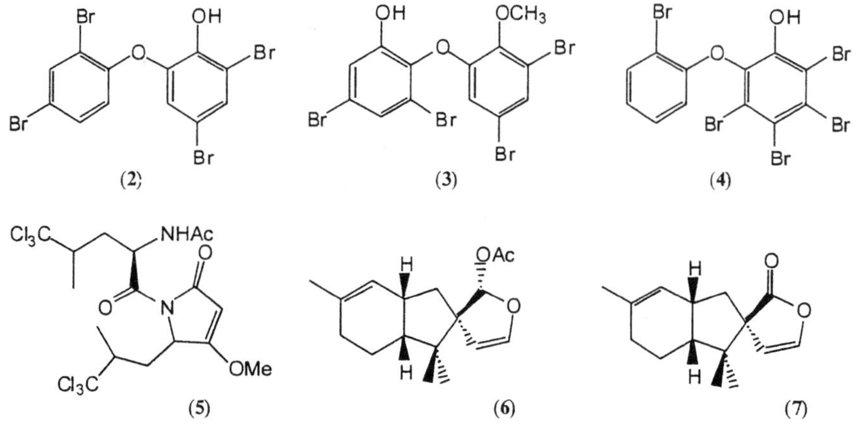

Figure 2. Selected metabolites isolated from *Asteronotus cespitosus*

Hence, *Asteronotus cespitosus* appears to acquire certain potent chemicals from this sponge, sequestering sesquiterpenes in its mantle tissue, while eliminating others such as bromophenols or chlorinated alkaloids that may be too potent to incorporate into body tissue. This is the first time the characteristic halogenated metabolites of *Dysidea herbacea* have been reported in a carnivorous mollusc, a result that strongly suggests a dietary origin as opposed to *de novo* synthesis. Furthermore, since the bromophenols, alkaloids and terpenes found in *A. cespitosus* arise from distinct biosynthetic pathways, it is highly unlikely that this mollusc could be capable of *de novo* synthesis of all these metabolites.

3.2 Phyllidid Nudibranchs and "Isocyanide" Sponges

A common structural motif in marine terpene chemistry is the presence of an N_1-C_1 functional group (-NC, -NCS, -NHCHO are common; the rarer – SCN, –NCO and –N=CCl$_2$ groups are also known)[7]. When "isocyanide" sponges and their associated molluscs are collected from our site at Mooloolaba, we encounter some interesting structural diversity. The nitrogen-containing functionality in the extracts is inferred by GC-MS study since the isocyanide (with M$^+$ 231) and isocyanate (M$^+$ 247) sesquiterpene metabolites elute ahead of the isothiocyanate and thiocyanate metabolites (M$^+$ 263). A ^{13}C NMR study then aids functional group confirmation, provided an extended pulse delay of >10 sec is used to maximise signal intensity for the N_1-C_1 carbon. The chemical shifts are typically: -NC 156-

170 ppm; -NCS 126 – 141 ppm; -SCN 111-114 ppm; and –NCO 121 – 125 ppm.

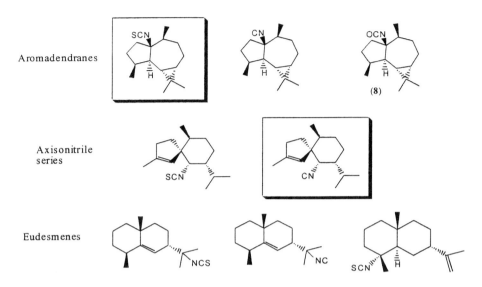

Figure 3. Selected metabolites isolated from *Acanthella cavernosa*. Metabolites also isolated from *Phyllidia ocellata* are boxed.

Specimens of *A. cavernosa* collected at Mooloolaba contained several known[8] sesquiterpenes (Figure 3), two of which were also found in the nudibranch *Phyllidia ocellata* that feeds on this sponge. A new isocyanate (**8**) was found in trace quantities in the sponge.

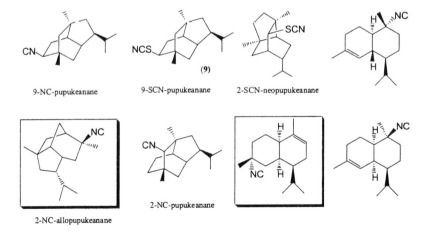

Figure 4. Selected metabolites isolated from the molluscs *Phyllidia varicosa* and the sponge *Axinyssa* n.sp. Boxed structures are found in both animals.

In contrast, the mollusc *Phyllidia varicosa* concentrates a range of sesquiterpene metabolites from its sponge prey, *Axinyssa* n.sp., which contains both tricyclic and bicyclic metabolites[9] (Figure 4). A new thiocyanate metabolite (**9**) was isolated from the nudibranch, but not yet from the sponge. When tested in the *H. curvata* assay, the various sponge sequiterpene fractions inhibited development, although purified *Acanthella* or *Axinyssa* metabolites were not individually effective. The reasons why the phyllidid molluscs selectively accumulated certain sponge chemicals, and the ecological implications of these two specific sponge-nudibranch associations, require further study.

4. CONCLUSION

Many of the sponge metabolites encountered in these chemical stories have been frequently isolated by marine researchers, subjected to bioassay, and then shelved as insufficiently active for detailed pharmacological study. But their biological effects are finely tuned, and maybe these chemicals have not yet been subjected to assays that might show their true potency. Through ecological study, we now appreciate that they act effectively to disturb the growth or survival of neighbouring organisms, and thus enhance the survival of the producer sponge or predatory nudibranch. The mechanisms by which these animals themselves deal with such toxic chemicals also now need our attention. There are potential applications of the research in human medicine (implants) and in the development of anti-fouling paints for use in the construction, aquaculture and transport industries.

ACKNOWLEDGEMENTS

Our research is funded by the Australian Research Council. I thank my senior co-workers Bernie Degnan and Greg Skilleter, and the enthusiastic postgraduates (sponge/ascidians Richard Clark, Kathryn Green, Kerry Roper; sponge/nudibranchs Shireen Fahey, Sharna Graham, Jamila Rogic, and Bronwin Stapleton) who have contributed to these emerging stories in marine chemical ecology.

REFERENCES

1. Garson, M.J., Clark, R.J., Webb, R.I., Kim L. Field, Charan, R.D., McCaffrey, E., 1999, *Mem. Qld. Mus.*, **44**, 205-13.

2. Clark, R.J., Field, K.L., Charan, R.C., Garson, M.J., Brereton, I.M., Willis, A.C., 1998, *Tetrahedron*, **54**, 8811-26.
3. Green, K.M., Russell, B.M., Clark, R.J., Jones, M.K., Skilleter, G.A., Garson, M.J.,Degnan, B.M., 2001, *Mar. Biol.*, (in press).
4. Fahey, S.A.,Garson, M.J., 2001, *J. Chem. Ecol.*, (submitted).
5. Cameron, G.M., Stapleton, B.L., Simonsen, S.M., Brecknell. D.J., Garson, M.J., 2000, *Tetrahedron*, **56**, 5247-52.
6. Unson, M.D, Rose, C.B., Faulkner, D.J., Brinen, L.S., Steiner, J.S., Clardy, J., 1993, *J. Org. Chem.*, **58**, 6336-42.
7. Garson, M.J., Simpson, J.S., Flowers, A.E., Dumdei, E.J., 2000, Cyanide and thiocyanate-derived functionality in marine organisms – structures, biosynthesis and ecology. In *Studies in Natural Products Chemistry* (Rahman, A.-ur., ed.), Vol. 21, Part B, Elsevier, Amsterdam, pp. 329-73.
8. Clark, R.J., Stapleton, B.L. Garson, M.J., 2000, *Tetrahedron*, **56**, 3071-6.
9. Simpson, J.S., Flowers, A.E., Garson, M.J., 2000, *Aust. J. Chem.*, **50**, 1123-7.

Anti-Cancer Metabolites from Marine Sponges

DENİZ TAŞDEMİR[*,#]
[*]*University of Utah, Department of Medicinal ChemistrySalt Lake City, Utah 84112, USA,*
[#]*Present address: Hacettepe University, Faculty of Pharmacy, Department of Pharmacognosy, 06100, Ankara, Turkey*

1. INTRODUCTION

The search of anti-cancer leads has been the mainstream of the marine natural product research. For many years, our research group at the University of Utah has focused on novel sponge metabolites with cytotoxic/antitumor activity. Here I will summarize the preliminary results of three major projects that involved isolation, structure elucidation and evaluation of anti-cancer activity of the secondary metabolites of the marine sponges, *Stylissa massa*[1], *Rhabdastrella globostellata*[2] (previously identified as a *Stelletta* sp.) and *Xestospongia* sp[3]. These studies were performed during my post-doctoral training with Dr. Chris M. Ireland at the University of Utah. These projects were supported by a National Cooperative Drug Discovery Group Grant (NCDDG), which was funded by the US National Cancer Institute (NCI). The majority of the sponge material came from the Philippines, as part of the collaboration within NCDDG between the University of the Philippines (Marine Science Institute) and the University of Utah. The biological investigations were carried out and are still underway at the Oncology Department of Wyeth Ayerst Research (NY) and at the Department of Pharmacology and Toxicology of the University of Utah.

Cancer research over the last two decades provided a rich and complex body of knowledge. Particularly, the identification of oncogenes that are activated, and tumor supressor genes that are lost, inactivated, or mutated in tumors brought a new dimension to this topic. Cells enter the cell cycle and perform the DNA synthesis in response to growth factors and extracellular stimuli. These signals are transmitted to the nucleus by signal transduction pathways that occur through phosphorylation of substrates via protein kinases. Disruption of cell cycle is the hallmark of the cancer. Improper kinase activity has also been implicated in numerous human diseases including cancer. Effectors of signal transduction are linked to cell cycle regulators and the pathways connecting these effectors and regulators have started to become apparent[4].

Eucaryotic cell cycle is composed of four distinc phases. DNA replication occurs in S (synthesis) phase and replicated DNA is distributed to the daughter cells in M (mitosis) phase. Between these two phases, there are two gap phases, G_1 and G_2. G_1 is the phase in which the cell is very sensitive to mitogenic signals and decides to proceed, pause or exit the cell cycle. Cell cycle transitions are controlled by cyclin-dependent kinases (CDKs) that require a cyclin regulatory unit for acivity. As shown in Figure 1, various cyclin/CDK pairs function during each phase of the cell cycle. CDK activity is tightly controlled by multiple mechanisms. For example, two families of CDK inhibitors, INK4 and Cip/Kip, oppose the activities of the various cyclin/CDKs (Figure 1). p21 ($p21^{WAF1/Cip1}$) is the first identified CDK inhibitor of Cip/Kip family that arrests the cell cycle by inhibiting specifically CDK2, and to a lesser extent, CDK3, 4, and 6[5]. There is strong evidence that p21 is a downstream target of the tumor suppressor gene p53, the most frequently mutated gene in human cancers. However, p21 expression may also be regulated in a p53-independent manner. In case of DNA damage, p21-dependent checkpoint control stops the progress of the cell cycle to allow for cellular repair, thus inhibit oncogenic proliferation[6].

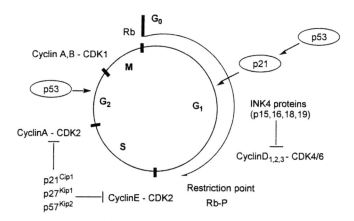

Figure 1. Cell cycle regulatory components

Signal transduction is a complex process that often involves the coordinated activation of cytoplasmic kinases. Ras/Raf/MEK/MAPK cascade is one of the best-characterized signaling pathways. Mitogen-Activated Protein Kinases (MAPKs), also called as ERKs (Extracellular Signal Regulated Kinases) are serine/threonine kinases that provide the transmission of the extracellular stimuli to the nucleus. Ras, the best-known family of oncogenes, triggers this cascade. A wide variety of hormones, growth factors, differentiation factors and tumor-promoting substances activate Ras proteins. Activated Ras binds to Raf kinases and causes their translocation to the cell membrane where they are activated by phosphorylation. Activated Raf-1 phosphorylates and activates MEK (MAP kinase kinase), which in turn phosphorylates and activates MAPKs. MEK has two isoforms, MEK-1 and MEK2, and Ras/Raf signaling complex formation favors MEK-1 activation[7]. Activated MAPKs then translocate to the nucleus and phosphorylates transcription factors and some cyclins to regulate proliferation, differentiation, apoptosis and cell survival. Ras is structurally altered in about 30% of all human cancers. Therefore, targeting of Ras and its downstream mediators became quite valuable for therapeutic intervention[8] and opened up new insights for medicinal chemists.

2. RESULTS AND DISCUSSION

2.1 *Stylissa massa (=Stylotella aurantium)*

Recently, our collaborators at Wyeth Ayerst Research laboratories developed a Raf/MEK-1/MAPK cascade ELISA assay[9] that uses activated Raf to activate MEK1, which in turn activates MAPK. Activation of MAPK through phosphorylation of Threonine and Tyrosine (202 and 204 of human MAPK) at the sequence TEY by a phospho-specific monoclonal antibody makes up the endpoint measurement of the ELISA. In order to discover small-molecule inhibitors of Ras/Raf/MEK/MAPK cascade, we have screened hundreds of marine sponge extracts. The $CHCl_3$ and MeOH-solubles of a Philippine sponge, *Stylissa massa*, showed activity at ng levels in this assay. Bioactivity-guided separation of these extracts gave eight known pyrrole alkaloids: aldisine (**1**), 2-bromoaldisine (**2**), 10Z-debromohymenialdisine (**3**), 10*E*-hymenialdisine (**4**), 10Z-hymenialdisine (**5**), hymenin (**6**), oroidin (**7**) and 4,5-dibromopyrrole-2-carbonamide (**8**). Table 1 illustrates the inhibitory activity of these compounds on the Ras/MAPK cascade. 2-Bromoaldisine (**2**, IC_{50} 539 nM), 10Zdebromohymenialdisine (**3**, IC$_{50}$ 881 nM) and hymenin (**6**, IC$_{50}$ 1288 nM) exhibited moderate to low inhibitory activity. Aldisine (**1**), oroidin (**7**) and 4,5-dibromopyrrole-2-carbonamide (**8**) were found to be inactive. The most potent compounds among all isolates were 10*E*-hymenialdisine (**4**) and 10*Z*-hymenialdisine (**5**) that showed the IC$_{50}$ values of 3 and 6 nM, respectively. It is interesting that both isomers showed almost the same activity. It was shown that 10*E*-hymenialdisine (**4**) converts to its 10*Z* derivative upon standing in DMSO[10], the solvent that we used as a vehicle in ELISA assay. Therefore, the activity of 10*E*-hymenialdisine (**4**) appears to be insignificant. We observed that geometrically pure **4** was replaced by a 1:1 mixture of both *E*/*Z* isomers during storage (- 20 °C) in a dry state.

R	R		
1 H	**3** H	**4**	**6**
2 Br	**5** Br		

7   **8**

Secondary assays were performed to identify the specific target of cascade inhibition by using either MEK1 or MAPK phospho-specific antibodies. All compounds showed the identical IC$_{50}$ values in the MEK to MAPK assays as in the Raf/MEK1/MAPK cascade ELISA, whereas none of the compounds exhibited activity in the Raf to MEK1 assay (Table 1), indicating them to be selective inhibitors of MEK-1. We included the IC$_{50}$ values of a general kinase inhibitor (staurosporine), and a specific MEK inhibitor (PD98059)[11], in both Raf/MEK1/MAPK ELISA and secondary assays on Table 1, for comparison.

Compounds **1-8** were tested for their ability to inhibit growth of two human colon tumor cell lines, LoVo and Caco-2. Only 10*E*-hymenialdisine (**4**) and 10*Z*-hymenialdisine (**5**) moderately inhibited growth of LoVo cells with the IC_{50}'s 586 and 710 nM, respectively. Compounds **4** and **5** showed much weaker activity towards Caco-2 cells (IC_{50}'s 3867 and 7799 nM, respectively).

Table 1. Kinase Inhibitory Activity of Compounds **1-8** (IC_{50} nM)

Compound	Raf/MEK/MAPK ELISA assay	Raf to MEK1 assay	MEK1 to MAPK assay
Aldisine (**1**)	>2500	2500	>2500
Bromoaldisine (**2**)	539	2500	539
10Z-Debromohymenialdisine (**3**)	881	2500	824
10E-Hymenialdisine (**4**)	3	2500	6
10Z-Hymenialdisine (**5**)	6	2500	9
Hymenin (**6**)	1288	2500	1288
Oroidin (**7**)	>2500	2500	>2500
4,5-dibromopyrrole-2-carbonamide (**8**)	>2500	2500	>2500
Staurosporine	2.5	2.5	2.5
PD98059	2800	>10000	2800

This study led the identification of a widespread bromopyrroloazepinone alkaloid, hymenialdisine, as a potent MEK-1 inhibitor. Despite its relatively modest potency on Lovo cells, hymenialdisine might provide a platform for development of more potent MEK1 inhibitors. Targeting of kinases is a new and exciting area for drug discovery community, although it sounded impossible before. We are aware that hymenialdisine is not specific inhibitor of MEK, since this compound has recently been reported to inhibit some other kinases, such as GSK-3β, CDK1, 2, 5, CK1[12], Chk1 and Chk2[13]. We are currently trying to assess the kinase activity of hymenialdisine on a very broad basis to find out if hymenialdisine is a non-specific kinase inhibitor, or, if it inhibits a family of related kinases. This will probably help us to understand cross talkings between Ras/Raf/MEK/MAPK cascade and other signaling systems.

2.2 *Rhabdastrella globostrella* (*Stelletta* sp.)

Isomalabaricane triterpenes are cytotoxic yellow pigments composed of a tricyclic terpenoid core and a conjugated side chain. They are rarely found in nature and have so far been described from four marine sponge genera, *Stelletta*, *Jaspis*, *Rhabdastrella* and *Geodia*. Isomalabaricanes are photosensitive compounds. Upon exposure to light, they rapidly undergo to a 1:1 equilibration of 13*E*/13*Z* isomers.

The specimen of *Rhabdastrella globostellata* (referred as a *Stelletta* sp. in the abstract) was collected from Mindanao, the Philippines. The crude MeOH extract was subjected to a standard solvent partitioning scheme. Bioactivity-guided isolation using the wild type (WT, $p21^{+/+}$) and p21 deficient ($p21^{-/-}$) HCT-116 cell lines in which the p21 gene was disrupted through homologous recombination was carried out on the $CHCl_3$ extract. Employment of C-18 flash CC followed by SiO_2 HPLC afforded two new isomalabaricanes, stellettin H (**9**) and I (**10**), and the known compounds, stellettins A (**11**), B (**12**), C (**13**), D (**14**), the optical antipode of stellettin E (**15**) and rhabdastrellic acid-A (**16**).

9 R=OCOCH$_3$, R$_1$=H
16 R=R$_1$=O

10 R=OCOCH$_3$, R$_1$=H
15 R=R$_1$=O

11 R=R$_1$=O
13 R=OCOCH$_3$, R$_1$=H

12 R=R$_1$=O
14 R=OCOCH$_3$, R$_1$=H

During the process of this project, we avoided light by wrapping all glassware with aluminum foil and stored all extracts, fractions and pure compounds at - 80 °C. Further, we acquired the NMR data of the isolates on 3 mm micro inverse-detection probe (Nalorac MDBG500-3 mm) immediately after their isolation. The use of conventional 3 mm micro probe allowed the sample to be dissolved in very small amount of (80-100 µl) deuterated solvent, and shortened the measurement time tremendously by increasing NMR sensitivity[14].This was not only extremely helpful in reducing isomer signals during measurement and having fairly cleaner spectra, but also allowed us to determine the bioactivity of the isolates before isomerization process became significant.

Compounds **9-16** were tested in a panel of isogenic colorectal cancer cells, wild type HCT-116 and the corresponding $p21^{-/-}$ HCT-116 cell line. Only stellettin B (**12**) and (-)-stellettin E (**15**) showed selective toxicity towards the $p21^{-/-}$ HCT cell line (IC_{50}'s of 43 and 39 nM, respectively). These compounds were also tested in the corresponding WT ($p53^{+/+}$) and $p53^{-/-}$ HCT-116 cell lines, but showed no differential. These results are quite interesting and may either suggest that stellettin B and (-)-stellettin E act in a p21 dependent manner, or that p21 deficient cells are more sensitive to the DNA damaging effects of **12** and **15** than WT cells and lead them to go through apoptosis. We are currently performing further investigations to understand the mechanism of action of these compounds. It is interesting that both stellettin B (**12**) and (-)-stellettin E (**15**) contain a keto function at C-3 and adopt Z geometry at Δ^{13}. Due to their instability, isomalabaricanes have been tested as isomeric mixtures in the past. The obvious differences observed concerning the bioactivity of 13Z and 13E isomers may indicate that we were able to test geometrically pure isomers.

2.3   *Xestospongia* species

Pyridoacridine alkaloids are a small class of cytotoxic marine-derived heteroaromatic pigments obtained from members of Porifera, Urochordata, Cnidaria and Mollusca. Our research group recently published the isolation of neoamphimedine (**18**) from two tropical *Xestospongia* species. Neoamphimedine is a new topoisomerase II inhibitor[15]. In order to obtain quantitative amounts of neoamphimedine for further bioactivity studies, several *Xestospongia* species from the Philippines and Palau were worked-up. These studies furnished amphimedine (**17**), neoamphimedine (**18**) and a new pyridoacridine alkaloid, deoxyamphimedine (**19**).

Amphimedine (**17**)   Neoamphimedine (**18**)   Deoxyamphimedine (**19**)

The animal materials were macerated with MeOH and MeOH:CHCl$_3$ mixtures. After the standard solvent partitioning procedure, pyridoacridines were purified by a combination of Flash CC and C-18 HPLC.

Deoxyamphimedine (**19**) showed cytotoxic activity against HCT-116 (IC$_{50}$ 335 nM). It was also tested in Chinese hamster ovary cells, AA8 and EM9, to exhibit the IC$_{50}$'s of 25 and 6 μM, respectively. In the DNA cleavage assays,[16] **19** showed the highest degree of cleavage. Interestingly, it did not require topoisomerase I or II for this activity (Figure 2). Since DNA cleavage required only a reducing agent and aerobic conditions, we hypothesize that deoxyamphimedine damages DNA through the production of ROS which results from quinone redox cycling or Fenton reaction.

1. DNA only
2. DNA + TopoII
3. DNA + TopoII + 1%DMSO
4. DNA + TopoII + 50μM etoposide
5. DNA + 100 μM **17**
6. DNA + TopoII + 100 μM **17**
7. DNA + 100 μM **18**
8. DNA + TopoII + 100 μM **18**
9. DNA + 100 μM **19**
10. DNA + TopoII + 100 μM **19**

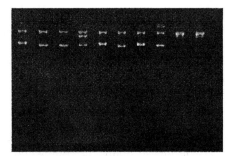

Figure 2. Pyridoacridines on ethidium bromide agarose gel (all incubated in Topo II 2 reaction mix buffer for 30 min, then SDS/Proteinase K was added for 60 min).

Neoamphimedine (**18**) is new topoisomerase inhibitor with a novel mechanism of cytotoxicity: it stimulates topoisomerase II to catenate DNA to a high molecular weight complex[15]. Neither deoxyamphimedine nor amphimedine exhibits this distinc activity. Detailed investigation of the effects of neoamphimedine on chromosomes and cell cycle machinery is in progress. A drug that can induce catenation of tumor cell DNA can serve as an effective anti-cancer agent. Therefore, we started doing some *in vivo* experiments in tumor-implanted mice. The preliminary results of the ongoing studies indicate that neoamphimedine could be a very useful tumoricidal drug.

3. CONCLUSION

Cancer drug discovery has already reached a new dimension as a direct consequence of the passionate study of the underlying genetic alterations associated with the cancerous state. This presents numerous opportunities to us for the development of new therapeutic agents. However, we need to understand better how the intracellular regulatory pathways work in each type of cancer. The ability of determining the genetic changes in a tumor cell genome will lead early cancer diagnosis and the employment of a potent and selective "new age" anti-tumoral drug or drug combinations.

ACKNOWLEDGEMENTS

I would like to thank Dr. Chris M. Ireland (University of Utah) for providing excellent facilities and for allowing me to present my studies. M.K. Harper, J.N.A Hooper and R.W.M.s Soest are acknowledged for taxonomical identifications. I am indebted to Dr. G.P. Concepcion and G.C. Mangalindan, (University of the Philippines) for their faithful collaboration. I thank the oncology research group at Wyeth Ayerst Research, as well as Dr. L.R. Barrows, K.M. Marshall and S.M. Verbitski (University of Utah) for performing biological assays. I am also grateful to Dr. D.J. Faulkner (Scripps Institution of Oceanography) for providing the Palauan sponge *Xestospongia* cf. *carbonaria*. Thanks are also due to Dr. Elliot Rachlin and Dr. Vajira Nanayakkara (University of Utah) for recording mass spectra and Dr. Bert Vogelstein at John Hopkins University for providing the isogenic HCT-116 cells. The research projects were supported by NIH Grants CA 67786 and CA 36622 (C.M.I). Partial funding for the Varian Unity 500 spectrometer was provided by NIH Grant RR 06262.

REFERENCES

1. Taşdemir, D., Mallon, R., Greenstein, M., Feldberg, LR., Kim, S.C., Collins, K., Wojciechowicz, D., Mangalindan, G.C., Concepción, G.P., Harper, M.K., Ireland, C.M., 2002, *J. Med. Chem.*, **45**, 529-32.

2. Taşdemir, D., Mangalindan, G.C., Concepción, G.P., Verbitski, S.M., Rabindran, S., Miranda, M., Greenstein, M., Hooper, J.N.A., Harper, M.K., Ireland, C.M., 2002, *J. Nat. Prod.*,(in press).
3. Taşdemir, D., Marshall, K.M., Mangalindan, G.C., Concepción, G.P., Barrows, L.R., Harper, M.K., Ireland, C.M., 2001, *J. Org. Chem.*, **66**, 3246-8.
4. Roussel, M.F., 1998, *Adv. Cancer Res.*, **74**, 1-24.
5. Johnson, D.G., Walker, C.L., 1999, *Annu. Rev. Pharmacol. Toxicol.*, **39**, 295-312.
6. Sielecki, T.M., Boylan, J.F., Benfield, P.A., Trainor, G.L., 2000, *J. Med. Chem.*, **43**, 1-18.
7. Jelinek, T., Catling, A.D., Reuter, C.W., Moodie, S.A., Wolfman, A., Weber, M.J., 1994, *Mol. Cell. Biol.*, **14**, 8212-8.
8. Barbacid, M., 1987, *Ann. Rev. Biochem.*, **56**, 779-822.
9. Mallon, R., Feldberg, L.R., Kim, S.C., Collins, K., Wojciechowicz, D., Hollander, I., Kovacs, E.D., Kohler, C., 2001, *Anal. Biochem.*, **294**, 48-54.
10. Eder, C., Proksch, P., Wray, V., Steube, K., Bringmann, G., van Soest, R.W. M., Sudarsono, Ferdinandus, E., Pattisina, L.A., Wiryowidagdo, S., Moka, W., 1999, *J. Nat. Prod.*, **62**, 184-7.
11. Pang, L., Sawada, T., Decker, S.J., Saltiel, A.R ,1995, *J. Biol. Chem.*, **23**, 13585-8.
12. Meijer, L., Thunnissen, A.-M.W.H., White, A.W., Garnier, M., Nikolic, M., Tsai, L.-H., Walter, J., Cleverley, K.E., Salinas, P.C., Wu, Y.-Z., Biernat, J., Mandelkow, E.-M., Kim, S.-H., Pettit, G.R., 2000, *Chem. Biol.*, **7**, 51-63.
13. Curman, D., Cinel, B., Williams, D.E., Rundle, N., Block W.D., Goodarzi, A.A., Hutchins, J.R., Clarke, P.R., Zhou, B.-B., Lees-Miller, S.P., Andersen, R.J.,Roberge, M. 2001, *J. Biol. Chem.*, **276**, 17914-9.
14. Russell, D.J., Hadden, C.E., Martin, G.E., Gibson, A.A., Zens, A.P., Carolan, J.L., 2000, *J. Nat. Prod.*, **63**, 1047-9.
15. de Guzman, F.S., Carté, B., Troupe, N., Faulkner, D.J., Harper, M.K., Concepción, G.P., Mangalindan, G.C., Matsumoto, S.S., Barrows, L.R., Ireland, C.M., 1999, *J.Org.Chem.* **64**, 1400-2.
16. Matsumoto, S.S., Haughey, H.M., Schmehl, D.M., Venables, D.A, Ireland, C.M., Holden, J.A., Barrows, L.R., 1999, *Anti-Cancer Drugs*, **10**, 39-45.

Altitudinal and Latitudinal Diversity of the Flora on Eastern and Western Sides of the Red Sea

AHMAD K. HEGAZY and WAFAA M. AMER
Department of Botany, Faculty of Science, Cairo University, Giza 12613, Egypt

1. INTRODUCTION

The Red Sea is 1932 km long, and average of 280 km in width. At its widest, in the south near Massawa, it is reaches 354 km wide and this narrows to 29 km at Strait of Bab al Mandab and to about 180 km before it branches to the Gulf of Suez and the Gulf of Aqaba. The latitudinal range of the Sea lies between 30°N and 12° 30'N, it has a long history in association with man's activities, but the degree of human impact and exploitation has until recently been negligible.

Though it is not classified under a "key environment" region (*sensu*)[1], the Red Sea has very rich and varied environment. Compared with many other tropical and subtropical seas, the high susceptibility to misuse and the obvious vulnerability of the Red Sea qualifies it for "key environment" status. Concerning its coastal flora, both sides of the Red Sea encompass a highly spectacular flora and interesting plant communities. The region is valued for its unique environment, high diversity, great scientific and ecological importance. With its north-to-south and sea-to-land gradients of the physical conditions, it offers a great scope for floristic richness and diversity.

The Red Sea region is interesting from the floristic and phytogeographic point of view because of its relation to the neighbouring regions of Africa and Asia. The rugged topography and inaccessibility of the mountainous escarpments of this region have resulted in a paucity of extensive or intensive floristic studies in general and its altitudinal diversity in special. Being diverged from the same origin by the Red Sea trough, the eastern and western sides are described as adjacent, and hence the observed continuity of the floristic elements.

Both sides of the Red Sea comprise three principal habitat types that include (a) The littoral habitats with its diverse coral edges, sandy beaches, mangrove wetlands and salt marshes. (b) The coastal plains that lie between the littoral zone and the mountain escarpment. These plains consist of sandy and/or stony deposits mainly derived from the mountain zone. A sub-coastal plains may occur and are central between two hill masses in an intermediate range of hills. (c) The mountain escarpment which can be differentiated into hill massif and mountain range. The hill massif may begin from the sea shore or from up to 80 km from the sea as in the Jizan region, south-west Saudi Arabia. The hill massif rises to about 500 m a.s.l. The mountain range extends parallel to the coastal with elevations from around 500 m to more than 3000 m a.s.l. in the highest peaks. The continuous range of the mountain escarpment forms a natural divide between seaward drainage to the Red Sea and landward drainage to the mainland. The three habitat types are traversed by wadis (drainage system) that are deeply incised into the coastal plains and their flood waters seldom reaches the sea, as they are gradually absorbed by the sandy substratum.

The continuous south-to-north extension of the coastal plains and mountain ranges on both sides of the Red Sea represent an important link which strengthen the floristic relationships in the region. The purpose of this study is to analyse the species diversity, chorology, and floristic relations at both altitudinal (sea landwards) and latitudinal (south to north) levels on both sides of the sea. The study provides an analysis into which further more detailed ecological and taxonomical studies may be fitted.

1.1 Study area

Six major localities (Map 1) are selected for this study including:
1- Ghedem mountain group on the eastern Eritrean coast (15° 20`- 15° 40` N), with maximum elevation 3054 m a.s.l.
2- Asir mountain group, South-West Saudi Arabia (16°-19° N), with maximum elevation exceeds 3000 m a.s.l.
3- Hijaz mountain group, West Saudi Arabia (21° 30` – 24° 30`N), with maximum elevation 2500 m a.s.l.
4- Elba mountain group at Sudano-Egyptian border (22°-23° N) with maximum elevation 2216 m a.s.l.
5- Shaieb El-Banat mountain group, Egypt (26° 30`- 27° 30` N), with

maximum elevation 2184 m. a.s.l.
6- South Sinai mountain group between Gulf of Suez and Aqaba, Egypt (27° 45`- 28° 45` N), with maximum elevation 2642 m. a.s.l.

Data of Hijaz, Ghedem and Elba mountain groups are complied from the literature, while data of the remaining three localities are based on field and literature surveys. The field work was carried out during the years 1997-2001.

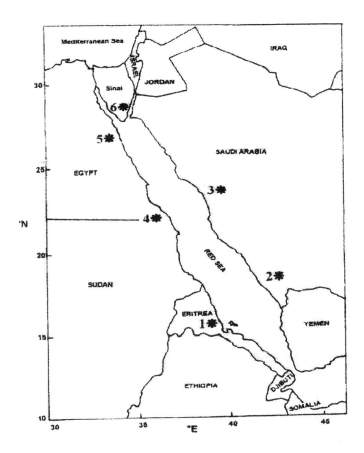

Map 1. Location of the transects, on both sides of the Red Sea.
1. Ghedem, 2. Asie, 3. Hijaz, 4. Elba, 5. Shaieb and 6. South

1.2 Geology

The Red Sea lies between the African and Arabian plates and is essentially a product of their divergence. The Sea cuts across Pre-Cambrian (700 Ma old) basement rocks of the Sudan and Saudi Arabia which were once united. It was established as a linear trough about the Oligocene period (38 Ma ago). Deposition during the Miocene period (25-5 Ma ago) was dominated by salt deposits in the central region, with transitions to marginal carbonates and siliciclastic deposits derived by erosion from the rising mountain fronts[2]. Rift movements appear to have been reduced during the Miocene but they began again about 5 Ma ago with the margins of the depression moving apart at about 0.9 cm/ year per flank. The fault-bounded escarpments formed have exposed Miocene salt deposits and black shales rich in heavy metals, Pliocene, Pleistocene and Recent sediments from a relatively thin cover along the margins within which lavas and igneous intrusions are common. The coastal margins are characterized by multilayered Pleistocene and Recent sediments. These rang from screes and wadi deposits, through alluvial fans and siliciclastic plain deposits, flowing from the mountain front, to reefs and associated bioclastic deposits.

Detailed description of the study localities are found in the references[3-8].

1.3 Climate

The climate varies from very hot and dry in the littoral and coastal plains to wet and cold and mostly foggy at high altitudes in the mountain belts. The distance between the sea and the mountain escarpment is an important factor affecting the climate in the different localities. There are some seasonal climatic differences between the southern and the northern parts, but both eastern and western sides are overwhelmingly arid.

The Red Sea region experiences some of the hottest and most arid conditions on earth. To the west stretches the almost rainless North African desert. Eastwards and north-eastwards desert and semi-desert extend even father through Arabia to central Asia. To the north lies the Mediterranean with its winter rain and summer drought. To the south, the copious summer rainfall of the Ethiopian highlands remains distant, and only alternating summer and winter monsoons, barely penetrates to the southern extremities of the Red Sea basin.

Rainfall over the region is sparse, sporadic and very localized. A particular location may receive no rain for years, then to experience a brief heavy rainfall which may then not be repeated for a similar lengthy period. In the Gulf of Suez and Aqaba, rain amount to about 25 mm/year. The western side, from Hurghada to about 22° N is virtually rainless, with any specific area only receiving few millimetres at intervals of several years. This is probably also true for the eastern side-south of 22° N at Dungunab the mean annual rainfall is about 40 mm, but Port Sudan it is about 100 mm and at Suakin, at 19° N, it reaches about 180 mm. Jiddah on the eastern side, averages about 50 mm/year with some heavy outbreaks of rain south of Jiddah. Further south in the western side from Massawa towards the Strait, the average rainfall is about 180 mm/year. This is true from the coastal belt on the eastern side from Jazan region towards the Strait. Much more rainfall is expected at the high elevations in the mountain belts of the eastern and western sides. Orographic precipitation is more pronounced on the slopes of the high mountains. For more environmental settings of the region were given in the literature[9].

2. METHODS

For the field survey of Asir, Shaieb El-Banat and South Sinai, the study sites were selected along a transect from the Red Sea coast and extending landward through the littoral habitats and the coastal plains to the mountain escarpment. Five altitudinal belts were recognized: 0-1, 1-500, 500-1000, 1000-1500, 1500-2000 and >2000m a.s.l. Altitudes were determined by altimeter readings adjusted for regular daily fluctuations of temperature and air pressure in every locality. A floristic list was recorded for every altitudinal belt. Plant specimens were collected and identified in Cairo University (CAI) and Agricultural Museum (CAIM) herbaria. Voucher specimens were deposited in CAI.

Data for the three sites namely Hijaz, Ghedem and Elba mountain groups were compiled from the literature and flora of the concerned countries and its surroundings[10-30]. The phytogeographical treatment of the floristic elements are followed by researchers[30-33]. Species identification are followed in the references[10, 12-14,22,24,27,28,31-42]. Floristic richness and species diversity in the different localities were analyzed. Species diversity index was determined in the different altitudinal belts of every locality. The beta diversity index (β_t) was calculated from Wilson & Shmida's measure[43] by adding the number of the gained species g(H) encountered along the altitudinal transect, to the number of species lost l(H) over the same transect, and the standardization by the mean species richness (α) according to the following formula:

$$\beta_t = [g(H)+l(H)]/2\alpha$$

The fewer species that the different communities or habitats (gradient position) share, the higher the beta diversity.

3. RESULTS

3.1 Floristic richness

A total of 1310 species were recorded from the six studied mountain transects. Figure 1 shows the number of families, genera and species. The total number attained the highest values in the southern transects namely Ghedem and Asir. The number of species decreased towards the middle sector which is represented by Elba, Hijaz and Shaieb mountain groups, and increased again in the northern sector represented by South Sinai mountain transect. In all transects the dicot species, genera and families were higher than monocots.

In the southern Red Sea sector, monocots are highly represented by the family Gramineae (e.g. *Aeluropus lagopoidus*, *Cenchrus ciliaris*, and *Lamarckia aurea*). Dicot families are dominated by Compositae (e.g. *Psiada arabica*, *Euryops arabicus* and *Bidens bipinnata*); Leguminosae (e.g. *Agrolobium arabicum*, *Acacia mellifera* and *Acacia enhrenbergiana*) and Euphorbiaceae (e.g. *Andrachne aspera*, *Euphorbia schimperiana* and *Jatropha villosa*).

The flora of the middle Red Sea sector (Figure 1) as represented by Elba and Hijaz transects contain low number of monocot species compared to the southern sector. The southern transects are dominated by family Gramineae (e.g. *Panicum turgidum*, *Stipagrostis plumosa* and *Lamarckia aurea*). Dicots are moderately represented; and appear higher in Elba transect than Hijaz transect and families dominated by Leguminosae (e.g. *Astragalus vogelii*, *Acacia tortilis* and *Indigofera spinosa*); Compositae (e.g. *Pulicaria crisipa*, *Iflago spicata* and *Centaurea aegyptiaca*) and Cruciferae (e.g. *Zilla spinosa*, *Farsetia longisiliqua* and *Anastatica hierochuntica*).

While the northern Red Sea sector (Figure 1) as represented by Shaieb and South Sinai transects shows the lowest number of monocot species among the investigated transects. South Sinai transect contain higher dicot species than that in Shaieb. Monocots are dominated by Gramineae (e.g. *Panicum turgidum*, *Lasiurus scindicus* and *Stipagrostis pulmosa*). Dicots are dominated by Leguminosae (e.g. *Acacia tortolis*, *Astragalus spinosus* and *Retama raetam*); Compositae (e.g. *Pulicaria crispa*, *Launaea spinosa* and *Seriphidium herba-alba)* and *Cruciferae* (e.g. *Zilla spinosa*, *Diplotaxis harra* and *Matthiola logipetala*).

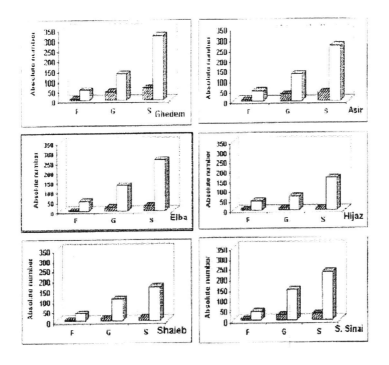

Figure 1. Floristic richness of the flora in the studied transects.
F. Family, G. Genera and S. Species.
Dicot   Monocot

Figure 1

Figure 2 outlines the families, genera and species number collectively on the eastern and western sides as well as in South Sinai. The number of species in the investigated transects in the western side reached 745, while in the eastern side and South Sinai transects amounted 434 and 231; respectively. The number of genera reached 212 in the western Red Sea side, 195 in the eastern side and 196 in South Sinai transect. The number of families was almost equal, in western side, eastern side and South Sinai transect and ranged between 54 in South Sinai and 53 in western side of the sea. Species diversity is in the similar pattern.

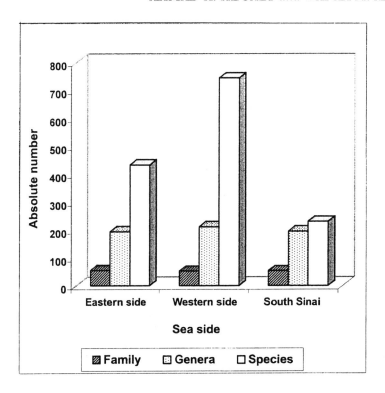

Figure 2. Total values of the floristic richness on both of Red Sea sides.

3.2 Species diversity

Analysis of species diversity along the altitudinal gradient in the study localities is shown in figure 3. Considering the overall site specific diversity, the two southern most sites namely Ghedem and Asir and the southern Sinai mountains show higher diversity than the remaining three sites. The altitudinal species diversity demonstrated an irregular pattern of variation among the different altitudinal belts in all study localities. The lowest species diversity values are found in the littoral habitats around the sea level (0-1 m a.s.l.). The highest species diversity is recorded in the altitudinal belt of 1500-2000 m a.s.l. in Ghedem and Asir mountains, while the coastal plains in the altitudinal belt 1-500 m a.s.l. in the remaining four study localities attain the highest species diversity values in the range of 0.79-0.95. Plant communities in the littoral habitats share a few number of species with the coastal plain habitats (1-500 m a.s.l.), while plant communities at altitudinal belt higher than 500 m a.s.l. share high number of species. This hold, true for all study localities.

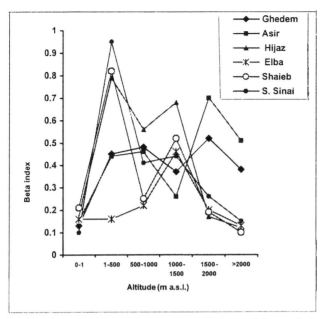

Figure 3. Species diversity along the altitudinal gradient of the studied transects locality. (Wilson and Shmida measure of beta diversity).

3.3 Endemism

The percentages of endemic species in the flora of the studied transects is shown in figure 4. Ghedem (most southern transect) and South Sinai (most northern transect = tropical corridor) show the higher percentages of endemism amounted (7 % & 6 % ; respectively). In Ghedem endemic species mostly belong to four genera: *Keniochloa* (Gramineae), *Oreophyton* (Cruciferae), *Haplosciadium* (Umbelliferae) and *Dianthoseris* (Compositae). Other endemic plants such as *Caralluma penicillata* and *Achillea arabica* are also among the endemics. In South Sinai endemic species are represented by 24 including: *Arabidopsis kneuckeri, Silene schimperiana, Bufonia multicepes, Phlomis aurea* and *Astragalus camelorum*.

Shaieb transect is represented by the lowest percentage of endemic species (0.24 %). This percentage includes the two species

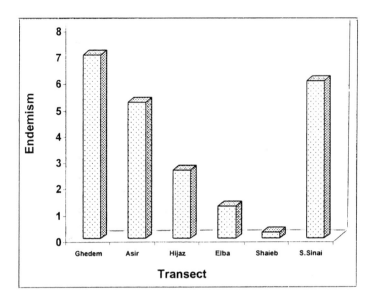

Figure 4. Percentages of endemism in the studied transects localities.

Colchicum cornigerum and *Crepis aegyptiaca. Biscutella elbensis* (Cruciferae), the only endemic species in Elba mountain transect, belongs to the Egyptian sector. Percentage endemism in Asir transect amounts 5.2 %. This value is represented by 11 species, among them: *Caralluma penicillata, Achillea arabica, Senecio odorus* and *Lavandula citriodora*. Two endemic species were traced in Hijaz mountain transect that represent 2.6 % of the total recorded species.

3.4 Floristic relations

Results in Table 1 indicate that the uniregional Sudano-Zambezian chorotype dominates the southern Red Sea sector as represented by Ghedem and Asir transects. This chorotype comprises 58.5 % and 36.99 % of the total species in each transect. Ghedem flora is associated with 11% Afro-Montane elements while flora of the slightly northern transect (Asir) is associated with the biregional Saharo-Sindian+Sudano-Zambezian chorotye, which represent, 13.71 % of the total species in the investigated transect.

The middle sector of Red Sea as represented by Hijaz and Elba transects are dominated by Sudano-Zambezian chorotype that includes 30.57 % in Hijaz and 28.47 % in Elba transect. Hijaz flora is codominated by the Saharo-Sindian+ Sudano-Zambezian chorotype (14.04 %), while Elba flora is codominated by Saharo-Sindian chorotype (17.14 %).

The northen sector of Red Sea as represented by Shaieb and South Sinai transects are dominated by Saharo-Sindian chorotype. Shaieb flora is characterized by 37.26 % Saharo-Sindian elements and about 11 % of the biregional Saharo-Sindian+Mediterranian elements with Saharo-Sindian+Sudano-Zambezian elements. South Sinai (the most northern transect) shows the dominance of Saharo-Sindian chorotype (31.48%) and codominance of Mediterranean +Saharo-Sindian chorotype (14.10 %). True Mediterranean (3.28 %) elements and Saharo-Sindian+Irano-Turanean elements (9.04 %) are also represented in the floral skeleton of S. Sinai transect.

The floristic affinity of the Red Sea vegetation shows the dominance of monoregional chorotype over the other chorotypes. Sudano-Zambezian chorotype dominats the southern Red Sea vegetation. This type decreased northwards and is replaced by the Saharo-Sindian chorotype till it reaches the Mediterranean and Irano-Turanean regions in more northern parts.

4. DISCUSSION

Ecologically and floristically the present work fairly defines the Red Sea region into three phytogeographical sectors: (a) the southern sector that is distinct with copious summer rain ; (b) the middle sector, virtually rainless or receives a few millimetres in some localities ; and (c) northern sector that is affected by the Mediterranean with its rainy winter and dry summer. The continuous south- to- north extension of the principal habitat types on both sides that include the littoral habitats, coastal plains and mountain escarpment represent an important link that strengthen the floristic relationship in the region. The south- to- north extension of both sides of the sea acted as a migration route between the Mediterranean and Tropical flora[31].

4.1 Floristic richness

Floristic analysis demonstrated the existence of very strong relationship between the flora of the two sides of the Red Sea. The species shared in the eastern and western sides constitute about 70 % of the total flora. About 5o % of the total flora on both sides are found in South Sinai among of them: *Panicum turgidum, Stipagrostis plumosa Blepharis edulis, Aerva javanica, Aizoon canariense, Calotropis procera, Heliotropium arabinese, Citrullus colocynthis, Anastatica hierochuntica, Lavandula coronopifolia, Acacia enhrenbergiana, Acacia tortilis, Echinops spinosus, Iflago spicata, Pulicaria undulata* and *Forsskaolea tenacissima*. This strong floristic relationship between the two sides seems to be attributed to the previous possible same origin as exhibited by the part played by the breaking off the land continuity that existed previously. It is a fact that the Arabian Peninsula was geologically and geographically part of Africa[6,44].

Table 1. Phytogeographical groups as percentages of the total number of species

Phytogeographic region	Transect locality					
	Ghedem	Asir	Hijaz	Elba	Shaieb	S. Sinai
Uniregonal						
Afro-Montane	11	1.50	-	-	-	-
Guineo-Congo	2.0	-	-	-	-	-
Mediterranean	-	-	-	0.95	3.10	3.28
Irano-Turanean	-	-	0.82	0.95	-	0.75
Saharo--Sindian	1.0	12.38	5.87	17.14	37.26	31.48
Sudano-Zambezian	58.8	36.99	30.57	28.47	3.34	1.0
Total uniregional	72.50	50.87	37.26	47.51	43.70	36.82
Biregional						
Med+Sah-Sind	-	11.06	5.78	5.71	11.18	14.10
Sah-sind+Sudano-Zamb	5.30	13.71	14.04	11.42	11.8	6.12
Med+IR-Tur	-	2.65	4.95	0.95	2.48	8.23
Med+Euro-sib	-	-	-	0.47	-	0.75
Sah-Sind+IR-Tur	-	3.53	9.91	8.57	14.90	9.04
Total biregional	5.3	30.95	34.68	29.52	40.36	38.24
Pluriregional						
Med+Sah-Sind+IR-Tur	-	0.44	4.13	2.85	5.59	7.30
Med+IR-Tur+Euro-Sib	-	-	-	0.47	-	1.25
Total pluriregional	-	6.19	9.91	10.47	13.04	14.1
Paleotropic	9.0	5.3	4.13	4.28	1.24	1.51
Pantropic	2.0	0.34	2.47	2.38	-	1.05
Cosmopolitan	3.0	2.21	4.13	1.9	1.40	2.5
Endemic	7.0	5.20	2.6	0.50	0.24	6.0
Others	1.2	0.03	4.82	3.44	0.02	0.08

Floristic richness revealed that the number of species in the western side is greater than that of the eastern side. Considering the restricted and common families to both sides of the Red Sea we find Gramineae, Leguminosae and Compositae are the major families met in both of the Red Sea sides as well as South Sinai. Aristolochiaceae and Dipsacaceae are restricted to the eastern side. Actiniopteridaceae is restricted to the western side. Aspleniaceae and Globulariaceae are restricted to South Sinai. As a common observation the number of salt tolerant species is 57 in the western side, 40 in South Sinai and 19 in the eastern side of Red Sea. The presence of salt tolerant species is linked with the width of the coastal plain that increases with the increase of the coastal width.

Halopyrum mucronatum (Gramineae) has an extensive distribution on maritime sands from East Africa northward to Somaliland, Eritrea, Sudan and Egypt to about 30 km north of Mersa Halaib at $22°$ N^{45}.

Mangrove vegetation in the western Red Sea side is dominated by *Avicennia marina*. It dominates the Eritrean coast, northward *A. marina* is associated with *Rhizophora mucronata* South of Suakin ($19°$ $15'$-$19°$ N) in the Sudanian coast. *Avicennia marina* dominates the more northern coasts till mixed with *Rhizophora mucronata* near Halaib at $23°$ N on the Sudano-Egyptian border. The northern limit of *A. marina* is Moys Hormos Bay near Hurghada at $27°$ $14'$ N. In addition to *A. marina* stand reported in Gulf of Aqaba (tropical corridor), South Sinai at $27°$ $40'$ N. In the eastern coast *Avicennia marina* pure communities dominant the coast of Yemen, and northward in Saudi Arabia to Jiddah; except individual trees of *Rhizophora mucronata* are associated with *A. marina* in Jizan swamps. Northern of Jiddah scattered shrubs are noticed at 90 km, 160 km and 520 km north of Jiddah. The northern limit of mangrove in the eastern Red Sea side is at $27°$ N, opposite to Moys Hormos in the western side$^{5,6,45-47}$.

The investigated localities are compared; we note in Ghedem transect, Gramineae, Compositae and Euphorbiaceae are the families having the highest species richness; *Senecio, Lobelia* and *Alchemilla* are the common genera. In Asir and Elba transects Gramineae, Leguminosae and Compositae are the families with the highest species richness. *Acacia, Euphorbia* and *Solanum* are genera with high number of species in both of Asir and Elba transects. Hijaz and Shaieb transects showed lower floristic richness compared to the previous transects. Again Compositae, Leguminosae and Gramineae are the families with the highest species numbers. The common genera in Hijaz are *Heliotropium, Acacia* and *Fagonia,* while in Shaieb *Cleome, Astragalus* and *Fagonia* are the common genera. The same three families are represented with high numbers in South Sinai, while *Astragalus, Silene* and *Erodium* are the genera represented with high species numbers.

The relatively high number of species in South Sinai may be attributed to its position as a tropical corridor and as a point of intersection for four biogeographical regions: Mediterranean, Irano-Turanean, Saharo-Arabian, and Sudanean[48]. These regions give South Sinai its rich floristic and biological diversity.

Also the mountain area receives sporadic precipitation about 50 mm/year and extra rain up to 300 mm occurs in higher altitudes due to orographic influence and snow falls occasionally at altitudes above 2000 m a.s.l.[49]. Nearly every shower becomes available to the plants growing on mountain crevices and soil pockets[50] (Danin, 1986). Runoff water enhancing dense wadis vegetation, lowlands of 0-1 m a.s.l. characterized by halophytic vegetation. Among the tropical species to the South Sinai are *Acacia tortilis, Avicennia marina, Hyphaene thebaica, Moringa peregrina, Salvadora persica* and *Suaeda monoica*.

Among the common species recorded in the six studied transects are: *Panicum turgidum, Blepharis edulis, Aerva javanica, Aizoon canariense, Calotropis procera, Heliotropium arabinese, Citrullus colocynthis, Anastatica hierochuntica, Lavandula coronopifolia, Acacia enhrenbergiana, Acacia tortilis, Stipagrostis plumos, Caylusea hexagyna* and *Forsskaolea tenacissima*.

The floristic richness in the eastern and western Red Sea sides as based on the studied transects reveals about 110 species common to both sides of which are: *Abutilon fruticosum, Acacia tortilis, Anastatica hierochuntica, Capparis deciduas, Lavandula pubescens, Leptadenia pyrotechnica, Moringa peregrina, Panicum turgidum, Pistacia khinjuk, Stipagroatis plumosa, Ruellia patula, Asphodelus africanus, Echium longifolium* and *Cometes abyssinica*.

4.2 Species diversity

Altitudinal variation is a major factor affecting the distribution of plant species and communities[8,10]. Vegetation studies of the six selected transects revealed that plant communities changed from one altitudinal belt to the next with broad transitional areas and overlap between low and high altitude vegetation. Similar results and observations were described by Kassas[50] in Sudan, Vesey-Fitzgerald[51] in Saudi Arabia, Kassas[52] in Egypt, Brooks & Mandil[53] in Saudi Arabia, Ghazanfar[54] in Oman, Hegazy, *et al.*[8] in Saudi Arabia, and Vetaas[55] in Sudan.

Wilson and Shmida measure of beta diversity reveals that the highest species diversity are found in the altitudinal range from 500-2000 m a.s.l. Similar observation was also found by Hegazy, *et al*[8] in south-west Saudi Arabia. This observation explained by Arcotech[49], who pointed out that the intermediate ridges are characterized by deep alluvium and loose deposits covered by fissured rocks, cobbles and stones. Deep deposits have a greater capacity for water storage than the terraces, providing better chance for vegetation growth.

This study revealed that the species diversity attained the highest values at the intermediate elevations decreased towards highlands (>2000 m a.s.l.) and lowlands (0-1 m. a.s.l.). The highest species diversity in southern Red Sea transects represented by Ghedem and Asir is in the altitudinal range of 1500-2000 m a.s.l. While in northern transects represented by Hijaz, Shaieb and S. Sinai attained the highest species diversity in the altitudinal range of 1-500 m. a.s.l. except Elba transect, the highest species diversity is in the altitudinal range of 1000-1500 m. a.s.l.

4.3 Endemism

Ghedem transect belongs to Afr-Oriental domain of Sudano-Zambezian region. The flora of this transect characterized by 7 % endemic species, it is the highest percentage of endemism among the studied transects. The flora of Asir transect attained 5.2 % endemic species. This transect belongs to South Arabian domain which is also a subregion of Sudano-Zambezian region. Sudano-Zambezian region is characterized by a few endemic families and a few endemic genera such as *Keniochloa* (Gramineae), *Oreophyton* (Cruciferae), *Haplosciadium* (Umbelliferae) and *Dianthoseris* (Compositae) and a large number of endemic species such as *Caralluma penicillata* and *Achillea arabica*.

Hijaz and Shaieb transects lie in the core of Saharo-Sindian region, both of transects containing low percentages of endemic species (2.6 % and 0.24 %; respectively). The low percentages of endemic species in these transects is referred to Wickens[31], who suggests that the Saharo-Sindian region is characterized by the absence of endemic families and the presence of a small number of endemic genera and a few endemic species. Sinai flora containing about 36 endemic species, most of them are confined to the mountain region[39]. Danin[17] mentions that the number of endemic species and subspecies in Sinai is 28 that comprises 3.2 % of the total Sinai flora. While, Boulos[36] notes that the number of endemism in Sinai is 33 species; four of them are sub-endemic species (Known from other regions in Egypt). Gibali[41] cited that the number of endemic species in Sinai is 33 species 9 of them is endemic to North Sinai and 24 in South Sinai. Sinai endemic species comprises 60.7 % of the total Egyptian endemic species among of them: *Arabidopsis kneuckeri, Silene schimperiana, Bufonia multicepes, Phlomis aurea* and *Astragalus camelorum*.

The high endemic species in Ghedem, Asir and South Sinai transects is supported with Boulos[56], who cleared that the endemic flora is highly represented in Islands, Peninsulas and mountain chains. Labiatae, Leguminosae, and Compositae are families with highest number of endemic species in South Sinai transect. Asclepiadaceae, Liliaceae and Euphorbiaceae are families with high endemic species in Asir transect. One crucifer is the only endemic species in Egyptian part of the Elba mountain group.

4.4 Chorological analysis

The southern transects namely Ghedem and Asir are dominated by monoregional Sudano-Zambezian chorotype elements. It is represented by 58.5 % and 36.99 % ; respectively of the total species. The dominance of Sudano-Zambezian chorotype species is supported by White & Leonard[33] suggestion about the extension of Sudano-Zambezian region to South Arabia known as South Arabian domain. Also Ghedem belongs to the same phytogeographic region and lies in the core of Afr-Oriental domain of the region[31].

Middle transects namely Elba and Hijaz transects are also dominated by Sudano-Zambezian chorotype: 30.57 % in Hijaz and 28.47 % in Elba. Elba flora is codominated by Saharo-Sindian chorotype (17.14 %). Hijaz flora is codominated by the Saharo-Sindian+Sudano-Zambezian biregional chorotype.

The northern transects Shaieb and South Sinai are dominated by monoregional Saharo-Sindian chorotype: 37.26 % in Shaieb and 31.48 % in South Sinai. The high percentage of Saharo-Sindian elements is explained by the presence of Egypt in the middle of Saharo-Sindian region which extends from Morocco to South Iran and Iraq[3].

The percentage of Sudano-Zambezian elements decreases northward from Elba to Shaieb and is replaced by Saharo-Sindian elements. These results are supported by the results obtained of Hassan[42], who showed that the Sudano-Zambezian elements decreased in the Nubian desert where Saharo-Sindian elements dominate. Further north, Mediterranean elements are apparently well represented[57]. This gradient corresponds to the pattern of precipitation[58], which gradually changes from predominantly winter rains in the north (Mediterranean type of climate) to predominantly summer rain (tropical type of climate) in the southern sector of the Red Sea. Conclusions

The Red Sea is defined ecologically and floristically into three major sectors (southern, middle and northern sector). The floristic richness decreased from south to north direction and relatively increased in south Sinai. About 70 % of the flora are shared between the eastern and western sides. South Sinai is considered a tropical corridor for many tropical species. Some species are restricted to each side and each sector, while others are common and recorded in all studied localities.

The altitudinal gradient showed three main zones *viz* low, intermediate and high altitude. Vegetation in the intermediate altitude attains the highest species diversity.

Endemic species are rich in the southern Red Sea sector and decreases to minimum in the middle sector and again increases in the northern sector namely South Sinai transect which represents the tropical corridor. Southern sector is characterized with a few endemic families and large number of endemic species, for example, *Caralluma penicillata*, *Achillea arabica* and *Pulicaria schimperi*. Middle sector is characterized by absence of endemic families and presence of a small number of endemic species. The endemic species in the northern sector showed Mediterranean and Irano-Turanean affinities. Among the endemic species are: *Arabidopsis kneuckeri, Silene schimperiana, Bufonia multicepes, Phlomis aurea* and *Astragalus camelorum*.

In general the Red Sea is defined ecologically and floristically into three major sectors:

Southern sector of the Red Sea

This sector is represented by Ghedem and Asir. The two representative transects showed the dominance of Sudano-Zambezian elements which is explained by the position of this sector within the Sudano-Zambezian phytogeographic region. This sector is characterized by the genera *Commiphora, Boswellia, Euphorbia* and *Dracena* genera and *Cometes abyssinica, Andrachne aspera* and *Argyrolobium arabicum* species.

Middle sector of the Red Sea

The sector is represented by Elba and Hijaz transects. The flora is characterized by the dominance of Saharo-Sindian elements. Among the characteristic species are *Calotropis procera, Panicum turgidum, Cornulaca monacantha* and *Moltkiopsis ciliata*.

Northern sector of the Red Sea

The sector is represented by Shaieb and South Sinai transects. The flora is characterized by the dominance of Saharo-Sindian elements. Dominant species as the middle sector. In addition to the presence of some Mediterranean elements among of them are *Asphodelus fistulosus*,

Astragalus cretaceous and Olea europaea subsp. cuspidate, while Irano-Turanean elements represented by Reseda stenostachya.

The Coastal vegetation is almost similar on both sides of Red Sea with minor differences over several degrees of latitude, but the differences rapidly become apparent as one moves away from the sea and its immediate influence.

ACKNOWLEDGMENT

The authors thank Prof. Dr. M. Kassas and Prof. Dr. L. Boulos for their valuable comments on the manuscript.

REFERENCES

1. Head, S. M., 1987, Introduction. In *Red Sea* (Edwards, A. J., Head, S.M. ,eds.), Pergamon Press, Oxford, pp.1-21.
2. Braithwaite, C.J.R., 1987, Geology and palaeogeography of the Red Sea region. In *Red Sea* (Edwards, A. J., Head, S.M., eds.), Pergamon Press, Oxford, pp. 22-44.
3. Nebert, K., 1970, Saudi Arabian Directory General of Mineral Resources Bulletin, No.4.
4. Liddicoat, W.K., 1971, The North Samran exploration area, Saudi Arbian Directory General of Mineral Resources, Technical Record, TR 1971-1.
5. Zahran, M.A., 1983, Introduction to plant Ecology and Vegetation types in Saudi Arabia. King Abdulaziz University, Jeddah, p.142.
6. Edwards, F.J., Head, S. M., (eds.),1987, Red Sea, Pergamon Press, Oxford.
7. Springuel, I., 1997, Vegetation, land use and conservation in the South Eastern Desert of Egypt, In *Reviews in Ecology: Desert conservation and development* (Barakat, H.N., Hegazy, A.K. eds.), Printed by Metropole, Cairo, pp. 177-206.
8. Hegazy, A.K., El-Demerdash, M.A., Hosni, H.A.,1998, *Journal of Arid Environments,* **38**, 3-13.
9. Edwards, F.J., 1987, Climate and oceanography. In *Red Sea* (Edwards, A. J., Head, S.M. eds.), Pergamon Press, Oxford, pp.45-70.
10. Abd El-Ghani, M. ,1996, *Journal of Arid Environments,* **32**, 289-304.
11. Boulos, L., 1975, The Mediterranean element in the Flora of Egypt and Libya, In *La florae du bassin méditerranéen: essai de systématique synthétique*, Colloques Internationaux du C.N.R.S. No. 235, Paris, pp. 119-124.
12. Boulos, L., 1999, Flora of Egypt, Vol. 1, Azollaceae-Oxalidaceae, Al Hadara Publ. Cairo, p. 417.
13. Boulos, L., 2000, Flora of Egypt, Vol. 2, Geraniaceae-Boraginaceae, Al Hadara Publ. Cairo, p. 352.
14. Burger, W.C. , 1967, Families of flowering plants in Ethiopia, Oklahoma, p. 231.

15. Bussmann, R.W., 1994, The forests of Mount Kenya. Vegetation, ecology, destruction and management of a tropical mountain forest ecosystem dissertation, University. Bayreuth, Vol. 1. p. 177., Vol 2. Appendices, p. 53., Vol. 3. p.15, Forest destruction and Management.
16. Collenette, S., 1999, Wild flowers of Saudi Arabia, NCWCD, Riyad, pp. 799.
17. Danin, A., 1986, Flora and vegetation of Sinai, *Proceeding of the Royal Society of Edinburgh*, **89B**, 159-65.
18. El-Hadidi, M.N., Hosni, H.A., 1996, *Biodiversity in the Flora of Egypt*. In *The Biodiversity of African Plants*, Van der Maesen, L.J.G. et al., eds.), Proceeding XIVth AETEAT Congress, Kluwer Academic Publ. Dordrecht, pp. 785-7.
19. Hedberg, O., 1951, *Svenk Bot. Tidsky.*, **45**, 140-202.
20. Hedberg, O.,1965, *Webbia*, **19**, 519-29.
21. Hedber, I., Edwards, S. (eds) ,1989, Flora of Ethiopia, Vol. 3, Pittosporaceae-Araiaceae. Addis Ababa and Asmara, Ethiopia, Uppsala, p. 659.
22. Hepper, F.N., Friis, I. ,1994, Flora Aegyptiaca-Arabica, The plants of Pehr ForssKål's. Royal Botanic gardens, Kew, U.K, p. 400.
23. Jackson, J.K. ,1956, *Journal of Ecology*, **44**, 341-74.
24. Migahid, A.M., 1988-1990, Flora of Saudi Arabia, 3rd ed.,. Vol 1, p. 251. (Equisetaceae-Neuradaceae); Vol 2. p. 282 . (Leguminosae-Compositae), Vol 3. p. 150. (Hydrocharitaceae-Orchidaceae), King Saud University, Saudi Arabia.
25. Ozenda, P., 1977, Flora Du Sahara, Centre National de la Recherche Scientifique, Paris, p. 621.
26. Quézel, P. , 1978, *Annuals of the Missouri Botanic Garden*, **65**, 479-534.
27. Täckholm, V., 1974, Student's Flora of Egypt, Ed. 11 Beirut, Cairo University Press, p. 888.
28. Teketay, D., 1995, *Mountain Research and development*, **15**(12), 183-6.
29. White, F., 1950, *Forest Society Journal*, **3**, 5.
30. Zohary, M., 1972, Flora Palestina. 2. Israel Academy of Science and Humanities, Jerusalem, p. 489.
31. Wickens, G.E., 1976,. *Kew Bulletin*, Additional Series V. London: HMSO, p.199.
32. Hosni, H.A., Hagazy, A.K. , 1996, *Candollea*, **51**, 169-202.
33. White, F., Leonard, J., 1991, *Flora de Vegetatio Mund*, **9**, 229-46.
34. Abulfatih, H.A., 1979, Vegetation of higher elevations of Asir, Saudi Arabia, *Proceeding of Saudi Biological Society*, **3**, 139-48.
35. Boulos, L. , 1985, *Journal of Scientific Research*, **3**, 67-94.
36. Boulos, L. , 1995, Flora of Egypt Checklist. Al Hadara Publ. , Cairo, p. 283.
37. Chao, C., Renvoize, S.A., 1989, **44**(2), 349-68.
38. Cope, T.A., Hosni, H.A., 1991, A Key to Egyptian Grasses. Royal Botanic Gardens, Kew London, p.75.
39. El-Hadidi, M.N., 1969, *Bulletin LA Societe de Geographic D' Égypt*, **T. XI**, 142-55.
40. Fayed, A.A., Zayed, K.M., 1989, *Arab Gulf Journal for Scientific Researches*, **7**, 97-117.
41. Gibali, M.A., 1988, Studies on the Flora of Northern Sinai. M. Sc. Thesis, Cairo University, Cairo, p.403.
42. Hassan, L.M., 1987, Studies on the flora of Eastern Desert, Egypt. Ph. D. Thesis, Cairo University, Cairo, p. 515.
43. Wilson, M. V., Shmida, A., 1984, *Journal of Ecology*, **72** 1055-64.
44. Furon, R., 1963, Geology of Africa. Oliver & Boyd, London, p. 377.
45. Kassas, M., Zahran, M.A.,1967, *Ecological monographs*, **37**, 297-315.
46. Hemming, C.F., 1961, *Journal of Ecology*, **49**, 55-78.

47. Zahran, M.A., 1977, *Africa A. wet formations of the African Red Sea Coast*. In Wet Coastal ecosystems, Ecosystems of the world 1 (Chapman, V. J. ed), Elsevier, Scientific Publ., New York, pp. 215-8.
48. Hegazy, A.K., 1999, Deserts of Middle East, In *Encyclopedia of Deserts Mares* (M. A., ed.), University of Oklahoma Press, Norman, pp. 360-4.
49. Arcotech, 1994, Feasibility study on St. Catherine protectorate and Bardawil lagoon protected area; Sinai Egypt. Prepared for European communities, commission, pp. 13-9.
50. Kassas, M., 1955, *Journal of Ecology*, **44**, 180-94.
51. Vesey-Fitzgerald, D.F., 1955, *Journal of Ecology*, **43**, 477- 89.
52. Kassas, M., 1957, *Journal of Ecology*, **45**, 187-203.
53. Brooks, W.H., Mandil, K.S.D., 1983, *Journal of Arid Environments*, **6**, 357-62.
54. Ghazanfar, S.A.,1991, *Journal of Biogeography*, **28**, 299-309.
55. Vetaas, O.R., 1992, The interaction between biotic and abiotic factors controlling temporal and spatial dynamics of arid vegetation in Erkowit, North-Eastern Sudan. Ph. D. Thesis, University of Bergen, Bergen.
56. Boulos, L., 1997, *Endemic flora of the Middle East and North Africa*. In Reviews in Ecology: Desert conservation and development. (Barakat, H.N., Hegazy, A.K. eds.), Printed by Metropole, Cairo, pp. 229-45.
57. El-Hadidi, M.N., 1993, *Natural vegetation*. In The Agriculture of Egypt (Craig, G.M., ed.), Oxford University Press, Oxford, pp. 39-62.
58. Ali, M.M., Badri, M.A, Hassan, L.M., Springuel, I.V., 1997, *Ecologie, t.* **28** (2), 119-28.

Biodiversity and Free Market Mechanism

KENAN OK
İstanbul University, Faculty of Forestry, Department of Forest Economics 80895 Bahçeköy İstanbul, Turkey

1. INTRODUCTION

Questions about what, how and for whom goods and services must be produced in an economy are the main problems, targeted by economic mechanisms to be enlightened. Mechanisms which are designed to answer these questions are usually classified as *market mechanism, command mechanism* and *mixed economies*.

A command mechanism is a method based on the authority of a ruler or ruling body while a market mechanism is a method based on individual choices co-ordinated through markets.

Neo-classical macroeconomists particularly stress the efficiency of the market mechanism. They believe that if some conditions are met, market mechanism can solve many economic problems. According to Adam Smith, an ***individual intends only for his own gain***. A producer wants to maximise his profit while a consumer tends to maximise his utility level. Market mechanism is based on profit-maximising producers and utility-maximising consumers. Targets of individuals co-ordinate an immense number of decisions like an invisible hand in an economy. The market is a co-ordination mechanism because it pools together the separate plans of all the individual decision makers who try to buy and sell any particular good[1]. In their opinion, market mechanism which is free from government intervention will automatically lead to an optimal allocation of production factors, full employment and the highest level of welfare.

However, not all economists agree on the qualities and advantages of the market mechanism. According to the market theory, utility- maximising behaviour of consumers and profit- maximising behaviour of producers will lead to efficiency in production and consumption. For this to be true the following conditions have to be met: *perfect competition, absence of public goods, absence of externalities*[2].

While debates on the results of the market mechanism are not yet concluded, new advocates who defend the market as an alternative to solve environmental problems rapidly increasing are seen in discussions. Anderson and Leal took over the main role with their textbooks titled Free Market Environmentalism (FME)[3]. According to Power and Rauber, Clinton gave the signs for market based environmental protection strategy with his sentence that we must "recognise that Adam Smith's invisible hand can have a green thumb" during his campaign[4].

According to adherents of the FME, the greatest hope for protecting environmental values lies in the empowerment of individuals to protect those environmental resources that they value (via a creative extension of property rights)[5]. The market mechanism can solve resource pollution, depletion of rain forest and global warming problems if the required conditions are generated[6]. On the other hand, FME and privateers are accused of distrusting democracy deeply, and are criticised for their over-simplistic, misleading and hyperbolic prescriptions[7].

In this study, the results of the free market mechanism, especially the aspects of biodiversity, are discussed. Firstly, the crowding out effect, which risen from producer and consumer types accordance with Smith's assumption is introduced in biodiversity subject. Then, activities aimed at creating free market conditions are discussed regarding international trade, conventions and treaties.

2. CROWDING OUT EFFECT OF THE MARKET IN BIODIVERSITY

The crowding out effect in economics is usually used as a term pertaining to fiscal policy, interest rate and investment level. Crowding out is the tendency for an increase in government purchases of goods and services to increase interest rates, thereby reducing investment[1]. Economists have known this effect for many years and they are interested in private investors discharged from the market by fiscal policies. However, even if government does not intervene in the market, a similar effect has also happened in biodiversity by changing actors.

As mentioned above, expected results of the market mechanism require some conditions. The first condition is *perfect competition*. Perfect competition occurs in a market where there are many firms, each selling an identical product, there are many buyers, there are no restrictions on entry into the industry, firms in the industry have no advantage over potential new entrants, firms and buyers are completely informed about the prices of the products of each firm in the industry[1]. Advocates of the market mechanism believe that the market, which is far from perfect competition conditions enumerated above, is the source of problems such as unemployment, inflation, low national wealth, and degraded natural resources.

However, a hypothetical market which has perfect competition conditions can be envisaged to understand the crowding out effect in biodiversity. For that purpose, let us assume that there are many apple producers on their own field for the market. They have no capacity to affect the market with their individual decisions. There is no restriction on entry into the apple production industry for new producers. Producers can easily learn and follow apple prices and other information needed for the apple market. Furthermore, they can compare advantages of each apple species. And lastly, fields allocated to apple production are owned privately.

According to Smith's assumptions, an apple producer will tend to his own gain and want to allocate his field for apple species which can maximise his own profit level. For this propose, the producer will prefer or select the apple species which will maximise his profit level. Prices, costs, consumer preference and other variables which must be taken into account for decision making at the local scale will not be the same variables needed for decisions at the national or global scale. When the market broadens from local to global scale, producers must to add new variables about some characteristics of candidate species such as suitability for storage and transportation. On the other side, consumers will demand apple species which can maximise his/her utility level concerning price, colour, form, taste, aroma. Therefore, the price of species preferred by consumers will increase. After that decision process, the producer will allocate his field for apple species which are preferred by the maximum number of consumers, has maximum price, advantages about productivity, storage and transportation and other species will be crowded out by producer.

Indeed, the species selection process under market rules, especially in agriculture, provoked superior species to pervade all cultivable lands, and other species could not find any field to survive. This result was observed in agricultural crops such as apple, grape, wheat, barley, corn and bean which have been cultivated for years in Anatolia. While present fields and vineyards were covered by species preferred by the market, other species were crowded out by that market from agriculture and then, they became extinct. The crowding out effect is not a result of consumer preferences. Some species are not offered to consumers because of their low profitability. While market is crowding out some species from cultivation lands, at the same time, it forced to consumer to buy same goods continuously. The market mechanism is as destructive a factor as forest fires, soil erosion and pollution, urbanisation and industrialisation of cultivable lands for extinction of these species. For this reason, market intervention must be used as a remedy for conservation of species crowded out by the market. In the apple example explained above, perfect competition occurred, there were no public goods, there was a property right system for producer and externalities in that production could be of ignorable scale. In other words, conditions which are demanded by free marketers were established but the market could not conserve biodiversity. On the contrary it destroyed biodiversity.

For all, free market environmentalists explain the species extinction problem by using property rights and assert that privately owned species are conserved automatically. According to Anderson and Show, "The wild and commonly owned passenger pigeon was hunted, killed, and transported to large cities, where it was eaten in homes and restaurants much the way that chickens are today. However, because there was no ownership ... the pigeons disappeared. In contrast, even though millions of chickens are eaten each year, private ownership ensures their continued existence[8]." If so, according to them, privatisation of all kinds of natural resources must be realised immediately. But, they ignore the chicken species crowded out by poulterer. Present poultry facilities are crowded in by one or two superior chicken species or subspecies under the auspices of Adam Smith's producer. The crowding out effect of the market mechanism can also be observed in animal husbandry. Essentially, agriculture is the best sector to observe that the expected results of the ownership philosophy advocated by free market environmentalists could not occur. There are many examples of degraded privately owned land as a result of misuse.

When genetically modified new plants or organisms which have some improved characteristics are introduced to the market, the crowding out effect of the free market mechanism will increase rapidly. "The privatisation of genetic resources that have been engineered and patented accelerates the trend toward monocultural cropping[9]." As seen in hybrid seed markets, privately owned enterprises will be controlled by monopolistic private corporations and fields will be covered by their plants or organisms.

The crowding out effect of the market can be seen in conversion of lands from one sector to another. Smith's profit maximising producers allocate their production factors, especially lands, for the sector which has minimum opportunity cost. Forest owners compare the candidate sectors with respect to their profit and if agriculture is more profitable than forestry, they want to cut their trees and to convert their land to agriculture, ignoring the biodiversity of the forest. That behaviour is in agreement with Smith's assumptions. As a matter of fact, Kumla Çiftliği Private Forest owned by The Sabancı Family, one of the bigger and famous entrepreneurs in Turkey, was converted to an industrial area by the same family. Under the direction of the market, northern forests of Anatolia, which contain important and rich biodiversity capacity, are invaded by tea and hazel plantations illegally. Maquis which is an important ecosystem for the Mediterranean region could not compete with banana plantations under market conditions. Besides forest and maquis, vineyards, olive and orange gardens were covered by tourism facilities in coastal regions of Turkey. In industrial regions, both forest and agriculture lands were allocated for industrial or settlement purposes.

As seen in the above explanation, the crowding out effect of the market on biodiversity may appear, even though there is no government intervention or fiscal policies. Under free market conditions, profitable species can survive and others become extinct. If land allocation for any sector or species selection in any sector is directed by free market mechanisms, forests together with biodiversity can not compete with other sectors or tend to monocultural forest. Conservation of forest resources is not a problem which can be solved by market mechanism. Even if the market reaches perfect conditions, the crowding out effect can occur on biodiversity. But, it is a reality that the same results can also be observed after government intervention. Conversion of forests to tea and hazel plantations were watched silently by governments concerned with their economic benefits.

Nations have different trade capacities determined by their natural, cultural, economic and technological characteristics. The theory of comparative advantages is based on these differences. Nations can change their economic and technological characteristics. Thus, they can compensate their disadvantages, especially in the long range. Species in biodiversity also have different tradable characteristics like national economies. But, they can not change their growth curves or improve new and productive assimilation techniques to increase their productivity or profitability rates. For this reason, the comparative advantages theory is not suitable to determine which species must survive or which species must not. Thus, some species crowded out because of their low profitability must be supported by government interventions to conserve biodiversity. Experiences show that some species which have biological advantages in the market may become an invasive species rapidly. However, While debates about the benefits and costs of the free market mechanism are still continuing, on the other side, international binding treaties are being signed to eliminate tariff and nontariff measurements (NTMs) that are important tools of government intervention in international trade in accordance with the free market spirit. For this

3. BIODIVERSITY AND TRADE LIBERALIZATION

The first GATT (General Agreement on Tariffs and Trade) was signed to improve international trade by decreasing of government interventions may be a barrier for trade in October 1947.

A main subject debated during GATT negotiations is tariffs. A tariff is a tax that imposed by the importing country. Quantitative restrictions, health and sanitary regulations, licensing, deposits, industrial standards, government procurement, subsidies, customs, formalities and requirements .. etc. are called as Non Tariff Measurements. If NTMs become a trade barrier, they are called non tariff trade barriers (NTBs) which are other important subjects negotiated in GATT progress.

Tariffs have been decreased by negotiation at continual rounds such as Kennedy (1967), Tokyo (1979), Uruguay (1988)[10,11]. But NTMs are argumentative subjects of negotiations. Usually, results of NTBs are discussed during negotiations in points of producer, importer, exporter, wealth and industrial level of importer countries. It is asserted that government intervention causes un-optimal resource allocation. However, optimization of resource allocation can be tested concerning the minimisation of the species extinction rather than profit maximisation. The market mechanism is focused on profit and is not aware of species extinction. In other words, it is interested in individual profit and loss but excludes social and ecological costs.

Tariffs and NTMs do not affect only the industrialisation level of the trading countries. International trade which is carried out legally or not changes the locations of many ecosystem components. Eggs of the insects which are transported by international trade activities may jeopardise the new ecosystems especially when there are no the predators of the transported insect in it. Risks which arise unwillingly from the transference of some organisms are at a maximum level in raw or less processed material transportation. For that reason, when the trading countries tend to sell more processed materials, they not only protect their economy and industry but also conserve ecosystems at global scale. Thus, contributions of the NTMs on ecosystem conservation must not be sacrificed to stimulate global trade. Actually, some NTMs are still used by developed countries. For example, The European Community requires all coniferous lumber to be barked, to avoid further infestations of the pine bark beetle[10]. Final processed goods such as furniture, paper can provide safer conditions than barked logs or fumigated lumber. Nation states can be free to protect local industries and agricultural activities from global invader firms and species. However, recent tendencies show that local markets are forced by international treaties to open up to for global firms together with their invasive and superior species. Credit demands of nations in financial crisis are dependent on some regulations which can allow free access to local markets.

As seen at The World Trade Organisation's (WTO) Seattle meeting in 1999, demonstrators in large number showed that they worry about environmental results of GATT regulations. In the world today, international regulations related to many subjects are increasing rapidly. Some accept it as globalisation but these regulations do not serve the same objective. As understood from Convention on Biological Diversity (CBD,1993) and Trade Related Aspects of Intellectual Property Rights Agreement (TRIPs,1995) examples, these international binding treaties are directed by contradictory mentalities. While CBD was signed to conserve endangered species, TRIPs tends to privatise them. There are at least three areas of outright contradiction in *their objectives*, *systems of rights* and *legal obligations*[12].

The convention is founded on the principle that local communities generate and are dependent on biodiversity and should continue to benefit from it. The WTO administers a global trading system, much of which is founded on the private monopoly rights of transnational corporations over biodiversity[12].

Even though there is an effort to improve the environmental aspect of GATT, TRIPs is based on free market environmentalism. Reflection of the free market environmentalism can be easily seen in TRIPs. However, TRIPs can not generate positive results on environmental problems, even if it has perfect conditions. The theory of free market environmentalism is founded on certain visions regarding **human nature, knowledge** and **processes**. And, free market environmentalism sees a much smaller knowledge gap between the experts and the average individual[3].

Essentially, difference among knowledge levels of individuals or nations is not small. The number of patents that are owned by nations is the indicator which can prove the difference between developed and developing countries. This situation is not in accordance with either assumptions of free market mechanism or the prescription of free market environmentalism about knowledge.

The traditional approach to knowledge is also changed by TRIPs. While knowledge is accepted as a public good in general, TRIPs accepts it as a private good. But, according to O'Neill, Einstein's Theory of Relativity is not a good which is owned by any private individual[13]. Also, Einstein did not produce the knowledge for "his own gain" especially in an economic manner. Many scientists produced knowledge and offered it to society in order to be used freely in their productions. If that experience was observed in human history, Is TRIPs' approach, which tends to privatise all of the biological resources including micro organisms, responsible to "mankind's nature"? TRIPs' approaches are in contradiction with the transparency principle of perfect competition in the market mechanism.

Actually, the transparency principle is applied arbitrarily in the present world. For example, NGOs publicised and shared their ideas about the Free Logging Agreement[14] which was an item in the agenda of the meeting held by WTO at Seattle, but, WTO did not open their documents. This can be found by searching in its website. They did not share their ideas or targets with people.

4. CONCLUSION

Biodiversity is becoming a more and more important subject because of its *option*, *bequest* and *existence values*. It is obvious that biodiversity contains many components of the ecosystems which should be conserved by mankind. As a result of its richness, many human activities affect it. When problem analysis is applied or a problem tree is illustrated about biodiversity conservation, many factors which can affect it such as urbanisation, industrialisation, agriculture, forestry can be determined easily. But, economic theories also affect biodiversity more implicitly with respect to other factors mentioned above. An economic decision, at one side, may cause pure economic results such as profit, wealth, wage, employment. On the other side, it may crowd out some species from production units. For this reason, conservation of biodiversity can not be achieved unless treaties which are imposed by the global organisations such as WTO, United Nations are not taken into account.

The mentality of the CBD does not agree with the WTO's ideology. Free market environmentalism is a result of some beliefs. Assumptions or prescriptions of the free market environmentalism and the real world are not well matched. If economic conditions changed, decreasing profit trends may increase or bankrupt enterprises may re-established. However an extinct gene source can not be found or recreated if it is lost. For this reason, management of biodiversity can not be entrusted to the hypothetical market models such as *free market environmentalism*.

REFERENCES

1. Parkin, M., 1990, Economics, Addison-Wesley Publishing Company.

2. Ierland, E. C. Van, 1993, Macroeconomic Analysis of Environmental Policy, Vol.2, Elsevier.
3. Anderson, T.L., Leal, D.R., 1991, Free Market Environmentalism. Pacific Research Inst. for Public Policy, USA.
4. Power, T. M., Rauber, P., 1993, The Price of Everything. Sierra, Nov/Dec 93, Vol.78. Issue 6, p. 86.
5. Smith, F. L., 1994, Sustainable Development- A Free Market Perspective, Boston Col. Environ, Affairs Law Review Winter 94, Vol. 21, Issue 2, p.297.
6. Ginsburg, J., 2000, Letting the Free Market Clear the Air, Business Week, Issue 3706, p.200.
7. Blumm, M.C., 1992, The Fallacies of Free Market Environmentalism, Harvard Journal of Law & Public Policy, Spring 92, Vol. 15, Issue 2, p.371.
8. Anderson, T. L., Shaw, J.S., 2000, Is Free-Market Environmentalism "Mainstream"? Social Studies, Sep/Oct 2000, Vol. 91, Issue 5, p. 227.
9. Schap,D., Young, A.T., 1999, Enterprise and Biodiversity: Do Market Forces Yield Diversity of Life? CATO Journal Vol. 19 Issue 1, p. 49.
10. Bourke, I.J., 1988, Trade in Forest Products: A Study of the Barriers Fared by the developing countries, FAO. F.P. 83.
11. Bourke, I.J., Leitch, J., 1998, Trade Restrictions And Their Impact on International Trade in Forest Products. FAO (in press).
12. GAIA, 1998, Global Trade and Biodiversity in Conflict, Grain, Issue no 1. April.
13. O'neill, J., 1998, The Market Ethics, Knowledge and Politics. (Translated in Turkish by Şen Süer Kaya and Ayrıntı Publications).
14. Menotti, V., 1999, Free Trade Free Logging, A Special Report by The International

Domestication and Determination of Yield and Quality Aspects of Wild *Mentha* Species Growing in Southern Turkey

[1]MENŞURE ÖZGÜVEN, [1]SALİHA KIRICI and [2]FİLİZ AYANOĞLU
[1]*Çukurova University, Faculty of Agriculture, Adana, Turkey*, [2]*Mustafa Kemal Univ., Faculty of Agriculture, Antakya, Turkey*

1. INTRODUCTION

Mentha species are cultivated in many countries commercially because of their valuable essential oil. Mint, which has anti-spasmatic, carminative, gastral, stimulating, and diuretic effects has been grown as a spice and curative plant. Mint oil is the richest natural source of menthol, which is widely used in medicine, food and cosmetics industry[1,2].

Seven *Mentha* species-*M.aquatica, M. longifolia, M. pulegium, M. spicata ssp.spicata, M. suaveolens, M. arvensis* and *Mentha x piperita*- are naturally grown in Anatolia[3]. According to a study it is determined that *M. longifolia* ssp. *thyphoides* has 1.8 % essential oil (47% piperitone), *M. pulegium* has 0.48-3.26% (27-28% pulegon), M. spicata ssp. spicata has 1.2-1.3% essential oil (68-69% carvone)[4].

According to the Turkish Codex, *M. piperita* has to contain 0.5 – 2.0 % essential oil and 0.7- 50.2 % menthol. Generally, *M. piperita* has yield of 250-500 kg of drug herbs/da and 100-200 kg of drug folia/da, and the rate of essential oil in *M. piperita* is normally 1-2 % and sometimes it reaches to 3.5 %. Good quality mint oil should contain 40-45 % menthol, less than 40 % menthone and less than 5 % menthofuran[5].

This study was carried out to determine drug yield, essential oil contents and the components of *Mentha* species naturally grown in Çukurova region under various ecological conditions.

2. MATERIALS AND METHODS

Çukurova region, extending from the Toros Mountains and its outskirts, in the north, the Amanos Mountains and Antakya in the south, Göksu Delta in the west, Kahramanmaraş and Ahırdağı in the east, have been scanned for *Mentha* species and different plant materials have been collected. Prof. Dr. Tuna Ekim, at the Department of Biology in Faculty of Arts and Sciences of Gazi University, did identification and nomenclature.

In the study, ten different *Mentha* species and ecotypes gathered from nature have been propagated and an experiment was set up in ecological conditions of Adana in autumn 1992. These species and ecotypes are 4.1 *M. spicata* L. *ssp. spicata*, 4.2 *M. longifolia* (L.) Hudson *ssp. typhoides var. typhoides*, 4.4 *M. spicata* L. *ssp. spicata*, 4.5 *M. spicata* L. *ssp. spicata*, 4.6 *Mentha x piperita*, 4.7 *M. longifolia* (L.) Hudson *ssp. typhoides var. typhoides*, 4.8 *Mentha x piperita*, 4.9 *M. longifolia* (L.) Hudson *ssp. typhoides var. typhoides*, 4.10 *M. longifolia* (L.) Hudson subsp. *typhoides* (Briq.) Harley and 4.25 *M. pulegium*.

The study was conducted for two years with four replication in 1993 and 1994. Plants were transplanted with 40 x 20 cm apart. Before planting 4 kg/da N and 4 kg/da P_2O_5 and after each harvest 4 kg/da N were applied.

In the first year (1993) of the experiment, plants were harvested at three different growth stages (pre-flowering, full flowering and post-flowering). In the second year (1994), different from the first year, the three different harvest stages applications have been terminated and all of the plants have been harvested at the beginning of flowering stage. Thus, under the ecological conditions of the experiments, it was possible to determine the total forms of the plants examined and in other words, the potential total yield per year was determined. Generally three harvests per year, was possible but the forth cut was also possible for 4.1 *M. spicata* and 4.2 *M. longifolia*. On the other hand, 4.25 *M. pulegium* harvested only twice.

During the study, drog herb yield (kg/da), drog leaf yield (kg/da), leaf rate (%), dry matter (%) (drying of fresh herba samples of 500 g, which is taken from each parcel for 24 hours at 105°C), essential oil content (%)(Neoclevenger), essential oil yield (l/da), essential oil components were investigated. Split plot (1993) and Randomly Block Design (1994) were used for statistical analysis.

The essential oil components were determined by gas chromatography (Carlo Erba, No. 1 GC 6000,series of Vega 2, column: FS-SE-54-DF-025; 50mx0.25mm <.D.; Entegration: 2mV, rotation rate 0.5 cm/minute-1,att.1:128). Experimental conditions of GC analyses; injector temperature 230°C, detector temperature 250°C, temperature program 70°C –5min., 3°C min^{-1} to 200 °C, flow rate of carrier gas (N) 1.5 ml. min^{-1}.

3. RESULTS AND DISCUSSIONS

3.1 Drog Herb Yields

In 1993, significant (p<0.01) differences have been determined as regard to drog herb yield among mentha ecotypes, harvest time and mentha ecotypes x harvest time interaction. Generally, drog herb yields of the plants were increased from pre-flowering harvests to post-flowering harvests. The highest drog herb yield was obtained from 4.7 *M. longifolia* in full-flowering and post-flowering harvests. Ecotypes 4.4 *M. spicata*, 4.2 *M. longifolia* and 4.1 *M. spicata* also had high drog herb yields (Table 1).

In 1994, drog herb yields of the plants were decreased from the first cut to the third and fourth cuts. The highest drog herb yield was obtained from 4.2 *M. longifolia* (1171 kg/da) in the first cut. For total drog herb yield 4.2 *M. longifolia* is superior to the other ecotypes with 2157 kg/da and, 4.1 *M. spicata*, 4.6 *M. x piperita* and 4.7 *M. longifolia* followed this with 2056 kg/da, 1954 kg/da and 1825 kg/da respectively (Table 1).

The values obtained from each three harvest in the first year and in each cuts or as total yield in the second year, were highly above the findings of some other researchers[5-9]. These results point out that collected material is very precious. However, drog herb yield of all ecotypes are very high in general, 4.1 *M. spicata*,,4.2 *M. longifolia*, 4.6 *M. x piperita* and 4.7 *M. longifolia* ecotypes are different from other species with their superior performance in both years.

3.2 Drog folia yield

In 1993, significant differences have been determined as regard to drog folia yields among the ecotypes, harvest stages (p<0.01) and ecotype x harvest stages interaction (p<0.05). High drog leaf yields were obtained generally from full flowering and pre-flowering harvests. Higher values were obtained from ecotype 4.7 *M. longifolia* in all three harvests (Table 2).

In 1994, in the first cut, the maximum drog folia yield was obtained from 4.7 *M. longifolia* with 465 kg/da and 4.6 *M x piperita* was 414 kg/da. In the second harvest, the highest drog leaf yield was obtained from 4.7 *M. longifolia* with 323 kg/da and, 4.9 *M. longifolia* followed this with 316 kg/da. The drog leaf yields obtained from 4.6 *M. x piperita*, 4.8 *M. x piperita* and 4.4 *M. spicata* were also high. In the third cut, maximum drog leaf yield was obtained from 4.7 *M. longifolia* with 222 kg/da, and it was followed by 4.6 *M.x piperita* with 190 kg/da. It was observed that 4.7 *M. longifolia* had the highest total yield with 1009 kg/da, 4.6 *M.x piperita* is the second with 903 kg/da.

The average drog leaf yields which were obtained from each growth stage harvests were highly above the values which was obtained from a cut of culture species cultivated in various locations[10]. In the second experiment year, early growth of the plants that survived the winter in spring have led to excess number of cuts[11], due to their early flourishment in spring, enabling to get higher yields. Total drog leaf yield of a vegetation season were also high.

Taking into consideration the results of the drog leaf yields, obtained in each two years, 4.7 *M. longifolia* and 4.6 *M.x piperita* ecotypes have been the most important.

3.3 Dry matter

In 1993, significant ($p<0.01$) differences have been determined as regard to dry matter among *Mentha* ecotypes and harvest time. Depending on the growth stages of the plants, dry matter increased from pre flowering to the post flowering stages and the highest values were obtained from post flowering period with 38.86%. Among the ecotypes the higher dry matter ratios were obtained from 4.7 *M. longifolia* with 34.50% and 4.1 *M. spicata* with 33.97 %. 4.4 *Mentha spicata*, 4.2 *M. longifolia* and 4.5 *Mentha spicata* had also high dry matter contents.

In 1994, the highest dry matter ratios were obtained from 4.1 *M. spicata* and 4.2 *M. longifolia* with 34.2 % in the first cut. Dry material ratio in *Mentha* culture species varies between 7.9-26.6 % depending on the place that they are reared[12]. In the experiment dry matter ratios of ecotypes increased in the first year from pre-flowering towards post flowering. This is due to the thickening and lignifications of the stalk of the plant. Besides, plant contains more water in the younger period. During the second year, dry matter ratio, which is high in the first cut, somewhat decreases in the second cut and tendency of increase is observed in the third cut. In all ecotypes, the ratios of dry material obtained every two years have been higher than *Mentha* culture species[5,13].

3.4 Essential oil content and essential oil yield

In 1993, significant ($p<0.01$) differences have been determined as regard to essential oil content among mentha ecotypes, harvest time and mentha ecotypes x harvest time interaction. If average values are considered, the values obtained in pre-flowering (3.30%) and full flowering (3.33%) were higher than the post flowering (2.55%). The essential oil content of plants varied between 1.35-4.43%. The highest levels of all harvests had been obtained from 4.10 *M. longifolia* and also 4.1 *M. spicata*, 4.2 *M. longifolia*, 4.5 *M. spicata* and 4.6 *M. x piperita* have indicated differences with their higher essential oil ratios than the other ecotypes. In 1994, essential oil contents of the plants were varied between 1.42-4.92 % and the high ratios were obtained from the ecotypes 4.10 *M. longifolia,* 4.6 *M. x piperita,* 4.8 *M. x piperita,* 4.25 *M. pulegium* and 4.1 *M. spicata* (Table 4).

The essential oil yields of the plants were varied between 1.26-12.90 l/da in 1993. For average essential oil yield full flowering and pre-flowering stages have indicated similar values, but after flowering was low. The highest value among the ecotypes was obtained from 4.6 *M. x piperita* (9.36 l/da). According to the growth stages, high essential oil yields were obtained from 4.7 *M. longifolia* and 4.25 *M. pulegium* in full flowering and from 4.1 *M. spicata* in pre flowering stages (Table 5).

In 1994, essential oil yields of the plants have varied between 5.91- 20.60 l/da in the first cut, 4.89-12.19 l/da in the second cut and 1.81-4.79 l/da in the third cut. When the total essential oil yield considered ecotype 4.6 *M. x piperita* takes the first place with 35.81 l/da, 4.7 *M. longifolia* takes the second place with 28.47 l/da and 4.8 *M. x piperita* takes the third place with 23.51 l/da (Table 5).

High oil yield per unit along with high essential oil ratio in *Mentha* is a desired quality. In the literature the volatile oil yield for the culture species is indicated as 4-10 kg/da[9,14,15,16,17,18,19]. When the total cuts were considered, very substantial results were obtained regarding essential oil yield. It was observed that all plants have the essential oil yield higher than 10 l/da, and also ecotypes such as 4.6 *M. x piperita,* 4.7 *M. longifolia,* 4.8 *M. x piperita,* 4.2 *M. longifolia,* 4.1 *M. spicata,* 4.4 *M. spicata* and 4.5 *M. spicata* have essential oil yields higher than 20 l/da. Although they are not culture forms, the essential oil contents of ecotypes in both years were much higher than the values in literature[7,8,11,20,21]. The obtained results confirm the opinions of the researchers who state that essential oil ratios increase in higher temperatures[20,22,23].

3.5 Essential Oil Components

In the first year of the study (1993), essential oil components of ten different *Mentha* species and ecotypes, at different development stages are indicated in the Tables 6, 7 and 8.

The piperitone was seen as the main component, in five *Mentha* ecotypes (4.1 *M. spicata,* 4.2 *M. longifolia,* 4.5 *M. spicata,* 4.8 *M. x piperita* and 4.9 *M. longifolia*). Cineole was the second major component in the ecotypes that were analyzed, but it was the main component in 4.10 *M. longifolia* (61.05%) only in full flowering stage. Carvone was only found and the main component in 4.6 *M. x piperita* and pulegon was the main component in 4.25 *M. pulegium* in all growing stages in 1993. Ecotypes 4.10 *M. longifolia.* 4.4

M. spicata, and 4.7 *M. longifolia* attract attention with their high menthol components in pre flowering and post flowering stages. Menthone was found at low levels, only 4.4 *M. spicata* in pre flowering (8.14%) and full flowering (7.18%) stages and 4.10 *M. longifolia* in post flowering stage (14.29%) have higher menthone content. Limonene was only found in 4.25 *M. pulegium* in pre flowering harvest. Jasmone was found in 4.1 *M. spicata,* 4.2 *M. longifolia* and 4.10 *M. longifolia* in post flowering harvest.

In the second year of the study (1994), essential oil components of ten different *Mentha* species and ecotypes, in two cuts are indicated in the Tables 9 and 10.

Piperitone was the main component of 4.1 *M. spicata*, 4.2 *M. longifolia*, 4.5 *M. spicata*, 4.6 *M. x piperita*, 4.8 *M. x piperita* and 4.9 *M. longifolia*. Cineole was the main component of 4.10 *M. longifolia*. Pulegon was the main component and only found in 4.25 *M. pulegium*. Menthol rate was the highest with 31.86% in 4.4 *M. spicata*, it was followed by 4.8 *M.x piperita* with 26.24% in the first cut and only found in the ecotypes 4.4 *M. spicata*, 4.7 *M. longifolia* and 4.10 *M longifolia* in the second cut. Menthone was found in some ecotypes but at low levels. Among the ecotypes the highest jasmone content was found in 4.7 *M. longifolia*. Limonene was not found in any ecotype.

In the study, in contrast to the researchers who indicate that the volatile oil composition is not affected by the cutting dates[24], it is found that there is a variation depending on the development stages of the plant[11,23]. In the culture species during the youth period the menthone, and as it matures the menthol ratio increases[9,25]. But, for the natural Mentha species and ecotypes it was not possible to make similar generalization. Because in the materials of the wild natured plants, in some ecotypes there was no or very low levels of menthone and menthol, and in some they were very high. This situation in addition to their genetic structures is also influenced by the climatic conditions, because the high temperatures along with the long and short day conditions influence the quality of the oil[10].

Carvone is known to be the main component in *M. spicata* L. ssp. *spicata*[2,4,26]. But, in the three different *M. spicata* L. ssp. *spicata* ecotypes taking place in the study no carvone was found (with the exception of 4.4 *M. spicata* in the second year). Instead of carvone, piperitone and menthol were determined as the main components. Similarly, in the naturally found *M. spicata* L. determination of three different chemotypes, one rich in carvone and limonene, the other two rich in piperitone epoxide[27] has indicated that non carvone contained ecotypes were a different chemotype than the *M. spicata* L. ssp. *spicata* found naturally generally in the West Mediterranean and Aegean regions.

As mentioned above in the 4.25 *M. pulegium*, there is a high ratio of pulegon, and its formation is linked to the ratio of flowering of the plant[9] and in order to obtain this component for artificial (DL) menthol[2] production, it was found that the pre-flowering and full-flowering periods were found to be the most appropriate periods.

In the study, for the four species of ecotype within *M. longifolia*, as mentioned in the literature[2,4,26], in the 4.2 *M. longifolia* and 4.9 *M. longifolia* piperiton was the main component in the two years. While changing according to the ecotype forms 4.7 *M. longifolia*, in general due to its main component menthone or menthol could be used for improvement of obtaining subspecies containing high menthol. Because 4.10 *M. longifolia* contains high level of cineol it could be named as "cineol chemotype".

Monoterpenes in Menthas is dependent on the carvone species, pulegon arising from the C3-p-menthan exposed to Oxygen or carvone arising from C6-p-Menthane[28]. The 4.3 *M. x piperita* in the research has taken place among C3-type monoterpenes group. Because in the first year 4.6 *M. x piperita* has totally taken place in carvone and in the second year's first cut in piperitone, in the second cut Menthone was the main component, this could be attributed to the unstable genetic structure of this ecotype and it is highly effected from ecological factors.

It is known that *Mentha* genus has an unstable and complex structure as regards its genetic and volatile oil components. As a result of the studies conducted on the *Mentha* species collected from the nature and the first analyses conducted on ecotypes as regards their volatile oil components and later during the development stage, the studies conducted on culture findings and the cuts made in different years clearly showed evident differences.

4. CONCLUSIONS

Ecotypes, 4.1 *M. spicata*, 4.4 *M. spicata*, 4.6 *M. x piperita*, and 4.8 *Mentha x piperita* developed well and showed the ability of regeneration after the cuts. Among the ecotypes, 4.2 *M. longifolia* and 4.1 *M. spicata* could be cut four times and therefore these ecotypes had the maximum drog herb yield. The yield of drog leaf of the ecotypes in the study was generally high. Especially, 4.7 *M. longifolia*, 4.6 *M. x piperita*, 4.4 *M. spicata*, 4.1 *M. spicata* and 4.8 *M. x piperita* attracted attention as regards drog leaf yield and decided as the promising ecotypes for cultivation.

Essential oil levels of the natural ecotypes, with a few exceptions, were higher than those given in codexes. If high amount of essential oil level and yield are aimed, the ecotypes of 4.6 *M. x piperita*, 4.7 *M. longifolia*, 4.1 *M. spicata*, 4.8 *M. x piperita* and 4.2 *M. longifolia* are the most promising ones.

Ecotypes, which are rich in menthone and menthol, are 4.4 *M. spicata*, 4.7 *M. longifolia*, also the low productive ecotypes of 4.10 *M. longifolia*.

4.25 *M. pulegium* contains high amount of pulegon so it can be used as a pulegon source. Finding high levels of piperitone in most of the ecotypes, indicates that this component has a natural character. It is known that *M. spicata* ssp. *spicata* contains high level of carvone which is especially important for chewing gum and tooth paste industry. In this experiment carvone was also found in 4.6 *M. piperita* and this ecotype can be used as carvone source. It is interesting to note that 4.10 *M. longifolia* had high level of cineole, which is especially used for cough syrups in the medicine industry. Also the ecotypes 4.6 *M. x piperita* and 4.8 *M. x piperita* are determined to be the other cineole ecotypes.

Considering that *Mentha* is consumed fresh and dried by public in our country, the 4.6 *M. x piperita* and 4.8 *M. x piperita* ecotypes could be suggested for these types of usages due to their both light colors and big leaves and also their good organoleptic characters as mild smell and taste. The experimented natural Mentha species and ecotypes have indicated a wide variation on essential oil components. For the selection of the appropriate plants for their usage in various sectors such as medicine, toothpaste, chewing gum, menthol Kleenex and cigarette, cosmetics, food and beverages, a good genetic pool has been established.

REFERENCES

1. Baytop, T., 1984, Türkiye'de Bitkiler ile Tedavi, İ.Ü. Yay. No: 3255, İstanbul, pp. 337-40.
2. Akgül, A., 1993, Baharat Bilimi ve Teknolojisi, Gıda Teknolojisi Derneği Yay. No: 15, Ankara, p. 451.
3. Davis, P.H., 1982, Flora of Turkey and East Aegean Island. Vol. VII, Edinburgh University Press, Edinburg, pp. 384-99.
4. Başer, K.H.C., 1993, *Acta Horticulturae*, **333**, 217-37.
5. Ceylan, A., 1987, Tıbbi Bitkiler II. (Uçucu Yağ İçerenler), E.Ü. Ziraat Fak. Yay. No: 481, Bornova-İzmir, p. 188.
6. Arslan, N., 1978, *Çiftlik Dergisi*, **219**, 17-8.
7. Ceylan, A., 1974, Ekolojik Faktörlerin Farklı Kökenli Mentha Çeşitlerinin Verimlerine, Uçucu Yağ Miktarı ve Bileşimlerine Etkisi, *Uluslararası Tıbbi Bitkiler Kollogiumu*, İzmir, pp. 68-75.
8. Ceylan, A., 1980, Nane (Mentha Spec.) Türlerinde Verim ve Ontogenetik Varyabilite Araştırması, *TÜBİTAK Bilim Kongresi*, Adana, (1980), pp. 217-38.
9. Court, W.A., Roy, R.C., Pocs, R., 1993, *Can. J. Plant Sci.*, **73**(3), 815-24.
10. Franz, Ch., Ceylan, A., Hšlzl, J., Všmel, A., 1984, *Acta Horticulturae*, **144**, 145-50.
11. Ruminska, A., Suchorska, K., Weglerz, Z., 1984, *Horticulture*, **12**, 33-9.
12. Kothari, S.K., Singh, V., Singh, K., 1987, *J. Agric. Sci. Camb.*, **108**, 691-3.
13. Özgüven, M., Kırıcı, S., Mengel, C., 1995, Nane (Mentha) Türlerinin Farklı Ekolojilerde Araştırılması, *Tıbbi ve Aromatik Bitkiler Workshop*, İzmir.

14. Ebert, K., 1982, Arznei-und Gew Ÿrzpflanzen, wissenschaftliche, Verlagsgesl. Stuttgart, pp. 221.
15. Saha, B.N., Borvah, A.K.S., Bordoloi, D.N., Mathur, R.K., Baruah, J.N., 1986, *Indian Perfumer*, **30**(2), 355-9.
16. Ky, L.D., Kirichenko E.B., 1988, *Byulleten, Glavnogo Botanicheskogo Sada*, **151**, 71-5.
17. Hadipoentyanti, E., 1990, *Pemberitaan Penelitian Tanamen Industri*, **16**(1), 18-22.
18. Shcherbakov, S.E., Zavyalova, L.E., Sidorenko, O.M.,1990, *Tekhnicheskie KulŌtury*, 1, 23.
19. Bouverat-Berniers, J.P., 1992, *Herba Gallica*, **2**, 1-15.
20. Duriyaprapan, S., Britten, E.J., Bastford, K. E., 1986, *Annais of Botany*, **58**,729-36.
21. Shalaby, A.S., El-Gamasy, A.M., El-Gengaihi, S.E., Khattab, M.D.,1988, *Egyptian Journal of Horticulture*, **15**(2), 213-24.
22. Singh, A.K., Naqvi, A.A., Singh, K., Thakur, R.S., 1988, *Current Sci.*, **57**(9), 480-1.
23. Gasic, O., Mimica-Dukic, N., Adamovic, D., 1992, *J. of Essential Oil Research*, **4**(1), 49-56.
24. Clark, R. J., Menary, R. C.,1984, *J. Sci. Food Agric.*, **35**, 1191-5.
25. Brun, N., Colson, M., Perrin, A., Voirin, B., 1991, *Can. J. Bot.*, **69**, 2271-8.
26. Wagner, H., Bladt, S., Zgainski, E.M., 1984, Plant Drug Analysis, Spring Verlag, Berlin, pp. 320.
27. Misra, L. N., Tyagi, B. R., Thakur, R. S., 1989, *Planta Medica*, **55**, 575-6.
28. Croteau, R., 1991, *Planta Medica*, Supplement Issue, **57**, 10-4.

Table 1. Drog Herb Yield (kg/da)

Ecotypes	1993				1994				Total Yield
	Pre-flowering	Full-flowering	Post-flowering	Mean	I. Cut	II. Cut	III. Cut	IV. Cut	
4.1 *M. spicata*	729 cd	807 c	1166 ab	901	1074 ab	508 bcd	211 c	263	2056
4.2 *M. longifolia*	579 cd	942 bc	1226 ab	916	1171 a	623 ab	155 de	280	2157
4.4 *M. spicata*	754 cd	895 bc	1347 ab	999	700 de	491 cd	200 cd	-	1391
4.5 *M. spicata*	673 cd	718 cd	921 bc	771	738 cd	401 d	331 b	-	1470
4.6 *M. x piperita*	853 bc	853 bc	1012 bc	906	920 bc	606 abc	428 a	-	1954
4.7 *M. longifolia*	744 cd	1399 a	1399 a	1180	817 cd	607 abc	401 a	-	1825
4.8 *M. x piperita*	569 cd	769 cd	1082 b	807	870 bcd	503 cde	289 b	-	1662
4.9 *M. longifolia*	535 d	776 cd	1061 bc	791	495 ef	648 a	234 c	-	1377
4.10 *M. longifolia*	253 e	345 de	476 de	350	291 f	448 d	136 e	-	875
4.25 *M. pulegium*	376 de	473 de	369 de	406	428 f	193 e	-	-	621
Mean	579	798	1006						
LSD 0.05	270				219	120			

Table 2. Drog Folia Yield (kg/da)

Ecotypes	1993				1994				Total Yield
	Pre-flowering	Full-flowering	Post-flowering	Mean	I. Cut	II. Cut	III. Cut	IV. Cut	
4.1 M. spicata	306 bc	255 bc	221 bc	261	260 d	222 bcd	77 f	136	695
4.2 M. longifolia	208 C	292 bc	186 cd	229	269 cd	258 abcd	53 g	77	657
4.4 M. spicata	273 bc	320 bc	295 bc	296	344 bc	261 abcd	107 d	-	712
4.5 M. spicata	246 bc	213 bc	176 cd	212	290 cd	193 cd	169 b	-	652
4.6 M. x piperita	272 bc	306 bc	249 bc	276	414 ab	299 ab	190 b	-	903
4.7 M. longifolia	335 ab	441 a	363 ab	379	465 a	323 a	222 a	-	1009
4.8 M. x piperita	275 bc	385 ab	189 cd	282	294 cd	272 abc	100 de	-	666
4.9 M. longifolia	236 bc	279 bc	204 c	240	211 d	316 a	130 c	-	657
4.10 M. longifolia	137 cd	81 d	73 d	97	125 e	190 d	80 ef	-	395
4.25 M. pulegium	222 bc	353 ab	61 d	212	219 d	108 e	-	-	327
Mean	251	292	202						
LSD 0.05	109				85	80	21		

Table 3. Dry matter ratio (%)

Ecotypes	1993			1994			
	Pre-flowering	Full-flowering	Post-flowering	Mean	I. Cut	II. Cut	III. Cut	IV. Cut
4.1 *M.spicata*	27.38	33.67	40.46	33.97	34.2 a	22.1 d	31.32 ab	29.6
4.2 *M. longifolia*	26.85	31.81	39.89	32.85 ab	34.2 a	29.7 a	30.60 ab	30.8
4.4 *M. spicata*	25.44	32.34	42.10	33.29 ab	31.7 abc	27.8 ab	25.20 cd	-
4.5 *M. spicata*	25.37	30.86	38.85	31.69 ab	33.0 ab	30.4 a	29.37 ab	-
4.6 *M. x piperita*	22.55	29.55	35.20	29.10 b	27.1 e	22.0 d	28.30 bc	-
4.7 *M. longifolia*	26.79	37.79	38.92	34.50 a	30.2 bcde	23.9 cd	24.90 d	-
4.8 *M. x piperita*	25.41	28.48	36.53	30.14 a	31.2 abcd	25.2 bc	32.32 a	-
4.9 *M. longifolia*	27.80	28.13	35.99	30.64 b	28.4 cde	27.7 ab	22.52 d	-
4.10 *M. longifolia*	24.88	26.14	41.85	30.95 b	27.7 de	24.2 cd	22.60 d	-
4.25 *M. pulegium*	24.79	26.68	39.17	30.21 b	28.2 cde	28.5 a		
Mean	25.72 c	30.54 b	38.86 a					
LSD 0.05	2.90 (E)		1.58 (H)		3.8	3.02	3.23	

Table 4. Essential oil content (%)

Ecotypes	1993				1994			
	Pre-flowering	Full-flowering	Post-flowering	Mean	I. Cut	II. Cut	III. Cut	IV. Cut
4.1 *M. spicata*	4.33 a	3.94 ab	2.22 cd	3.50	2.91 d	4.27 ab	3.63 ab	2.00
4.2 *M. longifolia*	3.98 ab	3.48 b	2.31 cd	3.25	3.30 cd	3.92 bc	3.42 b	2.40
4.4 *M. spicata*	2.93 bc	3.12 bc	2.19 cd	2.75	2.96 d	2.83 ef	2.71 c	–
4.5 *M. spicata*	3.47 b	4.43 a	3.76 ab	3.88	3.84 bcd	3.54 cd	1.95 d	–
4.6 *M. x piperita*	2.99 bc	3.21 bc	4.18 ab	3.46	4.87 ab	4.07 ab	1.59 e	–
4.7 *M. longifolia*	2.44 cd	2.89 bc	1.35 d	2.22	3.39 cd	2.43 fg	2.16 d	–
4.8 *M. x piperita*	3.25 bc	2.68 c	1.86 d	2.59	4.17 abc	3.15 de	2.70 c	–
4.9 *M. longifolia*	2.19 cd	1.92 cd	1.48 d	1.86	2.83 d	2.21 g	1.42 e	–
4.10 *M. longifolia*	3.93 ab	4.32 a	3.88 ab	4.04	4.92 a	4.42 ab	3.79 a	–
4.25 *M. pulegium*	3.49 b	3.40 bc	2.30 cd	3.06	2.82 d	4.58 a	–	–
Mean	3.30	3.33	2.55		1.43	0.51	0.35	
LSD 0.05		0.77						

Table 5. Essential oil yield (1/da)

Ecotypes	1993				1994				Total Yield
	Pre-flowering	Full-flowering	Post-flowering	Mean	I. Cut	II. Cut	III. Cut	IV. Cut	
4.1 M.spicata	12.67 ab	10.06 ab	4.85 c	9.19	7.53 de	9.41 bc	2.79 b	2.72	22.45
4.2 M. longifolia	8.08 bc	10.08 ab	4031 cd	7.48	8.96 cde	10.07 ab	1.81 c	1.84	22.68
4.4 M. spicata	7.92 bc	9.77 ab	6.40 bc	8.02	9.93 cde	7.36 cd	2.82 b	-	20.11
4.5 M. spicata	8.49 bc	9.16 b	6.59 bc	8.07	11.08 cd	6.87 de	3.29 b	-	21.24
4.6 M. x piperita	8.06 bc	9.61 ab	10.43 ab	9.36	20.60 a	12.19 a	3.02 b	-	35.81
4.7 M. longifolia	7.85 bc	12.90 a	4.99 c	8.57	15.74 b	7.94 bcd	4.79 a	-	28.47
4.8 M. x piperita	8.90 bc	10.10 ab	3.47 cd	7.48	12.23 bc	8.59 bcd	2.69 b	-	23.51
4.9 M. longifolia	5.14 c	5.44 c	2.96 cd	4.51	5.96 e	6.98 cde	1.85 c	-	14.79
4.10 M. longifolia	5.37 c	3.56 cd	2.78 cd	3.90	5.91 e	8.37 bcd	3.04 b	-	17.32
4.25 M. pulegium	7.75 bc	12.31 ab	1.26 ab	7.10	6.23 e	4.89 e	-	-	11.12
Mean	8.02	9.29	4.80						
LSD 0.05		3.56							

Table 6. Essential oil components (%) at pre-flower harvest in 1993

Ecotypes	α-Pinene	β-Pinene	Limonene	1.8-Cineole	Menthone	Menthofuran	Menthol	Pulegon	Piperiton	Carvone	Methyl acetate	β-Caryophyllene	Σ
4.1	0.93	1.32	-	23.76	2.05	0.38	1.63	-	54.93	-	0.25	3.59	88.84
4.2	0.86	1.19	-	23.54	2.22	-	1.36	-	54.95	-	0.63	4.44	89.19
4.4	0.56	1.16	-	19.60	8.14	1.42	41.08	-	-	-	0.52	6.06	78.54
4.5	0.58	0.89	-	17.93	0.36	0.19	0.31	-	45.76	-	0.78	3.75	70.55
4.6	0.36	0.53	-	20.56	0.35	0.42	1.11	-	-	-	0.35	0.73	68.82
4.7	0.43	0.88	-	9.34	0.46	0.28	28.12	-	-	64.11	0.72	3.55	43.80
4.8	0.32	0.45	-	24.39	0.53	0.47	1.51	-	63.42	-	0.54	1.30	92.93
4.9	0.59	1.03	-	9.65	0.54	-	0.63	-	67.98	-	0.71	3.68	85.21
4.10	-	-	-	23.59	3.22	-	56.06	-	-	-	-	-	82.87
4.25	0.24	0.20	0.62	1.17	3.55	2.38	0.89	83.74	-	-	0.82	0.27	93.88

Table 7. Essential oil components (%) at full-flowering harvest in 1993

Ecotypes	α-Pinene	β-Pinene	Limonene	1,8-Cineole	Menthone	Menthofuran	Menthol	Pulegon	Piperiton	Carvone	Methyl acetate	β-Caryophyllene	Σ
4.1	1.00	1.78	-	24.56	1.59	0.66	1.51	-	44.64	-	0.52	3.52	79.78
4.2	0.86	1.41	-	21.80	1.56	0.34	1.31	-	48.55	-	1.22	3.68	80.73
4.4	0.66	1.04	-	20.94	7.18	-	45.88	-	-	-	0.42	4.08	80.20
4.5	0.45	0.98	-	16.20	0.29	-	-	-	56.15	-	0.46	2.51	77.04
4.6	0.55	0.74	-	23.36	0.40	0.20	0.68	0.63	-	-	-	0.53	73.55
4.7	1.06	1.48	-	17.47	0.74	-	0.55	2.64	-	46.46	0.57	2.77	27.28
4.8	0.71	0.47	-	28.02	-	-	-	1.34	63.31	-	1.49	1.24	96.58
4.9	0.64	0.86	-	5.17	0.53	0.20	0.77	6.30	62.88	-	0.44	2.74	80.53
4.10	1.37	2.20	-	61.05	1.55	0.34	4.33	-	11.43	-	0.34	0.76	83.87
4.25	0.32	-	-	-	-	20.39	0.60	58.36	-	-	0.61	-	80.28

Table 8. Essential oil components (%) at post-flowering harvest in 1993

Ecotypes	α-Pinene	β-Pinene	1.8-Cineole	Menthone	Menthofuran	Menthol	Pulegon	Piperiton	Carvone	Methyl acetate	Jasmone	β-Caryophyllene	Σ
4.1	0.66	0.76	16.89	-	1.27	0.99	-	39.18	-	2.17	16.59	3.17	81.68
4.2	0.72	0.80	20.02	-	1.13	1.04	-	48.40	-	0.54	8.22	2.84	83.71
4.4	-	0.77	23.08	2.43	6.57	39.14	-	-	-	0.61	-	6.10	78.70
4.5	0.46	-	7.10	-	-	-	-	69.34	-	-	-	2.14	79.04
4.6	-	-	21.42	0.53	-	2.35	-	-	63.46	-	-	-	87.76
4.7	-	-	20.66	-	-	64.79	-	-	-	2.98	-	-	88.43
4.8	0.39	0.37	22.49	1.96	2.68	4.23	-	60.58	-	-	-	0.40	93.10
4.9	0.54	0.53	6.37	0.98	1.48	0.90	-	69.21	-	-	-	1.62	81.63
4.10	-	0.54	13.32	14.29	0.52	26.28	-	-	-	0.43	16.81	1.19	73.38
4.25	0.56	0.89	16.26	0.92	21.30	1.18	38.63	-	-	0.40	-	4.99	85.13

Table 9. Essential oil components (%) at the first cut in 1994

Ecotypes	α-Pinene	β-Pinene	1.8-Cineole	Menthone	Menthofuran	Menthol	Pulegon	Piperiton	Carvone	Methyl acetate	Jasmone	β-Caryophyllene	Σ
4.1	0.80	1.06	19.36	-	1.61	0.77	-	46.70	-	2.65	2.56	4.97	80.48
4.2	0.90	1.18	18.24	1.24	2.09	1.07	-	46.19	3.68	1.76	2.73	3.94	79.34
4.4	0.90	0.81	17.33	10.08	1.54	31.86	-	-	-	-	-	7.04	73.24
4.5	1.50	0.80	15.42	-	-	-	-	51.09	-	-	5.46	3.96	78.23
4.6	-	-	19.83	1.88	-	1.04	-	61.46	-	-	-	-	84.21
4.7	0.94	1.12	11.89	-	-	7.73	-	-	-	-	27.77	4.29	53.74
4.8	-	-	17.98	-	-	26.24	-	31.26	-	-	-	1.18	78.04
4.9	0.74	0.63	7.56	-	-	0.63	-	60.75	-	-	1.21	3.80	75.31
4.10	3.46	1.77	46.22	-	1.16	9.58	-	-	-	-	9.31	1.12	74.63
4.25	-	-	1.86	-	20.97	-	43.83	-	-	14.28	-	0.67	81.61

Table 10. Essential oil components (%) at the second cut in 1994

Ecotypes	α-Pinene	β-Pinene	Limonene	1.8-Cineole	Menthone	Mentho-furan	Menthol	Pulegon	Piperiton	Carvone	Methyl acetate	Jasmone	β-Caryoph-yllene	Σ
4.1	1.18	0.83	-	18.81	1.59	1.47	-	-	26.08	-	1.10	17.31	2.10	70.47
4.2	0.99	0.81	-	14.74	0.89	0.92	-	-	36.82	-	0.88	9.01	3.11	68.17
4.4	0.98	0.86	-	17.19	2.86	0.74	14.33	-	-	10.88	1.37	8.72	3.69	61.62
4.5	0.68	0.69	-	10.48	-	-	-	-	47.60	-	1.13	9.27	2.28	72.13
4.6	0.58	0.54	-	16.48	-	-	-	-	57.60	-	0.78	-	-	75.98
4.7	1.16	0.98	-	11.49	-	-	6.85	-	-	-	3.70	24.76	3.53	52.47
4.8	0.61	0.53	-	23.71	-	-	-	-	57.73	-	0.97	-	0.63	84.18
4.9	1.02	0.66	-	8.11	-	-	-	-	58.52	-	0.70	0.93	1.90	71.84
4.10	1.96	1.56	0.98	36.39	1.00	2.40	6.52	-	-	-	0.95	14.07	1.13	66.96
4.25	-	-	-	2.20	7.32	2.13	-	61.81	-	-	0.86	1.11	-	75.43

Some Ornamental Geophytes from the East Anatolia

MEHMET KOYUNCU
Ankara University, Faculty of Pharmacy, Department of Pharmaceutical Botany, 06100 Ankara, Turkey

1. INTRODUCTION

Turkey is rich in various of plants as well as the geophytes(bulb, tuber and rhizome bearing plants) of which more than 600 different species exist in Turkey[1-3]. Geophytes are important as ornamental plants with their splendid flowers and blossom in spring[4,5]. Besides their exportation as ornamental plants such as *Galanthus elwesii, Leucojum aestivum, Fritillaria imperialis, Cyclamen hederifolium* used in medicine. In the east Anatolia, especially around Van, Hakkari and Bitlis, leaves of some species e.g. *Allium vineale, A. schoenoprasum* are used in making cheese with herbs "otlu peynir". Fresh leaves of *Eremurus spectabilis, Ornithogalum narbonense* are sold in farmmarkets as vegetables. Bulbs of *Allium tuncelianum* are used instead of garlic. Tubers of many species of *Orchis, Ophrys* and *Dactylorhiza* are used to make a pleasant drink called "sahlep"[6]. In spite of the existence of other plants with beautiful flowers in the region, only *Fritillaria imperialis, F. persica* and *Sternbergia clusiana* are collected for exportation as ornamental plants. Among these, especially *Fritillaria imperialis* which exists around Semdinli, Bitlis, Siirt and Adıyaman is exported in large volumes. *Iris spuria* ssp. *musulmanica, Narcissus poeticus* and *Tulipa sylvestris* are used as ornamental plants in the parks and gardens in this region. Geophytes are used in the east Anatolia for various purposes and have an important economic value. In this study, we will concentrate on the species that are or can be used as ornamental purposes.

2. IMPORTANT GEOPHYTES OF EAST ANATOLIA

The east Anatolia region is one of the rich areas in terms of geophytes. The author of this work has collected number of species during field studies particularly around Van, Bitlis, Mus, Hakkari, Agrı and Erzurum districts. These species are classified according to the family as follows:

2.1 Amaryllidaceae

Ixiolirion tataricum (Pallas) Herbert
Narcissus poeticus L. ssp. *poeticus*
Sternbergia clusiana (Ker-Gawler) Ker-Gawler ex Sprengel
S.fischeriana (Herbert) M.J.Roem.

2.2 Berberidaceae

Bongardia chrysogenum (L.) Spach
Leontice leontopetalum L. ssp. *eversmannii* (Bunge) Spach

2.3 Fumariaceae

Corydalis rutifolia (Sibth. et Sm.) DC. ssp. *rutifolia*
Corydalis rutifolia (Sibth. et Sm.) DC. ssp. *Kurdica* Cullen & Davis

2.4 Geraniaceae

Geranium stepporum Davis
Pelargonium endlicherianum Fenzl
P.quercetorum Agnew

2.5 Iridaceae

Crocus cancellatus Herbert
C.kurduchorum Kotschy ex Maw
Gladiolus atroviolaceus Boiss.
G.kotschyanus Boiss.
Iris aucheri (Baker) Sealey
I.barnumae Baker et Foster
I.caucasica Hoffm.
I.iberica Hoffm. ssp. *elegantissima* (Sosn.)Takht. et Federov
I.paradoxa Stev.

I.persica L.
I.pseudocaucasica Grossh.
I.reticulata Bieb.
I.sari Schott. ex Baker
I.spuria L. ssp. *musulmanica* (Fomin) Takht.

2.6   Liliaceae

Allium affine Ledeb.
A.armenum Boiss. et Kotschy
A.aucheri Boiss.
A.cardiostemon Fisch. et Mey
A.hirtifolium Boiss.
A.kharputense Freyn et Sint.
A.noeanum Reuter ex Regel
A.scabriscabum Boiss. et Kotschy
A.schoenoprasum L.
A.szovitsii Regel
A.tuncelianum (Kollm.) culture
Bellevalia fominii Woronov
B.forniculata (Fomin) Deloney
B.longistyla (Miscz.) Grossh.
B.paradoxa (Fisch. et Mey.) Boiss.
B.pycnantha (C.Koch) A. Los.-Los.
B.rixii Wendelbo
Colchicum kotschyii Boiss.
C.szovitsii Fisch.et Mey.
Eremurus spectabilis Bieb.
Fritillaria alburyana Rix
F. aurea Schott
F.crassifolia Boiss. et Huet.
F. imperialis L.
F.michailovskyi Fomin
F.minima Boiss.et Noe
F.minuta Boiss.
F. persica L.
Merendera kurdica Bornm.
Muscari armeniacum Leichtlin et Baker
M.azureum Fenzl
M.coeleste Fomin
M.comosum (L.). Miller

M.tenuiflorum Tauch
Nectaroscordum tipedale (Trautv.) Grossh.
Ornithogalum arcuatum Stev.
O.oligophyllum E.D.Clarke
O.narbonense L.
O.platyphyllum Boiss.
Puschkinia scilloides Adams
Tulipa armena Boiss.
T.biflora Pallas
T.humilis Herbert
T.julia C.Koch
T.orphanidea Boiss. Ex Heldr.
T.sintenesii Baker
T.sylvestris L.

2.7 Orchidaceae

Dactylorhiza umbrosa (Kar. Et Kir.) Newki
Orchis simia Lam.

2.8 Ranunculaceae

Aconitum cochleare Woroschin
Paeonia mascula (L.) ssp. *aerietina* (Anders) Cullen et Heywood
Trollius ranunculinus (Smith) Stearn

3. CONCLUSION

As the list and pictures indicated, the east Anatolia is quite rich in terms of natural geophytes that can be used as ornamental plants. Words are insufficient to describe the joy of seeing and investigating these flowers in their original habitats. It is our wish to teach how to utilize these species without jeopardizing their natural habitats and threatening their population. Learning, protecting and wisely utilizing our natural resources and biological richness will play a key role on our way to being a developed country. It is our responsibility to take advantages of them without threatening their existence but on the other hand, it is not possible to protect a floral richness before learning and appreciating it. The most important purpose of this study is to help comprehend and protect the geophyte species that exist in Turkey.

REFERENCES

1. Davis, P.H., 1965-1988, Flora of Turkey and the East Aegean Islands, Vols. 1-9, Edinburgh University Press, Edinburgh.
2. Davis, P.H., Mill, R.R., Tan, K., 1988, Flora of Turkey and the East Aegean Islands (Supplement), Vol. 10, Edinburgh University Press, Edinburgh.
3. Güner, A., Özhatay, N., Ekim, T., Başer, K.H.C., 2000, Flora of Turkey and the East Aegean Islands (Supplement 2), Vol. 11, University Press, Edinburgh.
4. Mathew B., Baytop, T., 1984, *Bulbous Plants of Turkey*, p.10, B.T. Batsford Ltd., London.
5. Rix, M., Philips, R., 1983, The Bulb Book, Pan Books, Kent.
6. Sezik, E., 1984, Türkiye'nin Orkideleri, Sandoz Kültür Yayınları No.6.

Bioactive Molecules from *Cynodon dactylon* of Indian Biodiversity

NANJIAN RAMAN[1], A.RADHA[1], K.BALASUBRAMANIAN[2], R.RAGHUNATHAN[3] and R.PRIYADARSHINI[4]

[1]*Centre for Advanced Studies in Botany, University of Madras, Guindy campus, Chennai-600 025, India,* [2]*Department of Pharmacy, Baid Mehta College of Pharmacy, Thoraipakkam, Chennai-600, India,* [3]*Department of Organic Chemistry, University of Madras, Guindy campus, Chennai-600 025, India,* [4]*College of Pharmacy, Sri Ramachandra Medical College and Research Institute (Deemed University), Porur, Chennai - 600 116, India.*

1. INTRODUCTION

Cynodon dactylon (L.) Pers. (Gramineae), called Bermuda grass (English) is widely distributed in the tropics and the warmer areas of the temperate regions and used in traditional medicine for diarrhoea, leprosy, scabies, haemoptysis, haematuria, cephalalgia, catarrhal opthalmia, epileptic seizures, skin diseases, erysipelas, dropsy and anasarca and as a diuretic[1]. Eventhough the plant has been phytochemically investigated for the presence of flavonoids[2] triterpenoids[3] and streoids there are no reports on the presence of alkaloids (NAPRALERT). This is the first report on the isolation and structural elucidation of alkaloids (tryptamine, tyramine and gramine) from the aerial parts of *C. dactylon* of Gramineae from a 1996 collection. The structure of the alkaloids was proposed on the basis of UV, MS, IR and NMR (^{1}H and ^{13}C) spectral data. The antimicrobial activity of the crude extract and the alkaloids against human pathogenic bacteria and dermatophytes was reported for the first time.

2. MATERIALS AND METHODS

2.1 Extraction

Fresh and healthy aerial parts of *C.dactylon* were collected between September and October 1996 from Madipakkam area, Chennai, India. The shade dried powdered aerial parts (4kg) were exhaustively extracted with MeOH, concentrated and refrigerated overnight.The MeOH soluble fraction was partitioned with EtOAc and Hexane. The EtOAc soluble fraction was acidified with 0.01% H_2SO_4 and extracted with $CHCl_3$. The aqueous layer was basified with NH_3 and extracted with $CHCl_3$ to afford a mixture of alkaloids (4.92 g), after evaporation of solvent under vacuum. The alkaloid mixture (3 g) was subjected to column chromatography on neutral alumina, using $CHCl_3$, $CHCl_3$:MeOH and MeOH as eluents to afford tryptamine (72 mg), tyramine (58 mg) and gramine (62 mg).

2.2 Studied Activity

Antibacterial activity was checked by agar well-diffusion and disc diffusion method. Antifungal activity was studied by dilution tube technique.

2.3 Organisms Used

Human pathogenic bacteria - *Bacillus subtilis, Escherichia coli, Klebsiella pneumoniae, Proteus mirabilis, Salmonella typhi* 'H', *Shigella boydii* and *S. dysenteriae*.
Dermatophytes- *Epidermophyton floccosum, Microsporum gypseum, M. nanum, Trichophyton mentagrophytes* and *T. rubrum*.

3. RESULTS AND DISCUSSION

Fraction 1: Tryptamine, mp 116-118°C (uncorrected, EtOH);
Fraction 2: Tyramine, mp 163-165C° (uncorrected, boiling EtOH);
Fraction 3: Gramine, mp 137-139C° (uncorrected, EtOH);

The Spectral data of all the three compounds were identical with the authentic sample.

3.1 Antimicrobial Activity

The antibacterial property of CEE was evident when tested against human pathogenic bacteria. The antimicrobial activity of chloroform extract from the leaves of *C. dactylon* against Gram negative and positive bacteria has been reported[3] which may be due to the presence of aromatic acids such as 4-hydroxy benzoic acid, 3-methoxy-4-hydrozybenzoic acid, 2 (4'-hydroxyphenyl) propionic acid and 2 (3'-methoxy-4'-hydroxyphenyl) propionic acid. The aqueous extract of *C. dactylon* leaves in the form of 1% ointment has been reported to have wound healing activity.

The fungitoxic activity of CEE by dilution tube technique against dermatophytes was evident by the fact that even at a low concentration of 50 µg/mL complete inhibition of growth was observed in all the fungi. This suggests that the popular use of this plant as a remedy for skin infections and skin related diseases could be justified.

Tryptamine and tyramine individually did not exhibit any antibacterial activity. But a combination of the two significantly inhibited the growth of *Klebsiella pneumoniae* (7.7 and 8.8 %) at 500 and 1000 µg /mL concentration. This justified the use of *C.dactylon* against urinary inflammations, chest and wound infections. The alkaloid gramine effectively inhibited the growth of *Bacillus subtilis* and *Escherichia coli* (11.1%) at 1000 µg /mL. *Klebsiella pneumoniae* and *Proteus mirabilis* ((7.7 and 8.8%) at 500 and 1000 µg /mL and *Salmonella typhi* 'H' (7.77, 8.88 and 11.1%), at 100, 500 and 1000 µg /mL concentration. All the results were compared with standard antibiotic discs.

The growth inhibition of *P.mirabilis,* causal agent of serious urinary tract infection resulting in serious complications including cystitis and prostatitis confirmed the use of *C.dactylon* as an effective medicine in the treatment of prostatitis. The inhibition of growth of *E.coli* and *Bacillus subtilis* revealed the effective nature of *C.dactylon* in controlling dysentery and in the treatment of conjuctivitis.

The antifungal property of tryptamine was evidenced when tested against the growth of the fungi *Epidermophyton floccosum, Microsporum nanum, Trichophyton mentagrophytes and T. rubrum* at 1000 µg /mL concentration, where complete growth inhibition was observed when compared with the control and Griseofulvin. The compound inhibited the growth to 50% in *M.nanum, T.mentagrophytes* and *T.rubrum* at 50 and 100 µg /mL concentration.The growth of the fungus *E. floccosum* was inhibited up to 75% in 50 to 250 µg /mL conc. The compound was lethal to the growth of the fungus *M.gypseum* at all the concentrations. Tyramine effectively inhibited the growth of *E.floccosum, M.gypseum, M.nanum* and *T.mentagrophytes* completely (100% inhibition) at all the concentrations except *T. rubrum* where only 75% inhibition at 50 μg /mL conc. was observed. The potent antifungal activity of gramine against the dermatopytes was established by the fact that there was complete growth inhibition of the fungi at all the concentrations. For the first time the antifungal property of *C.dactylon* crude extract and alkaloids against dermatophytes is reported. This accounts for the use of *C.dactylon* in the use of skin diseases.

The present study clearly revealed that the antimicrobial activity of *C.dactylon* against human pathogenic bacteria and dermatophytes was due to the presence of the alkaloids tryptamine, tyramine and gramine. This indicated and justified the use of *C.dactylon* as a potent antimicrobial agent.

REFERENCES

1. Varier P.S., 1994, Indian Medicinal Plants - A Compendium of 500 species, Vol. 2, Orient Longman Ltd., pp. 282-92.
2. Millar R.P., *J.Agri.Res.,* 1967, **5**, 177-9.
3. Ohmoto T., Ikuse M., Natori S., 1970, *Phytochemistry,* **9**, 2137.

Phenylethanoid Glycosides with Free Radical Scavenging Properties from *Verbascum wiedemannianum*

İHSAN ÇALIS[1], HASAN ABOU GAZAR[1], ERDAL BEDİR[1,2] and IKHLAS A. KHAN[2]

[1]*Department of Pharmacognosy, Faculty of Pharmacy, Hacettepe University, 06100 Ankara, Turkey,* [2]*National Center for Natural Products Research, Research Institute of Pharmaceutical Sciences, Department of Pharmacognosy, School of Pharmacy, The University of Missisippi, University, MS 38677, USA*

1. INTRODUCTION

The genus *Verbascum* is represented by 228 species in the flora of Turkey. *Verbascum wiedemannianum* FISCH. & MEY. is among 192 endemic species with violet-purple corolla which widely spreads in Central Anatolia[1]. Steroids, oleanane-type triterpenic saponins, iridoids, phenylpropanoids and alkaloids have been reported from *Verbascum* species in previous studies[2]. In our comparative work on the chemical constituents of this species, we carried out a through examination of the aerial- and underground parts. Chromatographic studies afforded three new phenylethanoid glycosides, wiedemanniosides A-C (**1-3**), in addition to the four known glycosides, verbascoside, martynoside, echinacoside and leucosceptoside B.

2. MATERIAL AND METHODS

2.1 Plant material

Verbascum wiedemannianum Fisch. & Mey. was collected between Akmağden and Yildizeli, 4 km to Yildizeli, Sivas at June 2, 1998.

2.2 Extraction and Isolation

Air-dried powdered roots of *V. wiedemannianum* (350 g) were extracted with MeOH-H_2O (4:1) to yield 52 g of crude extract. The water soluble part of the methanolic extract was fractionated by a tactic combination of polyamide column chromatography (MeOH in H_2O; 0-100%), medium-pressure liquid chromatography (MPLC) on reversed-phase material (i-PrOH or MeOH in H_2O; 5-35% I-PrOH; 30-100% MeOH), and column chromatoraphy (CC) on normal-phase silica gel ($CHCl_3$-MeOH-H_2O mixtures) to afford three new phenylethanoid glycosides, wiedemannioside A-C (1-3), verbascoside, martynoside, echinacoside and leucosceptoside B (Table 1).

Reduction of DPPH Radical[3,4]

3. RESULTS AND DISCUSSION

All phenylethanoid glycosides were obtained as amorphous powders, whose UV spectra indicated their polyphenolic nature. Their IR spectra showed absorption bands for hydroxyls (3400 cm^{-1}); α,β–unsaturated esters (1690-1700 cm^{-1}); esters (for **1** and **2**: 1730-1740 cm^{-1}); olefinic double bonds (1635 cm^{-1}); and aromatic rings (1600-1520 cm^{-1}). The structures of the known glycosides were identified by comparison of their NMR data with those reported for verbascoside[5], martynoside[6], echinacoside[7] and leucosceptoside B[8].

Table 1. Phenylethanoid Glycosides isolated from *V. wiedemannianum* Fisch. & Mey.

Compounds	R_1	R_2	R_3	R_4	R_5	R_6
[1] **Wiedemannioside A**	Me	Me	H	H	H	Ac
[2] **Wiedemannioside B**	Me	Me	H	Ac	Ac	Ac
[3] **Wiedemannioside C**	H	Me	H	H	H	β-D-glucopyranosyl
Acteoside (= Verbascoside)	H	H	H	H	H	-
Martynoside	Me	Me	H	H	H	-
Echinacoside	H	H	H	H	H	β-D-glucopyranosyl
Leucosceptoside B	Me	Me	H	H	H	β-D-apiofuranosyl

The FAB-MS of wiedemannioside A [1] was compatible with the molecular formula $C_{33}H_{42}O_{16}$ (m/z 717 $[M+Na]^+$, Mol. Wt.: 694). The ^1H and ^{13}C-NMR data for 1 were extremely similar to those of martynoside, the major differences being the presence of a resonance for one acetoxyl signal. This observation was supported by the FAB-MS, which was 42 mass units higher than that of martynoside. The feruloyl and acetyl units were found to be located at the C-4' and the C-6' of the glucose moiety, on the basis of the strong deshielding of H-4' (δ 5.00 t, J = 9.5 Hz) and H_2-6' (δ 4.18 dd, J = 12.0 and 5.0 Hz; 4.09 dd, J = 12.0 and 2.0 Hz) signals of the glucose unit. The similar α-effect of esterification was also observed for the C-6' (δ 64.5) resonance of the glucose moiety. Thus, the structure of wiedemannioside A [1] was established as 6'-O-acetyl-martynoside.

The ^1H- and ^{13}C-NMR data of wiedemannioside B [2] were similar to those of 1 and martynoside. Additionally, the resonances arising from three acetoxyl signals were observed. The FAB-MS of 2 exhibited the pseudomolecular ion peaks at m/z 779 $[M+H]^+$, 801 $[M+Na]^+$, and 817 $[M+K]^+$ (calc. for $C_{37}H_{46}O_{18}$, Mol. Wt.: 778), which was 126 mass units higher than that of martynoside, confirming the presence of three acetoxyl functionalities. All structural assignments were substantiated by the results obtained from the 2D shift-correlated COSY, HSQC and HMBC experiments. The COSY experiment established the sites of four acylations: H-4' and H_2-6' of the glucose unit, and H-2" and H-3" of the rhamnose unit were shifted down-field. The HSQC experiment established direct C-H bondings. The HMBC experiment made clear the all intermolecular connectivities, where correlations were observed between H-1'/α-carbon of the aglycone moiety, H-1"/C-3', H-4'/carbonyl carbon of the feruloyl moiety, indicating that all connectivities were similar as in martynoside. Thus, the location of additional three acetyl units were found to be at the H-6' of glucose unit and the H-2" and H-3" of rhamnose unit. Consequently, the structure of wiedemannioside B [2] was established as 6',2",3"-tri-O-acetyl-martynoside.

The ^1H- and ^{13}C-NMR data of wiedemannioside C [3] were closely similar to those of echinacoside. Compound 3 contains an additional methoxyl signal as indicated from the ^1H- and ^{13}C-NMR data [δ$_H$ 3.89 s; δ$_C$

56.9 CH_3). The FAB-MS of **3** exhibited a pseudomolecular ion peak at m/z 823 $[M+H]^+$ (calc. for $C_{36}H_{48}O_{20}$, Mol. Wt.: 800), which was 14 mass units higher than that of echinacoside, confirming the presence of an methoxy-methyl group. The major differences were observed for the proton and carbon resonances of the acyl moiety. The chemical shifts of the signals assigned to the acyl unit were almost same with those of martynoside, wiedemanniosides A [**1**] and B [**2**] which have ferulic acid unit as acyl moiety. Based on these observations, the structure of wiedemannioside C [**3**] was elucidated as echinacoside-3-O''''-methylether.

The new compounds, wiedemanniosides A [**1**], B [**2**] and C [**3**] and the known phenylethanoid glycosides were found to have antioxidant properties, based on experiments with 2,2-diphenyl-1-picrylhydrazyl (DPPH), which indicated their ability to efficiently scavenge free radicals[3,4].

REFERENCES

1. Huber-Morath, A.,1978, *Flora of Turkey and the East Aegean Islands*, Vol. 6, (Davis, P.H. ed.) University Press, Edinburgh, pp. 461-603,.
2. Abou Gazar, H., 2001, Ph.D Thesis, Hacettepe University, Health Sciences Institute, Ankara.
3. Cuendet, M., Hostetmann, K., Potterat, O., Dyatmiko, W., 1997, *Helv. Chim. Acta,* **80**, 1144.
4. Takao, T., Kitatani, F., Watanabe, N., *et al.*, 1994, *Biosci. Biotech. Biochem.*, **58**, 1780.
5. Sticher, O., Lahloub, M.F., 1982, *Planta Med.*, **46**, 145.
6. Calis, I., Lahloub, M.F., Rogenmoser, E., Sticher, O.,1984, *Phytochemistry* , **23**, 2313.
7. Becker, H., Hsieh, W.C., Wylde, R., Laffite, C., Andary, C. ,1982, *Z. Naturforsch.*, **37c**, 351.
8. Miyase, T., Koizumi, A., Ueno, A., Noro, T., Kuroyanagi, M., Fukushima, S., Akiyama, Y., Takemoto, T.,1982, *Chem. Pharm. Bull.,* **30**, 2732-7.

Antioxidant Activity of *Capsicum annuum* L. Fruit Extracts on Acetaminophen Toxicity

BİLGEN ERYILMAZ[1], GÖKNUR AKTAY[2] and FUNDA BİNGÖL[1]
[1] *Gazi University, Faculty of Pharmacy, Department of Pharmacognosy, 06330,Ankara, Turkey,*
[2] *Ankara University, Institute of Forensic Medicine, Toxicology Unit, 06100, Ankara, Turkey*

1. INTRODUCTION

Parasetamol is very safe drug when used in therapeutic doses, it is known to cause lipid peroxidation and hepatotoxicity after overdose 1,2. Its toxicity has been attributed to the cytochrome P-450 catalyzed formation of a highyl reactive metabolic species, the N-acetyl-p-benzoqoinone imine, which is thought to bind covalently to protein in vivo. Paracetamol-induced lipid peroxidation (LPO) was inhibited by antioxidants such as vitamin E and vitamin C1.

Capsicum species (Solanaceae), known as "chilli pepper", "bell pepper" or "paprika" 3,4. The fruits have a characteristic green color and bitter flavor. The main chemical constituents of the *Capsicum* species are, capsaicin, vitamins, pigments, various volatile oils, proteins, sugars, minerals and organic acids[5-9]. The steady rise in consumption of paprika may be largely due to its high vitamin C content[7,8].

The aim of this study was to determine whether intake of water extracts of *Capsicum annuum* var. *frutescens*, *C. anuum* var. *longum* and *C. annuum* var. *grossum* could effect the level of lipid peroxidation in plasma, liver and kidney of acetaminophen-treated mice.

2. MATERIAL AND METHODS

Local breed male Swiss Albino mice, weighing approximately 20-25g were used in the present study.

Capsicum annuum fruit extracts were prepared from washed *Capsicum* fruits and equal weight of distilled water in a mixer (1:1 w/w), the contents were homogenized and centrifuged. The supernatant was filtered through filter paper[10].

The mice were divided into six groups. *Capsicum annuum* fruit extracts (0.1ml/10g of body weight/day for 7 days) were given in the experimental groups. The other group (control) was given distilled water and the ascorbic acid group was given ascorbic acid (100 mg/kg) in the same volume by gastric gavage for 7 days. Acetaminophen was administered 1 hr after the treatment of extracts and distilled water on 7th day in the experimental groups. The control group animals, however, received same volume of the NaCl %0.9, i.p. The oral administration was carried out once a day at 9:00-10:00 a.m. The animals were sacrificied on 7th day for examination after ether anesthesia. Blood was taken intracardiac. The liver and kidney were then removed, rinsed in ice-cold %0.9 NaCl, blotted dry, and weighed.

3. RESULTS

The analgesic drug acetaminophen is known to cause lipid peroxidation and hepatotoxicity after overdose[1]. The hepatotoxicity of acetaminophen is conventionally ascribed to metabolism by cytochrom P450 to N-acetyl-p-benzo-quinone imine and covalent binding to proteins[2].

The aim of this study was to determine whether intake of water extracts of *Capsicum annuum* var. *frutescens*, *C. anuum* var. *longum* and *C. annuum* var. *grossum* could affect the level of lipid peroxidation in plasma (nmol malondialdehyde /ml, nmol MDA/ml), liver and kidney (nmol MDA/g tissue) of acetaminophen-treated mice. The extracts administered with gastric gavage at a dose of 0.1ml/10 g body wt./day for 7 days. Acetaminophen at a single dose of 800 mg/kg on 7th day.

Acetaminophen, significantly increased the LPO in plasma, liver and kidney. When all parameters were compared, it was seen that *Capsicum annuum* extracts have an antioxidant activity in the plasma, liver and kidney.

Table 1. Antioxidant Activity of *Capsicum annuum* Fruit Extracts

	Plasma	Liver	Kidney
Control	2.71±0.327	297.4±7.044	131.1±5.116
Ascorbic Acid [b]	2.8 ±0.311**	298.8±3.146***	162.4±3.680***
Acetaminophen [a]	7.77±1.021***	443.7±9.258***	271.5±8.213***
C. annuum var. frutescens[b]	3.8 ±0.261**	252.5±13.571***	82.4 ± 9.887***
C. annuum var. longum[b]	6.8 ± 0.692 N.S.	339.1±15.919***	130.2± 6.512***
C. annuum var. grossum[b]	3.2 ± 0.518*	340.7±10.389***	179.1± 22.471**

*p<0.005 **p<0.01 *** p<0.001 N.S.: not significant
[a]: compare to control [b]: compare to acetaminophen group
(Mean ±SEM)

4. DISCUSSION

Acetaminophen by itself stimulates lipid peroxidation only when hepatic glutathione levels are lowered[11]. It was shown that in many studies, co administration of some antioxidants such as vitamin E, ascorbic acid, with acetaminophen protects the animals against hepatic toxicity[1].

When the effect of *Capsicum annuum* extracts were compared with acetaminophen group in plasma, liver and kidney we found a significant protective effect against acetaminophen-induced free radical injury.

According to our results and the other findings of different researchers, consumption of some vegetables such as *Capsicum annuum* may have beneficial effects against some diseases and may prevent acetaminophen toxicity without using any synthetic antioxidants. Further, chemopreventive effects of *Capsicum annuum* on the oxidative stress may depend on its structure, especially its vitamin C content.

REFERENCES

1. Fairhurst, S., Barber, D.J., Clark, B., Horton, A.A., 1982, *Toxicology,* 23, 249-59.
2. Straat, R.V., Bijloo, G.J.,Vermeulen, N.P.E., 1988, *Biochem. Pharmacol.,* 37(18), 3473-6.
3. Pnethi, J.S., 1976, Species and Condiments, National Bask Trust Press, New Delhi.
4. Suzuki, T., Iwai, A., 1984, The Alkaloids, Vol 23, Academic Press, New York.
5. Matsui, K., Shibata, Y., Tateba, H., Hatanaka, A., Kajiwara, T., 1997, *Biosci. Biotech. Biochem.,* **61** (1), 199-201.
6. Mende, P., Siddiqi, M., Preussmann, R.,Spiegelhalder, 1994, *Cancer Lett.,* **83**, 277- 82.
7. Somos, A., 1984,The Paprika, Akadémiai Kiadó, Budapest.
8. Peterson, M.A., Berends., H., 1993, *Z. Lebensm. Unters. Forsch.,* **197,** 546-9.

9. Atay, T., Değim, T., Büyükavşar, K., Akay, C., Cevheroğlu, S., 1993, XIIth International Symposium on Plant Originated Crude Drugs, 20-22 May 1993, Ankara-Turkey.
10. Komatsu, W., Yagasaki, K., Miura, Y., Funabiki, R., 1997, *Biosci. Biotech. Biochem.,* **61** (11), 1937-8.
11. Wendel, A., Feverstein, S., Konz, K.H., 1979, *Biocem. Pharmacol.,* **28**, 2051-9.

In Vitro Antileishmanial Activity of Proanthocyanidins and Related Compounds

HERBERT KOLODZIEJ[1], O. KAYSER[1], A.F. KIDERLEN[2], H.ITO[3], T. HATANO[3], T. YOSHIDA[3] and L.Y. FOO[4]

[1] *Institut für Pharmazie, Pharmazeutische Biologie, Freie Universität Berlin, Königin-Luise-Str. 2+4, D-14195 Berlin, Germany;* [2] *Robert Koch-Institut, Nordufer 20, D-13353 Berlin, Germany;* [3] *Faculty of Pharmaceutical Sciences, Okayama University, Okayama 700-8530, Japan;* [4] *New Zealand Institute for Industrial Research, P.O. Box 31-310, Lower Hutt, New Zealand*

1. INTRODUCTION

Species of the parasitic protozoa *Leishmania* are estimated to threaten some 350 million people world-wide with a broad range of disease[1]. In their mammalian hosts, protozoa of the genus *Leishmania* are obligate intracellular parasites of the monocyte-macrophage system. Despite the major adverse impact of *Leishmania* parasites on human populations, only a few new drugs are currently on clinical trial since the introduction of the pentavalent antimonials. As part of our research program to identify novel antileishmanial compounds, we have evaluated a series of proanthocyanidins and related compounds.

2. RESULTS AND DISCUSSION

The polyphenols tested according to Kayser et al.[2] are listed numerically and their structures are shown in Figure 1. Their in vitro antileishmanial activity against both extracellular promastigotes and intracellular amastigotes of *Leishmania donovani* are shown in Table 1. With IC$_{50}$ values ranging from 0.7–10.6 nM, all compounds were considerably leishmanicidal against amastigotes but catechin (**1**), cinchonain Ia (**6**) and 5',5'-bisdihydroquercetin (**17**) (> 41 nM), when compared with the IC_{50} value 10.6 nM of the clinically used drug, (Pentostam®). Compounds **10**, **14** and **16** exhibited the highest relative toxicity for intracellularly persisting *L. donovani* parasites with an IC_{50} of 0.7-0.9 nM. In contrast, none of the samples showed selective toxicity when tested against promastigotes.

Table 1. Antileishmanial activity [*L. donovani* amastigotes (AM) and promastigotes (PM)] and toxicity for RAW host cells of compounds **1 – 17** [IC_{50} values (nM)]

Compound	Leishmanicidal Activity		Toxicity for RAW cells
	PM	AM	
Flavan-3-ols and related compounds			
1 Catechin	> 86.2	50.3	67.9
2 Catechin 3-gallate	> 56.6	10.6	> 56.6
3 Epicatechin 3-gallate	> 56.6	10.6	> 56.6
4 Epigallocatechin 3-gallate	> 54.6	9.6	> 54.6
5 Phylloflavan	> 50.2	3.2	> 50.2
6 Cinchonain Ia	> 56.3	> 55.3	> 55.3
Proanthocyanidins			
7 Procyanidin B-1	> 43.0	6.5	> 43.0
8 Procyanidin B-2	> 43.0	2.9	> 43.0
9 Procyanidin B-3	> 43.0	1.4	> 43.0
10 Procyanidin B-4	> 43.0	0.8	> 43.0
11 Procyanidin C-1	> 28.8	6.7	> 28.8
12 Procyanidin C-2	> 28.8	7.7	> 28.8
13 Procyanidin undecamer	> 7.8	3.4	> 7.8
14 Procyanidin 3-gallate polymer	> 10.6	0.7	> 10.6
15 Prodelphinidin hexamer	> 13.6	4.2	> 13.6
Miscellaneous			
16 Pseudotsuganol	> 37.8	0.9	> 37.8
17 5',5'-Bis-dihydroquercetin	> 41.2	> 41.2	> 41.2
Pentostam	3.5	10.6	not det.

Regarding structure-activity relationships, marked leishmanicidal potency is associated with the presence of 3-O-acyl groups. For proanthocyanidins, 4α,8-coupled dimers (**9**, **10**) were more active than their 4ß,8-counterparts (**7**, **8**). Also, an increase in molecular weights, galloylation of constituent units or the presence of 2,3-*cis* flavanyl chain extender units enhanced leishmanicidal activity. Pseudotsuganol (**16**) showed pronounced antileishmanial activity. Since the structurally related compound **17** was inactive, the crucial structural element for activity of **16** may be due to the pinoresinol moiety rather than the taxifolin unit.

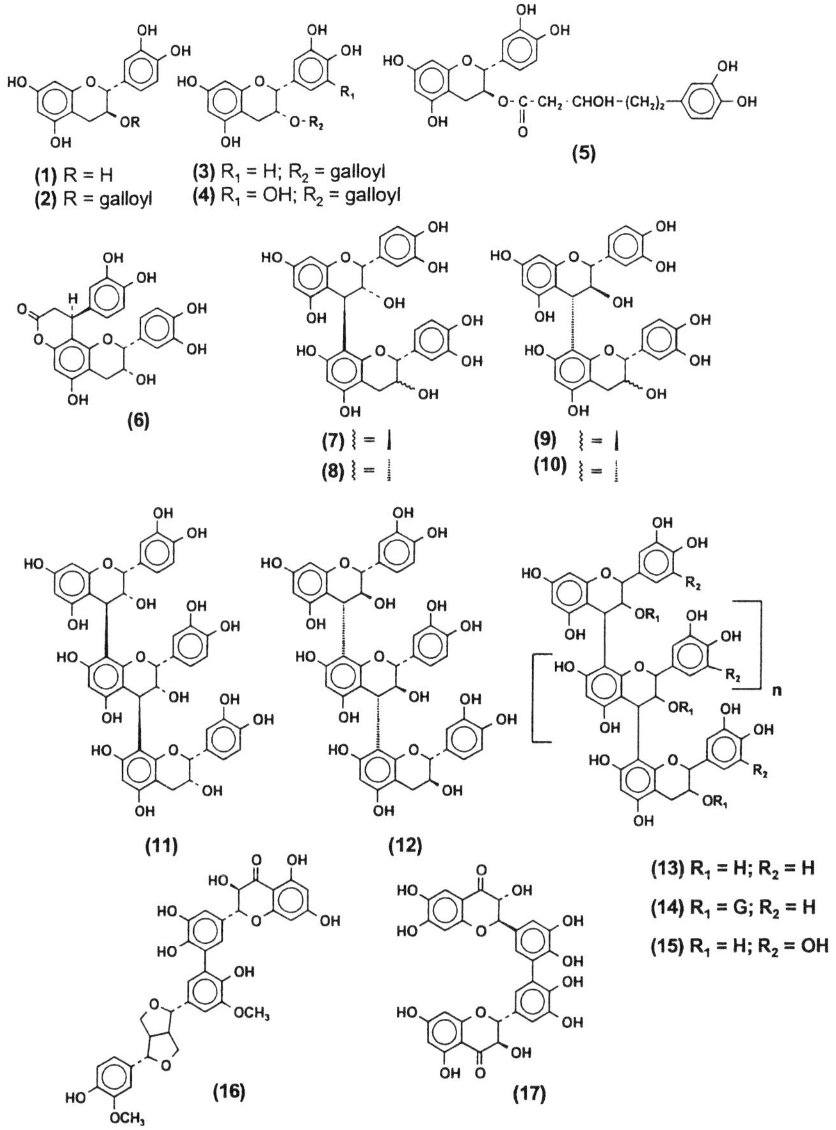

Figure 1. Structures of compounds **1-17**

REFERENCES

1. Ashford, R.W., Desjeux, P., DeRaadt, P.,1982, *Parasitol. Today,* **8**, 104.
2. Kayser, O., Kiderlen, A.F., Folkens, U., Kolodziej, H.,1999, *Planta Medica,* **65**, 316.

Evaluation of the Antileishmanial Activity of Two New Diterpenoids and Extracts from *Salvia cilicica*

NUR TAN, M. KALOGA, O.A. RADTKE and HERBERT KOLODZÏEJ
Institut für Pharmazie, Pharmazeutische Biologie, Freie Universität Berlin, Königin-Luise-Str. 2+4, D-14195 Berlin, Germany

1. INTRODUCTION

Leishmaniasis is a major public health problem. Given the limitations of the current treatments for this debilitating disease, there is an urgent need for the development of new therapeutics[1]. In traditional medicine, many plants have already provided valuable clues for potentially antiparasitic compounds including, e.g., phenols, flavonoids and terpenoids[2].

The cosmopolitan genus *Salvia*, which includes some 900 species abundantly distributed throughout the Mediterranean area, South East Asia and South America, is phytochemically characterised by the presence of phenols, flavonoids, essential oils, di- and triterpenoids. These constituents may well contribute to their biological activities: Salvia species are successfully employed to cure various inflammatory disorders and infectious conditions. Their reputed antimicrobial activities and the hitherto limited information on the chemical constituents of *Salvia cilicica*, a species that is endemic to Turkey[3], prompted the present study.

2. RESULTS AND DISCUSSION

2.1 Isolation and Characterization

The acetone extract of the roots of *S. cilicica* afforded the new diterpenoids (1) and (2), accompanied by oleanolic acid (3) and ursolic acid (4). Their isolation was achieved by column chromatography using a gradient system of petroleum ether/ethyl acetate (1:0 → 0:1). The content of fraction 200-1600 ml was subjected to preparative TLC with petroleum ether/toluene (1:1) to yield the new diterpenoid 1 (R_f 0.3). Similar purification of the same fraction using petroleum ether/toluene (4:1) as mobile phase gave compound 2 (R_f 0.4), representing another new natural product. Their structures have been established on the basis of spectroscopic evidence. The ^{1}H and ^{13}C NMR spectra of 1 and 2 indicated the presence of an abietane skeleton. The chemical shifts of the protons and carbons were assigned through one-bond and long-range C-H correlations using HETCOR and COLOC experiments.

The HR-MS of 1 revealed a molecular ion at *m/z* 326.1513 (calc. 326.1531), suggesting a molecular formula of $C_{20}H_{22}O_4$. Among the most relevant couplings observed in the NMR spectra of 1 are: the diagnostic proton signal for 15-H at δ 3.32 (*septet*, J= 7.0 Hz) with carbon atoms at δ 189.9 and 158.8 assigned to C-14 and C-12, respectively; methylene proton signals at δ 2.18 with both carbon atoms at δ 130.6 (C-2) and 27.4, assigned to the methyl functionality at C-4. Taking into account a signal for an enolic proton (δ 13.13, 7-OH) with showed correlations with carbon signals at δ 160.6 (C-7) and 111.8 (C-8), these spectral features collectively defined 1 as 7-hydroxy-12-methoxy-20-nor-abieta-1,5(10),7,9, 12-pentaen-6,14-dione.

The molecular formula of 2 was deduced to be $C_{20}H_{28}O_2$ from the HR-MS showing the [M]$^+$ at *m/z* 300.2075 (calc. 300.2089). The heteronuclear ^{1}H —^{13}C correlations indicated particularly the presence of an *ortho*-benzoquinone element. The key proton signal at δ 6.31 (12-H) indicated correlation with the carbon signals at δ 188.1 (C-11), 188.2 (C-14) and 26.3 (C-15). Based on the spectral evidence, compound 2 was identified as abieta-8,12-dien-11,14-dione (12-deoxy-royleanone).

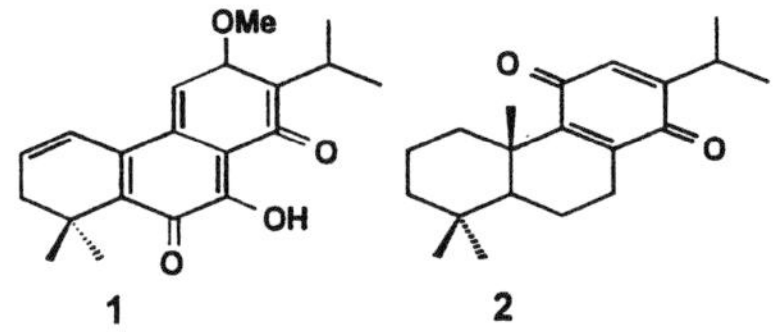

Figure 1. Structure and Compounds 1 and 2

The known compounds ursolic acid (3) and oleanolic acid (4) were characterised by comparing the chromatographic and spectral properties with those of authentic samples.

2.2 Leishmanicidal Activity

The antileishmanial activity of fractions and constituents (1) – (4) was assessed against both extracellular promastigotes and intracellular amastigotes of *Leishmania donovani* and *L. major* [3], and results are shown in Table 1 with Pentostam as reference. With IC_{50} values ranging from 23 to >100 µg/ml against both parasite species, the leishmanicidal activitiy of the samples was found to be only weak or moderate, when compared with the IC_{50} value 7.9 µg/ml of the antimonial drug, Pentostam. Interestingly, the fraction eluted 5.2–8.6 l exhibited pronounced activity against the extracellular promastigotes of *L. major*. Further purification of the active fractions resulted in the isolation of ursolic acid (3) as a potentially leishmanicidal constituent (IC_{50} 3.25 – 5.8 µg/ml).

Except for **3** (IC_{50} 7.1 µg/ml), none of the samples showed significant cytotoxicty (IC_{50} > 50 µg/ml) when tested against macrophage-like RAW 264.7 cells as a mammalian host cell control.

Table 1. Antileishmanial activity [amastigotes (AM); promastigotes (PM)] and toxicity for RAW host cells of fractions and constituents **1 - 4** of *S. cilicica* (IC_{50} values in µg/ml)

Sample	L. donovani		L. major		Toxicity for
	AM	PM	AM	PM	RAW cells
1	>100	62.5	>100	93.7	> 100
2	>100	36.5	>100	54.7	57.3
3	41.5	5.8	23.4	3.2	7.1
4	41.5	28.7	62.5	54.7	60.5
extract	23.5	>100	>100	>100	73.0
eluates (l)					
2.8 – 5.2	>100	>100	>100	40 to >100	>100
5.2 – 8.6	>100	40 to >100	40 to >100	17-30	>100
> 8.6	40 to >100	>100	60 to >100	40 to >100	83-85
Pentostam	7.9	2.6	2.7	Not determ.	not determ.

REFERENCES

1. Croft, S.L., 1997, *Parasitology*, **114**, 3-15.
2. Davis, P.H., 1982, Flora of Turkey and the East Aegean Islands, University Press, Vol. 7, Edinburgh, p. 400.
3. Corona, M.R.C., Croft, S.L., Phillipson, J.D., 2000, *Curr. Opinion Anti-infective Invest. Drugs*, **2**, 47-62.
4. Kayser, O., Kiderlen, A.F., Folkens, U., Kolodziej, H., 1999, *Planta Med.*, **65**, 316-9.

Antibacterial and Antifungal Activities of *Sedum sartorianum* subsp. *sartorianum*

M. KORAY SAKAR[1], M. ARISAN[1], M. ÖZALP[2], M. EKİZOĞLU[2], D. ERCİL[3] and H. KOLODZİEJ[3]
[1]*Hacettepe University, Faculty of Pharmacy, Department of Pharmacognosy, 06100,Ankara, Turkey,* [2]*Hacettepe University, Faculty of Pharmacy, Department of Pharmaceutical Microbiology, 06100,Ankara,Turkey,* [3]*Free University Berlin, Institute of Pharmacy,Pharmaceutical Biology,Königin-Luise-Str.2+4,D-14195 Berlin, Germany*

1. INTRODUCTION

Sedum sartorianum Boiss. subsp. *sartorianum* (Crassulaceae) is a small, glabrous, perennial, succulent plant[1]. *Sedum* species are traditionally used in many parts of the world for the treatment of a wide variety of diseases including inflammation, gastrointestinal disorders, microbial infections and haemorrhoids. *Sedum* species are also used as hypotensives, emetics and an effective corn removers[2]. In Turkey, *S. sartorianum* ssp. *sartorianum* is utilised as a wound healing agent.

We have previously reported our phytochemical studies about *Sedum* species on the polyphenolic compounds (flavan gallates, arbutin derivatives, dimeric prodelphinidin, flavonoid glycosides and phloroglucinol glucosides)[3-5]. In the present study, the chemical composition of the EtOAc extract and the antimicrobial potency of the EtOAc, *n*-BuOH, H_2O and MeOH extracts and isolated compounds SSAR1 and SSAR2 of the title plant have been studied.

2. EXPERIMENTAL

Aerial parts of *S. sartorianum* ssp. *sartorianum* (736 g) was collected in June 2000 (01-016) at Abant, Bolu, Turkey. The fresh material was chopped and extracted with MeOH (2.3 Lx3). and the combined extracts were evaporated to dryness at 45°C under reduced pressure to give a residue (31.62 g, 0.04%).

The lyophilized material was partitioned between EtOAc (500 mLx6) and water to yield 5.5 g crude EtOAc soluble material (0.07%). The remaining aqueous phase was extracted with *n*-BuOH (500 mLx4). Evaporation of solvents followed by lyophilisation produced 7.85 g (0.011%) and 13.82 g (0.018%), crude extracts respectively.

Preliminary phytochemical screening[6,7] indicated the presence of flavonoids, proanthocyanidins and organic acids in the parent MeOH extract. nitrogen-containing precursors from the EtOAC extract, flavan-3-ol monomers, lower oligomers and flavonoid monoglycosides and the more complex flavonoids from the EtOAc, aqueous and *n*-butanol extracts.

The EtOAc extract was applied on a Sephadex LH 20 column (40.0x3.0 cm) and eluated with EtOH (95%, $^v/_v$). Fraction volumes were about 25 mL each. Fractions 26-28 were combined and evaporated at 45°C under reduced pressure until dryness. Then dried fractions (65 mg, 0.009%) was chromatographed on VLC (RP-18 column, 20-45 μ, 15.5x2.2 cm) and eluted with MeOH/H_2O mixture. The elution was started with 10% MeOH and MeOH content in the mixture was increased stepwise. Fraction volumes were about 15 mL each.

After TLC, fractions RP_{36a}-RP_{68a} were evaporated at 45°C under reduced pressure until dryness (34 mg; 0.005%) and named as SSAR 1. Fractions 21-23 were combined and lyophilized. The dried fraction (100 mg, 0.014%) was dissolved in 10% MeOH and applied to the RP 18 column (20-45μ, 15.5x2.2cm) and eluated with MeOH/H_2O mixture (10-50%). After TLC, fractions RP_{34b}-RP_{62b} were combined and lyophilized (35 mg, 0.005%) and named as SSAR 2.

Two flavonol glycosides, gossypetin-4'-methylether-8-*O*-α–L-arabinopyranoside (SSAR1) and gossypetin-3'-methylether-8-*O*-α-L-arabinopyranoside (SSAR 2) were isolated from the EtOAc extract and their structures were elucidated by spectroscopic analysis (UV, ^{1}H-NMR, ^{13}C-NMR, HMBC, HMQC, NOESY and FABMS).

3. ANTIMICROBIAL ACTIVITY METHODS

The Broth Micro Dilution Method, recommended by National Comittee for Clinical Laboratory Standarts (NCCLS) was used to evaluate antimicrobial activities[8,9]. Mueller–Hinton Broth (MHB, Difco laboratories, Detroit, USA) was used for antibacterial studies. For testing yeast-like fungi, RPMI–1640 medium with L-glutamin (ICN-Flow, Aurora, USA) was used. The inoculum densities were approximately 5×10^5 cfu/mL and 0.5-2.5×10^3 cfu/mL for bacteria and fungi, respectively.

The EtOAc, n-BuOH, H_2O and MeOH extracts and the pure compounds (SSAR1 and SSAR2) were dissolved in sterile water. Two fold concentrations were prepared in the wells of the microtiter plates. The initial concentrations were 1250 μg/mL for n-BuOH and water, 250 μg/mL for EtOAc and 625 μg/mL for MeOH extracts and 10mg/mL for SSAR1 and SSAR2. Ceftazidime and Tobramycin were used as the reference antibiotics for bacteria and Fluconazole for yeast-like fungi, respectively (64.0-0.0625 μg/mL for all).

Incubation was performed at 35 °C for 18-24 h for bacteria and 48 h for yeast-like fungi. After the incubation period, minimum inhibitory concentration (MIC) values were defined as the lowest concentration of the EtOAc, n-BuOH, H_2O and MeOH extracts and SSAR 1 and SSAR 2 that inhibits the visible growth of microorganisms.

In this study, two gram-positive and two gram-negative bacteria as well as three yeast like-fungi were used as test microorganisms. These strains were chosen from the stock culture collections maintained in our laboratory.

4. RESULTS AND DISCUSSION

Different extracts of *S. sartorianum* ssp. *sartorianum* were evaluated for antibacterial activity against two gram-positive (*Staphylococcus aureus, Enterococcus faecalis*) and two gram-negative microorganisms (*Escherichia coli, Pseudomonas aeruginosa*) as well as for antifungal potency against three yeast like fungi (*Candida albicans, C. krusei, C. parapsilosis*), using the broth microdilution method. With MIC values of 62.5-312.5 μg/mL, all extracts were found to be only moderately active against the spectrum of the test bacteria, while the *n*-BuOH extract showed selective antifungal activity (MIC 9.75-19.5 μg/mL). The relatively highest antibacterial activity resided in the EtOAc extract (MIC 62.5-125 μg/mL).

Two major flavonoids, SSAR1 and SSAR2 were isolated from the EtOAc extract, showed relatively strong activity against all tested fungi and bacteria (MIC 0.0625–0.125 μg/mL) (Tables I and II). The demonstrated antimicrobial activity of *S. sartorianum* ssp. *sartorianum* lend support to the traditional uses of the plant as wound healing agent.

Table 1. Antibacterial activities of *Sedum sartorianum* Boiss. subsp. *sartorianum* different extracts and Gossypetin-4'-methylether-8-O-α-L-arabinopyranoside (SSAR1) and gossypetin-3'-methylether-8-O-α-L-arabinopyranoside (SSAR 2).

Microorganisms	MIC (µg/ml)							
	MeOH extract	EtOAc extract	n-BuOH extract	H$_2$O extract	SSAR1	SSAR2	Ceftazidime (reference)	Tobramycin (reference)
Gram Negative Bacteria								
E. coli (ATCC 25922)	n.t.*	125	156	312.5	≥0.125	≥0.125	0.5	0.25
P. aureginosa (ATCC 27853)	n.t.	62.5	78	156	≥0.125	≥0.125	2	0.25
Gram Positive Bacteria								
S. aureus (ATCC 25923)	n.t.	62.5	78	156	0.0625	0.0625	0.5	0.125
E. faecalis (ATCC 29212)	n.t.	62.5	312.5	312.5	n.t.	n.t.	n.t.	32

* n.t.: not tested

Table 2. Antifungal activities of *Sedum sartorianum* Boiss. subsp. *sartorianum* different extracts and Gossypetin-4'-methylether-8-O-α-L-arabinopyranoside (SSAR1) and gossypetin-3'-methylether-8-O-α-L-arabinopyranoside (SSAR 2).

Microorganisms	MIC (µg/ml)						
	MeOH extract	EtOAc extract	n-BuOH extract	H$_2$O extract	SSAR 1	SSAR 2	Fluconazole (reference)
Fungi							
C. albicans (ATCC 90028)	39	31.25	19.5	39	0.0312	0.0312	0.25
C. krusei (ATCC 6258)	9.75	n.t	9.75	19.5	0.0625	≥0.125	16
C. parapsilosis (ATCC 22019)	n.t.*	n.t	9.75	78	0.0625	0.0625	0.5

ACKNOWLEDGEMENT

This study was supported by Hacettepe University Research Fund. Project no: 99.02.301.002

REFERENCES

1. Davis, P. H., 1972, Flora of Turkey and the East Aegean Islands, Vol. IV, Edinburgh University Press, Edinburgh, p. 224.
2. Bremness, L., 1994, Herbs Dorling Kindersley, London, p. 214.
3. Sakar, M. K., Petereit, F., Nahrstedt, A., 1991, *Phytochemistry*, **33**, 171.
4. Sakar, M. K., Petereit, F., Nahrstedt, A., 1997, *Sci Pharm.*, **65**, 33.
5. Petereit, F., Sakar, M. K., Nahrstedt, A., *Pharmazie*, 53(4), 280-1 (1998).
6. Harborne J. B., 1994, Phytochemical Methods, Chapman and Hall, London.
7. Stahl, E., Schild, W., 1981, Pharmazeutische Biologie, Gustav Fischer Verlag, Stuttgart.
8. National Comittee for Clinical Laboratory Standarts (NCCLS),1997, Methods for dilution antibacterial susceptibility tests for bacteria that grow aerobaically, 4th ed., Approved Standart, M7-A4, Wayne, P.A.
9. National Comittee for Clinical Laboratory Standarts (NCCLS),1997, Methods for dilution antifungal susceptibility tests for bacteria that grow aerobaically, 4th ed., Approved Standart, M27-A, Wayne, P.A.

Blood Pressure Lowering Activity Of Active Principle from *Ocimum basilicum*

KHALID AFTAB
Islamabad Medical & Dental College, Department of Pharmacology, Islamabad. & University of Karachi, H.E.J. Research Institute of Chemistry, Karachi, Pakistan.

1. INTRODUCTION

Ocimum basilicum belongs to the family labiatae and is commonly known as Basil (Tulsi). It is a widespread plant cultivated in the world. It grows to a height of approximately 50 cm. The leaves are oval and slightly toothed, and the flowers are white or purple. *Ocimum basilicum* looks very similar, but grows a bit taller (50-80 cm). *Ocimum basilicum* is the most widely used. It is used in cosmetics, liqueurs, medicines and perfumes.

In Indo-China, the ashes of the roots are suggested as a remedy for skin diseases. The plant is used as aromatic, antimicrobial, astringent in dysentery, while the leaves are antipyretic. The seeds are laxative, particularly in case of habitual constipation. The Juice of the leaves and flowers is a treatment of cough. A decoction may be given after parturition as emmenagogue and febrifuge. The leaves are carminative, antispasmodic and sedative. Preparations of basil are used for supportive therapy for feeling of fullness and flatulence, for the stimulation of appetite and digestion, and as diuretic.

Reported constituents of *Ocimum basilicum* are mostly phenolic compounds (volatile oil) e.g. Linalool, chavicol methyl ether (estragole), eugenol, caffeic acid derivatives and flavonoids. Bioassay-directed fractionation of *Ocimum basilicum* has resulted in the isolation of Eugenol, which present in other plants as well. It is reported to have analgesic, anesthetic antiaggregating, anti-inflammatory, anti-oxidant, antiseptic, antitumor, anti-ulcer, carminative, cytotoxic, fungicide, sedative, spasmolytic activities. However, its effect on blood pressure has not been studied so far.

Eugenol; 2-methoxy-4-(2-propenyl)phenol, Molecular formula: $CH_2CH_2CH_2C_6H_3(OCH_3)OH$, Physical data; Appearance: Colorless liquid with a strong odor of cloves, Melting point: -9 C, Boiling point: 253° C, Density (g/cm^3): 1.06, Flash point 110°C, Water Solubility: negligible. Stable, Combustible, Incompatible with strong oxidizing agents. It may act as a skin or eye irritant. Toxicity data: ORL-RAT LD_{50} 2680mg/kg, ORL-MUS LD_{50} 3000 mg/kg, IPR-RAT LDLO 800 mg/kg, IPR-MUS LD_{50} 500 mg/kg.

2. RESULTS AND DISCUSSION

In anaesthetized rats, ethanolic extract, fractions and Eugenol (0.3-10.0 mg/kg) produced dose-dependent fall in blood pressure and heart rate [1]. These effects were not blocked by atropine (1 mg/kg) and Eugenol did not modify presser response of norepinephrine which rules out the possibility of cholinergic stimulation or $\propto$-adrenergic blockade. In spontaneously beating atria, Eugenol caused decrease in force and rate of atrial contractions. These effects remain unaltered in the presence of atropine. In rabbit aorta, Eugenol caused relaxation of norepinepherine and K^+- induced contractions in a concentration-dependant manner [2].

These results suggest that the direct relaxant activity of Eugenol on myocardium and blood vessels may be responsible for its hypotensive and bradycardiac effects observed in the *in vivo* studies.

Table 1. Effect of the crude extract, fraction and pure compound (Eugenol) of *Ocimum basilicum* on the mean arterial blood pressure (MABP) and heart rate (HR) in anesthetized rats.

Dose mg/kg	No. of obs.	% Fall					
		Crude extract		Fraction		Eugenol	
		MABP (mm Hg)	HR bts./min.	MABP (mm Hg)	HR bts./min.	MABP (mm Hg)	HR bts./min.
0.3	5					10.35 ± 05.50	
1	5	10.57 ± 02.70		16.24 ± 03.51	05.83 ± 04.35	20.24 ± 03.01	06.83 ± 01.36
3	5	20.00 ± 03.00	14.25 ± 02.50	30.59 ± 04.50	11.16 ± 03.30	50.59 ± 02.59	21.16 ± 03.37
10	5	34.43 ± 04.90	28.50 ± 03.33	42.87 ± 04.33	20.50 ± 04.84	69.87 ± 04.55	39.50 ± 02.84
30	5	48.70 ± 05.23	32.85 ± 05.45	63.55 ± 05.35	40.35 ± 05.37		
100	5	65.37 ± 04.21	40.66 ± 07.55				

Values shown represent mean ± standard error of the mean.

REFERENCES

1. Aftab, K., Shaheen, F., Mohammad, F. V., Noorwala, M., Ahmed, V. U. ,1996, *Adv. Exp. Med. Biol.*, **404**, 429-42.
2. Aftab, K., Atta-ur-Rahman, Usmanghani, K. ,1995, *Phytomedicine*, **2** (1), 35-40.

Chemical Variability in *Azadirachta indica* Growing in Tamil Nadu State of India

NUTAN KAUSHIK[1] and B GURDEV SİNGH[2]
[1] *Bioresources and Biotechnology Division, TERI, Darbari Seth Block, Habitat Place, Lodhi Road, New Delhi 110 003, India.* [2] *Institute of Forest Genetics & Tree Breeding Coimbatore, India*

1. INTRODUCTION

Neem (*Azadirachta indica*) is an evergreen multipurpose tree grown in Indian subcontinent and south-east Asian countries. Products derived from Neem have been used for centuries, particularly in India, for medicinal and pest management purposes. The seed contains many important limonoids and 30-50% oil. Among all the limonoids present in the seed, azadirachtin is the most important one and determines the value of the seed. Individual neem trees may vary in their chemical make-up as the oil content and limnoid content of neem tree is governed by genetic and environmental factors. Very few studies have been carried out so far, in India and abroad to find the existing variability of azadirachtin content in neem trees and even fewer on oil content variability[1-4]. Efforts are underway at TERI to assess the azadirachtin and oil content diversity in Neem seeds being collected from various parts of the country under the "National Network on Integrated Development of Neem" supported by National Oilseeds and Vegetable Oils Development Board, Ministry of Agriculture, Gurgaon. Studies carried out so far under the network show great potential for tapping the genetic variability of Neem tree for azadirachtin and oil content. Azadirachtin variability recorded in Tamil Nadu State of India is presented in this paper.

2. EXPERIMENTAL

The neem kernel powder (1 g) was heated with ethanol (10 ml) for two hours on a water bath maintained at 50°C. The contents were filtered and residue was washed with 2x10 ml of ethanol. The extracts thus obtained were pooled, filtered through 0.22 µ membrane and injected into waters LC-Module-1 HPLC via an autoinjector. Acetonitrile-water (40+60) @ 1 ml/min was used as a mobile phase. The peaks were monitored at 214 nm.

3. RESULTS AND DISCUSSION

In present study, variability of the azadirachtin content in seeds of 137 accessions collected from Tamil Nadu State of India was studied. The azadirachtin content ranged from 725-15112 ppm (µg/g of kernel) (Figure 1). Six accession viz V/6, V/4, V/14, V/15, V/19 and Karur 9 recorded azadirachtin > 10000 ppm. About 58 accessions recorded azadirachtin in the range of 5000-10000 ppm and about 79 sample recorded < 5000 ppm azadirachtin content. Only two accession V/3 and III/9 recorded azadirachtin < 1000 ppm. Thus a great variability in terms of azadirachtin content was observed. Accession collected from Ramananthanpuram in general were found to be having high azadirachtin. This variability can be exploited for the selection of elite tree for azadirachtin content. These elite trees need to be conserved and further propagated through tissue culture and utilized for plantation programme. This variability can also be used for biodiversity related studies and need to be correlated with morphological and molecular markers.

ACKNOWLEDGEMENT

The financial assistance from the National Oilseeds and Vegetable Oils Development Board (NOVODB), Ministry of Agriculture, New Delhi Government of India, is gratefully acknowledged.

Figure 1. Azadirachtin variability in Neem seed samples

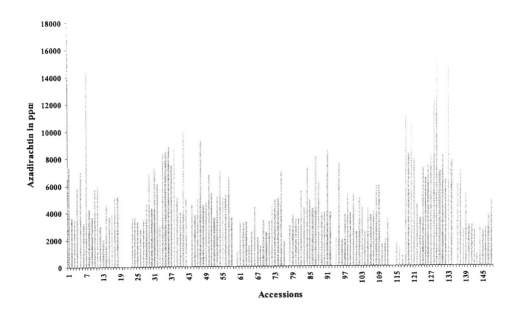

REFERENCES

1. Bally, I.S.E., Ruddle, L., Simpson, B., 1996, Azadirachtin levels in Neem seed grown in Northern Australia. Abstracts of the International Neem Conference, Feb 4-9, Queensland. Australia, p. 17.
2. Ermel, K., Pahlich, E.,Schmutterer, H., 1986, Azadirachtin contents of neem kernels from different geographical locations and its dependence on temperature, relative humidity and light. Proceedings of the International Neem Conference, Nairobi, Kenya, p. 171.
3. Kumar, M.G., Kumar, R.J:, Regupathy, A., Rajasekharan, B., 1995, *Neem Update*, **1**(1), 4.
4. Rengaswamy, S., Kaushik, N., Kumar, J., Kaul, U., Parmar, B. S., 1993, Azadirachitin content and bioactivity of neem ecotypes of India, Proceedings of the World Neem Conference, 24-28th Feb, Bangalore.

Pesticidal Activity of *Eucalyptus* Leaf Extracts against *Helicoverpa armigera* Larvae

NUTAN KAUSHIK
Bioresources and Biotechnology Division, TERI, Darbari Seth Block, Habitat Place, Lodhi Road, New Delhi 110 003, India.

1. INTRODUCTION

Plant biodiversity is of immense importance for finding a wide range of biomolecules. Plants have remained source for many important pesticides such as rotenoids, nicotine, pyrethroids, neem etc[1]. The discovery of many synthetic pesticides also find its origin from plant based chemicals. The need for newer pesticides remain ever persisting in order to combat the problem of resistance in the insects. *Helicoverpa armigera* is one of the major insect pest in India. It attacks many economically important crop species viz. cotton, pigeonpea, chickpea, tomato, sunflower, etc[2]. Currently, it is the most difficult species to control because of emergence of resistance to most of the commercially available insecticides. In search of newer molecules for pesticidal action, leaves of *Eucalyptus species* were tested against *Helicoverpa armigera* reared on artificial diet. The paper summarises the findings of this investigation.

2. EXPERIMENTAL

Leaf material of *Eucalyptus* species was collected for conducting bioassays and for preparation of extracts. Insect culture of *Helicoverpa armigera* was reared on artificial diet in a BOD at 27 ± 2°C, 70% RH and 10:14 LD photoperiod. Bioassays were conducted by mixing the test material with the dry portion of the artificial diets[3]. Ten 1st instar larvae per replication were released on treated diet and diet treated with solvent alone was kept as control. Ten replications were maintained per treatment. The larval development, molting, pupal weight and survival rate were recorded from the 1st instar larvae upto emergence of adults as performance variables.

3. RESULTS AND DISCUSSION

Preliminary bioassays by mixing the crude leaf powder (5% w/w basis) in the insect diet exibited promising results as slow growth and development of larvae was observed. Larvae were very small in size and none of the larve could convert into pupae (Figure 1). Further bioefficacy experiments were conducted with different levels of *Eucalyptus* leaf powder (2% and 1%) in the artificial diet. Slow growth and development of the larvae was observed at both the levels. The larvae could not survive beyond L_3 stage at 2% level. Among the various extracts of Eucalyptus leaves when tested for their bioefficacy against *H.armigera*, maximum activity was recorded in alcohol extract. About 88% growth inhibition was recorded. Although, no direct larval mortality was observed. However, subsequent increase in the larval period and moulting disruptions lead to the death of the larvae. Thus, suggests that *Eucalyptus* has growth inhibiting property. The final objective of the study is to characterize the active principle, preparation of formulations and field evaluation of the formulations.

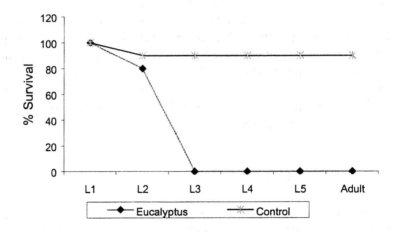

Figure 1. Survival of *H. armigera* larva at different stages with *Eucalyptus* leaf extract.

ACKNOWLEDGEMENT

Financial assistance from Department of Biotechnology, Ministry of Science and Technology, Government of India is duly acknowledge.

REFERENCES

1. Jacobson, M., 1971, Naturally Occurring Insecticides, Marcel Dekker Inc., New York.
2. Reed, W., and Pawar, C.S., 1982, *Heliothis*: A global problem. Proc. Int. Workshop on *Heliothis* management, 15-20 Nov. 1981, ICRISAT, pp. 9-14.
3. Singh, A.K., Rembold, H., 1992, *Insect Sci. Applic.*, **13**(3), 333-8.

Two New Lignans from *Taxus baccata* L.

NURGÜN ERDEMOĞLU and BİLGE ŞENER
Gazi University, Faculty of Pharmacy, Department of Pharmacognosy,06330 Ankara, Turkey

1. INTRODUCTION

Genus *Taxus* L. (Taxaceae) is widely distributed in the world which is represented by one species in Turkey, *Taxus baccata* L. (European yew)[1]. Many lignans have been isolated from *Taxus* species until now[2-4]. In our continuing researches for bioactive compounds, two new lignans, 3'-demethylisolariciresinol-9'-hydroxyisopropylether (**1**) and 3-demethyl isolariciresinol (**2**), have been isolated from the heartwood of *T. baccata*, along with a known lignan isolariciresinol (**3**). In this study, the isolation and structure elucidation of these compounds (Fig. 1 and 2) were described.

2. RESULTS

Three compounds were isolated from *Taxus baccata* L. Compounds **1** and **2** are new lignans isolated from the heartwood of *T. baccata* L. Compound **1** is the first example of a lignan containing a hydroxyisopropyl group at C-9'.

2.1 3'-Demethylisolariciresinol-9'-hydroxyisopropylether (1)

Compound 1 was obtained as fine yellowish crystals from chloroform; IR v_{max} (KBr) cm^{-1} : 3482 (OH), 2990, 2935 (CH), 1614 (C=C), 1513, 1447 (aromatic region), 1113, 1092 (C-O-C); EIMS m/z: 386 [M-H$_2$O]$^+$; DCI + (NH$_3$) m/z: 386 [M-H$_2$O]$^+$; M$^+$: 404, C$_{22}$H$_{28}$O$_7$.

On the basis of the spectral data (Table 1) compound 1 was identified as 3'-demethylisolariciresinol-9'-hydroxyisopropylether. The spectral data of 1 was found to be obvious similarities to our findings of the spectral data for isolariciresinol. Besides, the structure of compound 1 was also compared with the reported ^{1}H-NMR spectral data for isolariciresinol[5].

2.2 3-Demethylisolariciresinol (2)

Compound 2 was obtained as cream colour amorphous powder; IR v_{max} (KBr) cm^{-1} : 3386 (OH), 2360 (CH), 1610 (C=C), 1511,1446 (aromatic region), 1113, 1050 (C-O-C); EIMS m/z: 346 (M$^+$); DCI + (NH$_3$) m/z: 346, M$^+$: C$_{19}$H$_{22}$O$_6$.

On the basis of the spectral data (Table 2) compound 2 was concluded to be 3-demethyl isolariciresinol. We found that the spectral data of 2 are quite similar, except for a methoxyl group in isolariciresinol. In addition, the structure of 2 was also compared with the reported ^{1}H-NMR spectral data for isolariciresinol[5].

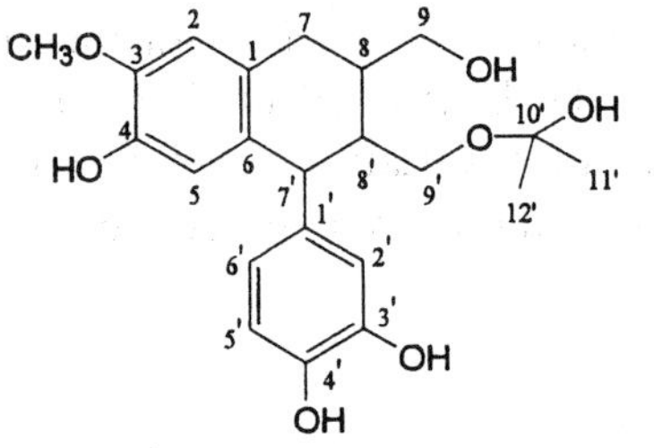

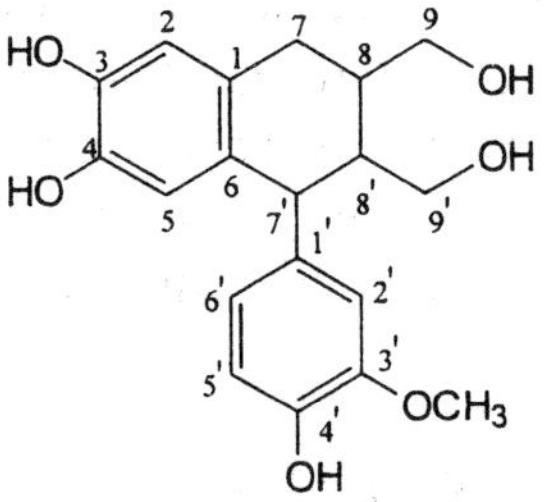

Figure 1. Compounds 1 and 2

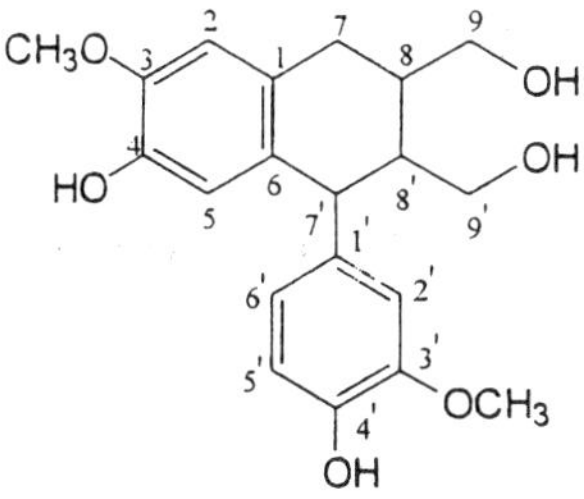

Figure 2. Compounds 3

2.3 Isolariciresinol (3)

Compound 3 was obtained as dark white small particules; IR v_{max} (KBr) cm^{-1} : 3386 (OH), 2360 (CH), 1610 (C=C), 1511,1446 (aromatic region), 1113, 1050 (C-O-C); EIMS m/z : 360 (M$^+$, 100); DCI + (NH$_3$) m/z: 360, M$^+$: $C_{20}H_{24}O_6$.

The structure of 3 was elucidated to be isolariciresinol by comparing of those reported ^{1}H-NMR spectral data[5]. The detailed spectral analysis of 3 was firstly given in this study (Table 3). It was earlier isolated from *T. baccata*[2], *T. cuspidata*[6] and other plants such as *Fitzroya cupressoides*[7], *Picea excelsa*[8], *Araucaria angustifolia*[9] and *Justicia tranquebariensis*[10].

3. MATERIALS AND METHODS

3.1 General

The IR spectra were taken in KBr pellet on a BRUKER VECTOR 22 FT-IR Spectrophotometer. The ^{1}H- and ^{13}C-NMR spectra were recorded on a JEOL JNM-ALPHA 500 FT-NMR Spectrometer in CD$_3$OD. The EIMS were measured on a HITACHI M-2500. The DCI spectra were recorded on a MS 80 MASPEC Spectrometer.

3.2 Plant Material

Taxus baccata L. (Taxaceae) was collected from the vicinity of Çamlıhemşin- Rize, Turkey, in June 1995. A voucher specimen (GUE 1560) was deposited in the Herbarium of Faculty of Pharmacy, Gazi University.

3.3 Extraction, Isolation and Purification

The air-dried and powdered heartwood (3078 g) was extracted with 95 % EtOH at room temperature. The ethanolic extract was evaporated under reduced pressure to give a reddish residue. The residue was partitioned between $CHCl_3$ and H_2O. The $CHCl_3$-soluble portion was evaporated under reduced pressure to give a residue (49 g). The residue was applied to a silica gel column and elution carried out with solvents of increasing polarity using hexane, acetone, $CHCl_3$ and CH_3OH to give seven main fractions (I-VII). Compound 1-3 were isolated from fraction VII which was obtained by CC eluting with $CHCl_3:CH_3OH$ (80:20, v/v) mixture. Fraction VII (5.65 g) was rechromatographed on a silica gel eluting with $CHCl_3$: CH_3OH (100:0→92:8, v/v) to give sixty four sub-fractions. The sub-fraction 11-13 was separated by PTLC with $CHCl_3$: CH_3OH (90:10, v/v) to give **1** (20.2 mg). The sub-fraction 14-22 was crystallized from $CHCl_3$ to afford 1 (42.6 mg). Thus, compound **1** was obtained 62.8 mg (0.0039 %). The sub-fraction 28-33 was crystallized from $CHCl_3$ to give **3** (41.3 mg, 0.0026 %). The sub-fraction 58-64 was subjected to a silica gel column eluting with $CHCl_3$: CH_3OH (100:0→80:20, v/v) to give fraction 4-14, which was purified by prep.TLC developed with $CHCl_3$: CH_3OH (80:20, v/v) to afford **2** (71.4 mg, 0.0044 %).

REFERENCES

1. Davis, P.H., Cullen, J., 1965, *Taxus* L. In *Flora of Turkey and the East Aegean Islands* (Davis, P.H., ed.), Vol.1, Edinburgh University Press, Edinburgh, pp.75-6.
2. Das, B.,Takhi, M, Srinivas, K.V.N.S.;Yadav, J.S., 1993, *Phytochemistry*, **33**(6), 1489-91.
3. Das, B., Takhi, M, Srinivas, K.V.N.S.,Yadav, J.S., 1994, *Phytochemistry*, **36**(4), 1031-3.
4. Shen, Y.C., Chen, C.Y., Lin, Y.M., Kuo, Y.H., 1997, *Phytochemistry*, **46**(6), 1111-3.
5. Abe, F., Yamauchi, T., 1989, *Phytochemistry*, **28**(6), 1737-41.
6. Erdtman, H., Tsuno, K., 1969, *Phytochemistry*, **8**, 931-2.
7. Buckingham, J., 1994, Dictionary of Natural Products, Chapman & Hall Data Base, Cambridge University Press, Cambridge.
8. Weinges, K., 1961, *Chem. Ber.*, **94**, 2522-33.
9. Fonseca, S.F., Campello, J.P., Barata, L.E.S., Ruveda, E.A., 1978, *Phytochemistry*, **17**, 499-502.
10. Raju, G.V.S., Pillai, K.R., 1989, *Indian J. Chem.*, **28B**, 558-61.

Table 1. The NMR Spectral data of 1

Position	^{1}H (J,Hz)	^{13}C (HMQC)	DEPT	HMBC
1	-	127.36	C	-
2	6.19 s	116.71	CH	C-1, C-7', C-3, C-4, C-6, C-5, C-7
3	-	146.51	C	-
4	-	144.52	C	-
5	6.63 brs	111.54	CH	C-6, C-7, C-4, C-3
6	-	133.11	C	-
7α	2.62 dd (15.87)	32.08	CH$_2$	C-1, C-6, C-8', C-8, C-5, C-9
7β	2.47 t (14.9)			C-1, C-8, C-6, C-9, C-8'
8	1.71 m	42.50	CH	C-8'
9a	3.73 t (11.9)	66.44	CH$_2$	C-10', C-8', C-8
9b	3.60 d (12.5)			C-10', C-8', C-8
1'	-	137.36	C	
2'	6.53 brs	116.11	CH	C-5', C-7', C-4', C-3'
3'	-	145.68	C	-
4'	-	144.10	C	-
5'	6.74 d (8.24)	115.34	CH	C-1', C-3'
6'	6.49 d (7.93)	121.13	CH	C-2', C-4', C-7'
7'	3.31 d (10.98)	48.75	CH	C-6, C-1', C-2', C-5', C-1
8'	1.68 d (12.82)	49.76	CH	C-9', C-8, C-7'
9'a	3.37 d (11.9)	64.92	CH$_2$	C-10', C-8
9'b	3.64 d (11.9)			C-10', C-8, C-8'
10'	-	101.56	C	-
11'	1.32 s	24.41	CH$_3$	C-8', C-9'
12'	1.32 s	24.32	CH$_3$	C-10', C-12'
3-OCH$_3$	3.81 s	55.58	CH$_3$	C-3

Table 2. The NMR Spectral data of 2

Position	^{1}H (J,Hz)	^{13}C (HMQC)	DEPT	HMBC
1	-	129.01	C	-
2	6.17 s	117.35	CH	C-3, C-1, C-7', C-4
3	-	147.21	C	-
4	-	145.28	C	-
5	6.64 s	112.38	CH	C-6, C-4
6	-	134.17	C	-
7	2.77 d (7.63)	33.58	CH$_2$	C-1, C-8, C-7', C-6, C-5, C-9, C-2
8	2.00 m	39.99	CH	-
9a	3.69 dd (4.89, 10.99)	65.96	CH$_2$	C-7
9b	3.64 d (6.72)			C-8'
1'	-	138.68	C	-
2'	6.66 brs	113.79	CH	C-5', C-4', C-7', C-3'
3'	-	149.03	C	-
4'	-	145.96	C	-
5'	6.60 d (7.93)	123.20	CH	C-4', C-2', C-7'
6'	6.73 d (7.93)	115.98	CH	C-3', C-1'
7'	3.80 d (10.07)	48.05	CH	C-1', C-6, C-2', C-5', C-8', C-1, C-9'
8'	1.76 t (10.07)	48.00	CH	-
9'a	3.66 d (4.27)	62.21	CH$_2$	C-8'
9'b	3.39 dd (3.97, 10.99)			C-8
3-OCH$_3$	3.80 s	56.39	CH$_3$	C-3

Table 3. The NMR spectral data of 3

Position	^{1}H (J,Hz)	^{13}C (HMQC)	DEPT	HMBC
1	-	128.98	C	-
2	6.20 brs	117.46	CH	C-1, C-3, C-7', C-4
3	-	147.15	C	-
4	-	145.24	C	-
5	6.63 brs	112.35	CH	C-6, C-4, C-7, C-3
6	-	134.32	C	-
7	2.74 d (3.36), 2.75 s (J=9.16 Hz)	33.56	CH$_2$	C-1, C-8', C-8, C-6, C-5, C-9
8	1.97 m	40.08	CH	-
9a	3.68 d (8.24)	66.04	CH$_2$	C-7'
9b	3.63 d (7.63)			C-7'
1'	-	138.76	C	-
2'	6.53 d (2.14)	117.27	CH	C-5', C-3', C-7', C-4'
3'	-	146.33	C	-
4'	-	144.70	C	-
5'	6.51 dd (2.14, 7.94)	122.04	CH	C-2', C-3', C-7'
6'	6.70 d (7.94)	116.06	CH	C-1', C-4', C-3'
7'	3.71 d (10.38)	47.79	CH	C-1', C-6', C-2', C-5', C-8', C-9', C-1, C-6
8'	1.72 dd (3.66,10.07)	48.10	CH	-
9'a	3.64 d (8.24)	62.38	CH$_2$	C-8', 3'-OCH$_3$, C-2'
9'b	3.41 dd (4.12, 11.29)			C-8
3'-OCH$_3$	3.79 s	56.40	CH$_3$	C-3'

Lignans from *Taxus baccata* L.

NURGÜN ERDEMOĞLU and BİLGE ŞENER
Gazi University, Faculty of Pharmacy, Department of Pharmacognosy, 06330 Ankara, Turkey

1. INTRODUCTION

Genus *Taxus* (Taxaceae) is represented by eight species and two hybrids in the world. However, *Taxus* has only one species growing in Turkey, *Taxus baccata* L. (European yew)[1,2]. Many compounds such as taxoids, lignans, flavonoids, steroids, diterpenoids and sugar derivatives have been isolated commonly in this genus[3,4].

As a part of our ongoing investigation on *Taxus baccata*, we have isolated two lignans lariciresinol (**1**) and taxiresinol (**2**) from the heartwood of this plant. In this study, the isolation and structure elucidation of **1** and **2** (Fig. 1) were described.

2. RESULTS

Continuing our researches on *T. baccata* growing in Turkey, two lignans lariciresinol and taxiresinol have been isolated from the heartwood of this plant. Compounds **1** and **2** have possessed the same skeleton.

2.1 Lariciresinol (1)

Compound **1** was obtained as a white fine crystalline compound from $CHCl_3$; IR v_{max} (KBr) cm^{-1} : 3372 (OH); 2937, 2885 (CH); 1606 (C=C); 1515, 1449 (aromatic region); 1154, 1113, 1033 (C-O-C); DCI + (NH$_3$) m/z: 360, M$^+$: $C_{20}H_{24}O_6$.

The structure of 1 was identified as lariciresinol by comparison with the reported ^{1}H- and ^{13}C-NMR spectral data[5,6] and its structure was also confirmed by 2D-NMR methods (Table 1). This is the first report of the isolation of 1 from the genus *Taxus*, however 1 was previously isolated from different plants such as *Araucaria angustifolia*[5], *Larix decidua*[7] and *Justicia tranquebariensis*[8].

2.2 Taxiresinol (2)

Compound 2 was obtained as scaly, pinky-orange coloured compound from CHCl$_3$; IR v_{max} (KBr) cm^{-1} : 3372 (OH); 2937, 2885 (CH); 1606 (C=C); 1515, 1449 (aromatic region); 1154, 1113, 1033 (C-O-C); EIMS m/z: 347 (M+1)$^+$; DCI + (NH$_3$) m/z: 346, M$^+$: C$_{19}$H$_{22}$O$_6$.

The structure of 2 was elucidated to be taxiresinol (3'-demethyllariciresinol) by comparing with those reported ^{1}H-NMR spectral data[9]. Moreover, its detailed spectral data was firstly determined in this study (Table 2). The presence of 2 has been previously reported in *T. baccata*[9] and *T. wallichiana*[10].

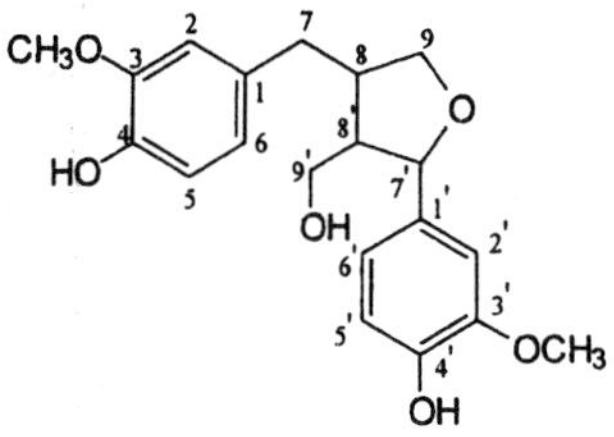

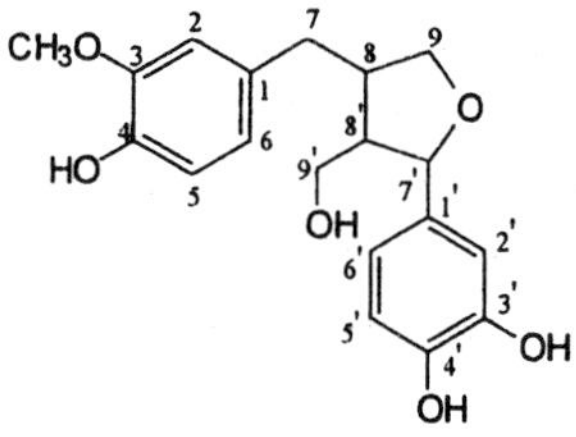

Figure 1. Compounds 1 and 2

3. MATERIALS AND METHODS

3.1 General

The IR spectra were taken in KBr pellet on a BRUKER VECTOR 22 FT-IR Spectrophotometer. The ^{1}H-, ^{13}C-NMR, DEPT, HMQC and HMBC spectra were recorded on a JEOL JNM-ALPHA 500 FT-NMR Spectrometer (500 MHz for ^{1}H- and 125 MHz for ^{13}C-NMR) in DMSO-d$_6$ and CD$_3$OD for 1 and 2, respectively. The EIMS of 2 was measured on a HITACHI M-2500.

The DCI spectra of **1** and **2** were recorded on a MS 80 MASPEC Spectrometer.

3.2 Plant Material

The heartwood of *Taxus baccata* L. (Taxaceae) was collected from the vicinity of Çamlıhemşin, Rize, Turkey, in June 1995. An authenticated voucher specimen (GUE 1560) was kept in the Herbarium of Faculty of Pharmacy, Gazi University.

3.3 Extraction, Isolation and Purification

The air-dried and powdered heartwood (3078 g) was extracted with 95 % EtOH at room temperature. The ethanolic extract was evaporated to dryness in vacuo and a reddish residue was obtained. The residue was diluted with

Table 1. The NMR Spectral data of **1**

Position	^{1}H (J,Hz)	^{13}C (HMQC)	DEPT	HMBC
1	-	131.76	C	-
2	6.65 brs	112.70	CH	C-4, C-6, C-3, C-7
3	-	147.46	C	-
4	-	144.57	C	-
5	6.58 brs	115.39	CH	C-1, C-3
6	6.48 dd (1.83, 7.94)	120.61	CH	C-4, C-2, C-7, C-3
7α	2.73 dd (4.88, 13.43)	32.16	CH$_2$	C-1, C-2, C-9, C-8, C-6
7β	2.33 t (12.0)			C-1, C-2, C-6, C-8, C-9
8	2.48 m	41.92	CH	-
9α	3.79 t (6.41)	71.84	CH$_2$	C-7', C-7, C-8'
9β	3.46 t (6.6)			C-7, C-7', C-8', C-8
1'	-	134.73	C	-
2'	6.74 brs	109.94	CH	C-6', C-3', C-7', C-5', C-1', C-4'
3'	-	147.38	C	-
4'	-	145.52	C	-
5'	6.61 brs	115.05	CH	C-1'
6'	6.60 d (4.88)	118.20	CH	C-2'
7'	4.58 d (6.1)	81.78	CH	C-9', C-2', C-6', C-8', C-1', C-9, C-8
8'	2.10 dd (7.02, 13.73)	52.44	CH	C-9', C-1', C-7', C-8, C-9
9'a	3.58 m	58.60	CH$_2$	C-7'
9'b	3.37 m			C-8
3-OCH$_3$	3.65 s	55.57	CH$_3$	C-3
3'-OCH$_3$	3.65 s	55.54	CH$_3$	C-3'

H$_2$O and then extracted with CHCl$_3$. The CHCl$_3$-soluble portion was evaporated under reduced pressure to give a residue (49 g), which was subjected to CC eluted with increasing polarities of different solvents (hexane→acetone→CHCl$_3$→CH$_3$OH) to give seven main fractions (I-VII) according to TLC.

Compound **1** and **2** were isolated from fraction VII which was obtained by CC eluting with CHCl$_3$:CH$_3$OH (80:20, v/v) mixture. Fraction VII (5.65 g) was rechromatographed on a silica gel eluting with CHCl$_3$: CH$_3$OH (100:0→92:8, v/v) to give sixty four sub-fractions. The sub-fraction 11-13 was separated by PTLC with CHCl$_3$: CH$_3$OH (90:10, v/v) to give compound **1** (45.7 mg, yield: 0.0028 %). The sub-fraction 47-52 was purified by PTLC using CHCl$_3$: CH$_3$OH (80:20, v/v) as the developing system to afford compound **2** (84.6 mg, 0.0052 %).

Table 2. The NMR Spectral data of **2**

Position	^{1}H (J,Hz)	^{13}C (HMQC)	DEPT	HMBC
1	-	133.55	C	-
2	6.82 brs	113.38	CH	C-5, C-4, C-3
3	-	148.95	C	-
4	-	145.73	C	-
5	6.67 m	122.17	CH	C-2, C-4
6	6.75 d (8.24)	116.12	CH	C-3, C-1, C-2
7α	2.96 dd (4.58, 13.4)	33.59	CH$_2$	C-9, C-1, C-8, C-5, C-2, C-8'
7β	2.49 t (12.36)			C-1, C-2, C-5, C-8, C-9, C-8'
8	2.75 d (6.0)	43.83	CH	C-8', C-7', C-9', C-1, C-9
9α	3.98 dd (7.6, 14.4)	73.43	CH$_2$	C-7', C-8', C-8
9β	3.74 dd (5.8, 7.9)			C-7', C-8', C-8
1'	-	135.80	C	.
2'	6.82 brs	114.16	CH	C-5'
3'	-	145.73	C	-
4'	-	146.30	C	-
5'	6.68 m	118.65	CH	C-3', C-2'
6'	6.76 d (8.24)	116.15	CH	C-4', C-1'
7'	4.71 d (7.02)	83.90	CH	C-9', C-5', C-2', C-8', C-9, C-1', C-8
8'	2.38 d (6.85)	54.05	CH	C-9', C-1', C-7', C-8, C-9
9'a	3.83 dd (8.23, 14.6)	60.41	CH$_2$	C-7', C-8', C-9, C-8
9'b	3.65 dd (6.4, 10.9)			C-8, C-7', C-8'
3-OCH$_3$	3.85 s	56.35	CH$_3$	C-3

REFERENCES

1. Van Rozendall, E.L.M., Kurstjens, S.J.L., Van Beek, T.A., Van Den Berg, R.G., 1999, *Phytochemistry*, **52**, 427-33.

2. Davis, P.H., Cullen, J., 1965, *Taxus* L. In *Flora of Turkey and the East Aegean Islands* (Davis, P.H., ed.), Vol.1, Edinburgh University Press, Edinburgh, pp.75-6.
3. Baloğlu, E., Kingston, D.G.I., 1999, *J. Nat. Prod.*, **62**, 1448-72.
4. Parmar, V.S., Jha, A., Bisht, K.S., Taneja, P., Singh, S.K., Kumar, A., Raijni Jain, P., Olsen, C.E., 1999, *Phytochemistry*, **50**, 1267-304.
5. Fonseca, S.F., Campello, J.P., Barata, L.E.S., Ruveda, E.A., 1978, *Phytochemistry*, **17**, 499-502.
6. Abe, F., Yamauchi, T., 1989, *Phytochemistry*, **28**(6), 1737-41.
7. Haworth, R.D., Kelly, W., 1937, *J. Chem. Soc.*, 384-91.
8. Raju, G.V.S., Pillai, K.R., 1989, *Indian J. Chem.*, **28B**, 558-61.
9. Mujumdar, R.B., Srinivasan, R., Venkataraman, K., 1972, *Indian J. Chem.*, **10**, 677-80.
10. Chattopadhyay, S.K., Kulshrestha, M., Saha, G.C., Sharma, R.P., Jain, S., Kumar, S., 1997, *J. Med. Aromat. Plant Sci.*, **19**(1), 17-9.

Heraclenol and Isopimpinellin: Two Rare Furocoumarins from *Ruta montana*

N. BENKIKI[1], M. BENKHALED[2], ZAHIA KABOUCHE[1] and C. BRUNEAU[3]

[1]*Université Mentouri-Constantine, Faculté des sciences, Département de chimie, Laboratoire d'Obtention de Substances Thérapeutiques (LOST), 25000 Constantine, Algérie,* [2]*Université de Batna, Faculté des sciences, Département de chimie, Batna, Algérie,* [3]*Université de Rennes 1, UMR 6509 CNRS, Campus de Beaulieu, 35042 Rennes Cedex, France.*

1. INTRODUCTION

Ruta montana is a common specie of the genus *Ruta* (Rutaceae) but still studied since very recently four new alkaloids have been isolated from this specie[1]. The presence of xanthotoxin, bergapten and other known furocoumarins have been reported from *Ruta montana*[2-5] but isopimpinellin[2] and heraclenol[3] are two rare furocoumarins reported here for the first time from the specie and the genus *Ruta*, respectively. Few plants contain heraclenol and isopimpinellin. recent pharmacological studies on isopimpinellin have shown it's antimicrobial activity[6], heraclenol have been reported for it's anti-inflammatory[7] and anti-mycobacterial[8] properties with helpful nmr high techniques, we succeeded to identify heraclenol and isopimpinellin, two furocoumarins isolated for the first time from *Ruta montana*.

2. RESULTS AND DISCUSSION

The compounds (**1-2**) and compound (**3**), found in the seventh fraction of the chromatographied chloroformic extract, gave a violet color reaction when treated with hydroxylamine then ferric chloride, the IR spectrum exhibited bands indicative of coumarin carbonyl (1720 cm^{-1}). The mass spectral fragmentations and ^{1}HNMR spectral patterns of these compounds (Table 1-2) showed furanic protons and aromatic protons characteristic of linear furanocoumarins. The resonance signals due to H-3 and H-4 appeared as doublets, J=9.5-9.6 Hz, the chemical shift of H-4 resonated relatively upfield at 7.69-8.06 which suggested a C-8 substitution. Except in compound (2) were it didn't appear, the proton H-5 was observed as a singlet at 7.26-7.36. The furano protons H-2' and H-3' appeared as doublet J =1.7-2.0 Hz and resonated at 7.56-7.60 and 6.74-6.90, respectively. In the compound (1), in addition to the molecular ion pic at m/z 216 indicating the molecular formula $C_{12}H_8O_4$ the methoxy signal at δ 4.21 and the other ^{1}HNMR and the ^{13}CNMR chemical shifts (Table 1-4) suggested the C-8 substituted psoralen. An authentic sample confirmed the 8-methoxypsoralen or xanthotoxin structure. The molecular ion pic of (2) at m/z 246 corresponding to the molecular formula for $C_{13}H_{10}O_5$ and the two methoxy signals at δ 4.11 and 4.10, respectively added to the chemical shifts of H-4 at δ 8.06 (Table 1-2) suggested the 5-8-dimethoxypsoralen structure or isopimpinellin which have been confirmed by the ^{13}CNMR signals (Table 3-4). In compound (3), in addition to the coumarin carbonyl band, the IR spectrum exhibited hydroxy bands at 3544 cm^{-1} and the molecular ion pic was observed at m/z 304 corresponding to the molecular formula $C_{16}H_{16}O_6$.

The ^{1}HNMR spectrum exhibited two doublets at 6.34 and 7.75 (J = 9.5 Hz) due to H-3 and H-4, respectively and two other doublets at δ 7.36 and 6.82 attributed to H-2' and H-3' (J = 1.7 Hz), respectively and a singlet at 7.36 due to H-5, two methyl groups appeared at δ 1.30 and 1.34, a CH$_2$ group at δ 4.41-4.75, a CHOH group at δ 3.89 (Table 1-2). Based on the detailed spectral analysis, (3) was identified as 8-(2,3-dihydroxy-3-methylbutoxy)psoralen or heraclenol which have been subsequently confirmed by the application of ^{1}H-^{1}H-COSY, ^{1}H-^{13}C-HMQC and HMBC techniques.

	R$_1$	R$_2$
(1)	OMe	H
(2)	OMe	OMe
(3)	(2,3-dihydroxy-3-methylbutoxy)	H

Table 1. ¹HNMR chemical shifts of (1), (2) and (3) (400 MHz, CDCl₃, δ values, coupling constants Hz)

Compound	H-3	H-4	H-5	H-8	H-2'	H-3'
(1)	6.31 (d, J 9.6)	7.70 (d, J 9.6)	7.29 (s)	-	7.62 (d, J 2.0)	6.75 (d, J 2.0)
(2)	6.31 (d, J 9.6)	8.06 (d, J 9.6)	-	-	7.56 (d, J 2.0)	6.93 (d, J 2.0)
(3)	6.34 (d, J 9.5)	7.75 (d, J 9.5)	7.36 (s)	-	7.69 (d, J 1.7)	6.82 (d, J 1.7)

Table 2. ¹HNMR chemical shifts of the side chain protons of (1), (2) and (3) (400 MHz, CDCl₃, δ values, coupling constants Hz)

Compound	H-1"a	H-1"b	H-2"	H-3"	H-4"	H-5"	H(OMe)
(1)	-	-	-	-	-	-	4.21 (s)
(2)	-	-	-	-	-	-	4.11, 4.10 (s), (s)
(3)	4.75 (dd, J 10.0, J 2.1)	4.41 (dd, J 10.0, J 7.7)	3.89 (dd, J 7.7 - J 2.1)	-	1.30 (s)	1.34 (s)	

Table 3. ¹³CNMR chemical shifts of (1), (2) and (CDCl₃, δ values)

Compound	C-2	C-3	C-4	C-5	C-6	C-7	C-8	C-9	C-10	C-2'	C-3'
(1)	160.5	115.3	144.8	112.9	126.0	147.6	132.7	142.9	116.4	146.6	106.7
(2)	160.7	112.9	139.3	144.4	114.9	149.8	128.9	143.7	107.1	146.0	105.1
(3)	160.5	114.6	144.4	112.4	126.1	147.6	130.8	142.9	116.4	146.6	105.08

Table 4. ^{13}CNMR chemical shifts of the side chain carbons of (1), (2) and (3) (CDCl$_3$, δ values)

Compound	C-1"	C-2"	C-3"	C-4"	C-5"	OMe
(1)	-	-	-	-	-	61.3
(2)	-	-	-	-	-	61.3, 60.0
(3)	75.5	76.0	71.6	25.0	26.5	-

3. EXPERIMENTAL

3.1 Plant material

Aerial parts of *Ruta montana* were collected from Mila, a region of north-east of Algeria. A voucher specimen has been deposited in the herbarium of the Laboratoire d'Obtention de Substances Thérapeutiques (L.O.S.T), Faculté des Sciences, Université Mentouri, Constantine (Rue NB01).

3.2 Extraction and purification

The aerial parts of *Ruta montana* (1 Kg) were pulverised and extracted with petroleum ether then filtered. The residue was continously extracted with methanol (80%), concentrated and acidified with HCl 2%. After filtration and extraction with CHCl$_3$ the obtained aqueous phase was basified with NaOH until pH-9, then extracted with CHCl$_3$ and fractionated on a silica gel column eluting with EtOAc-petroleum ether. The fractions obtained with the gradients 70:30 and 20:80 were separately rechromatographied with on silica gel column eluted with EtOAc-petroleum ether-MeOH and hexane-CHCl$_3$-MeOH gradients, respectively. Xanthotoxin (1) was found in several fractions while isopimpinellin (2) was found in the fraction 12 with the eluent petroleum ether-ethylacetate (90:10). Heraclenol (3) was isolated from fraction F-3 as following:

F-1: CC gradient: CHCl$_3$-n-hexane (50:50); TLC gradient: CHCl$_3$-n-hexane (5:5, 7:3, 8:2)

F-3: CC gradient: CHCl$_3$-n-hexane (80:20, 85:15, 0:100)

XANTHOTOXIN (1)-. Mp 249° (CH$_2$Cl$_2$/Et$_2$O) (lit. 251-252°) IR ν cm^{-1} 3290, 1712, 1605, 1498. ^{1}HNMR and ^{13}CNMR are given in the Tables 1-4.

[M]⁺ m/z 216.0336 (Calc. for $C_{12}H_8O_4$, 216.0432). EIMS m/z 216[M]⁺ 202, 201 (100), 185, 173, 145.

ISOPIMPINELLIN (2)-. Mp 151-152° (CH_2Cl_2/Et_2O) (lit. 150-151) IR ν cm⁻¹ 3320, 1723, 1608, 1455, 1376, 1374, 1083, 825. ¹HNMR and ¹³CNMR are given in the Tables 1-4. [M]⁺ m/z 246.1245 (Calc. for $C_{13}H_{10}O_5$, 246.1362). EIMS m/z [M]⁺ 225, 210, 202, 201 (100%).

HERACLENOL (3)-. Mp 118° (CH_2Cl_2/Et_2O) (lit. 117-118) IR ν max cm⁻¹ 3544, 1723, 1608, 1495, 1390, 1374, 875. ¹HNMR and ¹³CNMR are given in the Tables 1-4. [M]⁺ m/z 304.1002 (Calc. for $C_{16}H_{16}O_6$, 304.0946). EIMS m/z [M]⁺ 304, 289, 245, 202, 201 (100), 173, 145.

ACKNOWLEDGEMENTS

The authors thank the CMEP (France-Algeria) and ANDRS (National Health Research Agency, Oran, Algeria) for their financial support.

REFERENCES

1. Touati, D., Atta-ur-Rahman, Ulubelen, A., 2000, *Phytochemistry*, **53**(2), 277.
2. Sepulveda-Arques, J., Viguera-Lobo, J., Sanchez-Parareda, J., 1974, *An. Quim.*, **70**(12), 1020.
3. Rofaeel, S., El-Gengaihi, S., Shalaby, A.S., 1984, *Ann Agr. Sci Moshtohor*, **20**, 3.
4. Şener, B., Mutlugil, A., 1985, *GUEDE-Gazi Univ Eczacilik Fak Derg.*, **2**, 109.
5. Ulubelen, A., Güner, H., 1988, *J. Nat. Prod.*, **51**, 1012.
6. Manderfeld, M.M., Schafer, H.W., Davidson, P.M., Zottola, E.A., 1997, *J. Food Prot.*, **60**(1), 72.
7. Garcia-Argaz, A.N., Ramirez Apan, T.O., Parra Delgado, H., Velasquez, G., Martinez-Velasquez, M., 2000, *Planta Medica*, **66**(3), 279.
8. Rastogi, N., Abaul, J., Goh, K.S., Devallois, A., Philogene, E., Bourgeois, P., 1998, *Immun. Med. Microbiol.*, **20**(4), 267.

A Chemotaxonomic Study on the Genus *Ferulago*, Sect. *Humiles* (Umbelliferae)

EMİNE AKALIN[1], BETÜL DEMİRCİ[2] and K.HÜSNÜ CAN BAŞER[2]
[1]*İstanbul University, Faculty of Pharmacy, Department of Pharmaceutical Botany, 34452-Beyazıt-İstanbul, Turkey*, [2]*Anadolu University, Medical and Aromatic Plant and Drug Research Centre (TBAM), 26470- Eskişehir, Turkey*

1. INTRODUCTION

Ferulago is a genus of c.46 species distributed from Europe (W. Central, SW, S, SE, E) Asia (SW, Middle, the Caucasus), Africa (N, NW)[1] and it is represented by 31 species in Turkey of which 16 are endemic[2]. The genus *Ferulago* and its related genera such as *Ferula, Cachyrs, Glaucosciadium* and *Prangos* are known "Çakşır or Çağşır" as vernacular names and their roots are used as aphrodisiac, sedative, digestive, carminative, tonic, food, spices also in the treatment of intestinal worms and haemorrhoids[3].

2. MATERIAL AND METHODS

2.1 Plant Material

The plant materials were collected in Western Anatolia. Voucher specimens are kept at Herbarium of the Faculty of Pharmacy at Istanbul University, Istanbul (ISTE), Turkey. The drawings and measures are based on collected specimens. SM Lux trinocular stereo-microscope is used for drawings. Distribution maps are prepared based on all examined specimens.

2.2 Isolation of the Essential oil and Analysis

The essential oils were obtained by micro-distillation from cushed fruits of twelve species of *Ferulago* using an Eppendorf MicroDistiller®[4] Then the essential oils were analysed using a Hewlett-Packard G1800A GCD system. HP-Innowax FSC column (60 m x 0.25 mm inner diameter, with 0.25 μm film thickness). Helium (0.8 ml/min) was used as carrier gas. GC oven temperature was kept at 60°C for 10 min and programmed to 220°C at a rate of 4°C/min and then kept constant at 220°C for 10 min to 240°C at rate of 1°C/min. Mass range was recorded from m/z 35 to 425. Injections were applied both splitless or with a split ratio 50:1. Injection port temperature was at 250°C. MS were recorded at 70 eV. Relative percentage amounts of the separated compounds were calculated automatically from peak areas of the total ion chromatogram. n-Alkanes were used as reference points for the calculation of relative retention indices (RRI). Library search was carried out using both "Wiley GC/MS Library" and "TBAM Library of Essential Oil Constituents"[4].

2.3 Flavonoids

Paper and thin layer chromatography methods are used for separation. Five common substance-apigenin, luteolin, kaempherol, quercetin and isorhamnetin are applied for a marker[3].

3. RESULT AND DISCUSSION

3.1 Morphological Characteristics

Ferulago humilis Boiss., *F.macrosciadia* Boiss. et Bal., *F.sandrasica* Peşmen et Quézel and *F.idaea* Özhatay & E.Akalın are closely related four endemic species in Turkey. The genus *Ferulago* has been subdivided into 9 sections by L.P.Tomkovich and M.H.Pimenov[5] and according to their classification. *F.humilis* and *F.macrosciadia* are presented in Section *Humiles*, *F.sandrasica* is belonged in the Section Bernardia. After close and detailed examination of these species, Table 1 show that they are very similar and must be treated in same Section (Sect. *Humiles*). Also the new species is related to *F.humilis*[6] and it is presented in Sect. *Humiles*.

Table 1. Differences of morphological characters in four *Ferulago* species[3].

	F. humilis	F. macrosciadia	F. sandrasica	F. idaea
Stem	25-75 cm, straight, green	30-70 cm, straight, green	25-45 cm, straight, glaucescent	5-25 cm, decumbent, reddish green
Leaves (outline)	Narrowly-lanceolate, (9-)10-35 x 2-10 cm, green	Narrowly-lanceolate, 30-32 x 2.5 cm, green	Linear to narrowly-lanceolate, 8-27 x 0.6-3 cm, glaucescent	Linear, 6-18 x 0.5-2 cm, green
Leaves segments	Setaceous, 0.8-4 x 0.1-0.3 mm	Setaceous, 1-4(10) x 0.1-0.3 mm	Oblong-linear, 0.5-3 x 0,3-0.7 mm	Oblong-linear, 0,5-3 x 0,5 mm
Rays	(4-)5-9(-13)	4-11(-16)	3-12	(2-)3-7
Bracts	Linear	Linear	Ovate-lanceolate	Linear
Sepals	Lanceolate	Ovate-lanceolate	Ovate	Linear-lanceolate
Fruit shape	Elliptic-oblong, 7-12 x 4-8.5 mm,	Ovate-orbicular, 8-13 x 4.5-8.5 mm	Elliptic-oblong, 8.5-15 x 4.5-7.5 mm	Pyriform, 8-13 x 4-8 mm
Altitude	150-1030 m	20-1050 m	1680-2000 m	1750 m

3.2 Anatomical Characteristics

Fruit anatomical characteristics are useful with other properties for determination of the Ferulago species. Important characters are summarized in Table 2.

Table 2. Results of anatomical characters of fruit[3].

Species	General View (Cross-section of fruit)		Secretion canals number
	Dorsal wings	Lateral wings	A= Dorsal; B= Commissural; C= in mesocarp
F. humilis	Projections	Medium Size Wings	A= 35; B= 23; C= 23
F. macrosciadia	Distinct projections	Long wings	A= 27; B= 17; C= 22
F. sandrasica	Smooth	Short wings	A= 29; B= 21; C= 57
F. idaea	Projections	Short wings	A= 26; B= 20; C= 10

3.3 Distribution

F.sandrasica and *F.idaea* which are both named after their locality (Sandras Dağı and Ida/Kaz Dağı) and have very narrow distribution. *F.humilis* are grown from Northwest Anatolia to Southwest Anatolia. *F.macrosciadia* are distributed from Northwest Anatolia to further east.

3.3.1 Red data book categories

	Red Data Book (1989)	Red Data Book (2001)
F.humilis	Not listed	LR (lc)
F. macrosciadia	Not listed	LR (lc)
F.sandrasica	Not listed	EN
F. idaea	Not listed	Not listed

LR (lc): Lower Risk (least concern), EN: endangered.

3.4 Chemotaxonomy

Chemotaxonomy is a combinative discipline between chemistry and taxonomy. Mainly it is concerned with chemical properties of definite groups of plant. Chemotaxonomy tends to act very much as a support of classical taxonomy. Chemical characters of plants can often be used in classification. These four species are investigated by chemical survey of leaves flavonoids[3] and fruit essential oils[7]. The results of chemical studies are shown to as diversities and similarities among the related species. For example; isorhamnetin has been identified in *F.humilis*, but there is not in *F.macrosciadia*. Kaempherol presents in F.humilis, in spite of *F.idaea* have not. This situation is an important evidence for a new species. The best distinctive characters are among essential oils that every species have different main component and different amount main groups. E.g. Althouth (Z)-β-ocimen is main component of *F.humilis* (31 %), it is identified in *F.macrosciadia* very few amount (0.3 %). Also there is not any in *F. idaea* [7]. All results are given in Table 3.

Table 3. Main groups and main components of essential oils and flavonoids of four *Ferulago* species (Q: quersetin, K: kaempherol; Isorh.: isorhamnetin).

Species	Main Groups of Essential Oils[3](%)	Main Component of Essential Oils[3](%)	Flavonoids[1]		
			Q	K	Isorh.
F.humilis	Monoterpens (88.2)	(Z)-β-ocimen (31)	+	+	+
F. macrosciadia	Oxygeneted Monoterpens (78.5)	Carvacrol methyl ether (78)	+	+	-
F. sandrasica	Monoterpens (47.1)	α-pinene (40.8)	+	+	+
F. idaea	Oxygeneted Monoterpens (53.9)	P-ocimene (31.9)	+	-	+

REFERENCES

1. Pimenov, M.G., Leonov, M.V., 1993, The Genera of the Umbelliferae, A Nomenclator, Royal Botanic Gardens Kew.
2. Peşmen, H., 1972 *Ferulago* W.Koch. In *Flora of Turkey and the East Aegean Islands*, (Davis, P.H. Ed.), Vol.4, Edinburgh University Press, Edinburgh, pp.453-71.
3. Akalın, E., 1999, Taxonomical studies on the genus *Ferulago* in Western Anatolia; Ph.D Thesis, Istanbul University, Istanbul.
4. Briechle, R., Dammertz, W., Guth, R., Volmer, W., 1997, *GIT Lab. Fachz.*, **41**, 749-53.
5. Tomkovich, L.P., Pimenov, M.G., 1987, *Bot, Žourn.*, **72**(7), 964-71.
6. Özhatay, N., Akalın, E., 2000, *Botanical Journal of the Linnean Society*, **133**, 535-42.
7. Başer, K. H C., Demirci, B., Özek, T., Akalın, E., Özhatay, N., 2001, *Journal of Pharmaceutical Biology*, (in press).

Aromatic Biodiversity among Three Endemic *Thymus* Species of Iran

SEYED EBRAHIM SAJJADI
Isfahan University of Medical Sciences, Faculty of Pharmacy Department of Pharmacognosy, Isfahan, Iran

1. INTRODUCTION

The genus *Thymus*, which belongs to the Lamiaceae family, consists of about 400 species widespread throughout the world[1]. In Iran 14 species are present, among which four are endemic[2]. The oil of *Thymus* species have been traditionally used as anthelmintic, bacteriostatic, antiseptic and spasmolytic agents[3]. The antimicrobial properties are often due to their phenol content[4].

Due to the use of *Thymus* species or their essential oils in the food and drug industries, and because of the importance of thymol and carvacrol in these oils[4], it is decided to investigate the chemical composition of the essential oils of some *Thymus* species, which are endemic to Iran.

2. PLANT MATERIAL

Aerial parts of *T. persicus* (Ronniger ex Rech. f.) Jalas, *T. daenensis* Celak and *T. kotschyanus* Boiss. & Hohen. were collected during their flowering periods (June-July) near Hammadan (north of Iran), Zanjan (north of Iran) and Isfahan (center of Iran) respectively.

The plants were identified at the Botany Department of the Research Institute of Forests and Rangelands, Tehran, Iran and voucher specimens have been deposited in the Herbarium of Faculty of Pharmacy, Isfahan University of Medical Sciences, Isfahan, Iran.

The essential oils were isolated by hydrodistillation for 3h according to the *British Pharmacopoeia*[5]. The oils were dried over anhydrous sodium sulfate and stored in sealed vials at low temperature before analysis.

3. GC AND GC-MS ANALYSIS

The oils were analyzed on a Perkin-Elmer gas chromatograph Model 8500, equipped with a FID detector and a BP-1 capillary column (30 m× 0.25 mm; film thickness 0.25 µm). The oven temperature was programmed from 60°C to 280°C at 4°C/min. The carrier gas was helium with a flow rate of 2 mL/min. Injector and detector temperatures were 280°C.

GC-MS analysis was performed on a Hewlett-Packard 6890 mass selective detector coupled with a Hewlett-Packard 6890 gas chromatograph, equipped with a HP-5MS capillary column (30 m × 0.25 mm; film thickness 0.25 µm) and operating under the same condition as described above. The MS operating parameters were : ionization voltage ,70 eV ; ion source temperature, 200°C. Identification of components of the oils were based on GC retention indices relative to *n*-alkanes and computer matching with the WILEY275.L library, as well as by comparison of the fragmentation patterns of the mass spectra with those reported in the literature[6]. The relative percentages of the oils constituents were calculated from the GC peak areas.

4. CONCLUSION

The essential oils isolated from the aerial parts of *T. persicus*, *T. daenensis* and *T. kotschyanus* were obtained in yields of 0.2%, 0.7% and 1.1% respectively. The major components in the oil of *T. persicus* are thymol (42.3%), ρ-cymene (23.9%), γ-terpinene (9.8%), carvacrol (4.7%), linalool (2.3%) and borneol (1.9%).

In the oil of *T. daenensis* the main components are thymol (73.9%), carvacrol (6.7%), ρ-cymene (4.6%), β-bisabolene (1.5%) and terpinen-4-ol (1.4%). *T. kotschyanus* oil contain carvacrol (69.8%), thymol (6.8%), γ-terpinene (3.9%), ρ-cymene (3.2%), borneol (3.3%) and β-caryophyllene (1.3%) as major constituents.

The volatile oils of various species of *Thymus*[7-14] can be classified into two groups. The first group contains those in which aromatic alcohols (thymol and carvacrol) or their biosynthetic precursors are the predominant components, and the second includes the essential oils in which aromatic ring-containing components are scarce or altogether lacking.

The results of this study indicate that all of these three oils can be classified in first group. However there are also considerable diversities in the number and amounts of other components of these three *Thymus* species. For example in spite of the oil of *T. kotschyanus* which carvacrol is the main component, thymol is the major constituent of the oils of *T. persicus* and *T. daenensis*. The percentage of p-cymene and γ-Terpinene are much more in *T. persicus* in comparison to *T. kotschyanus* and *T. daenensis*. There are also some components which are exist in just one of the oils, for instance 1-octen-3-ol was only identified in the oil of *T. persicus* (2.6%).

REFERENCES

1. Evans, W.C., 1989, *Trease and Evans' Pharmacognosy*, 13th Edition, Bailliere Tindall, London , p. 217.
2. Mozaffarian, V., 1996, *A Dictionary of Iranian Plant Names,* Farhang Moaser, Tehran, p. 547.
3. Zafra-Polo, M.C., Blazquez, M.A. and Villar, A., 1989, *Fitoterapia,* **60**, 469-73.
4. Bauer, K., Garbe, D. and Surburg, H., 1997, *Common Fragrance and Flavor Materials*, Wiley-VCH , Weinheim, p. 214.
5. *British Pharmacopoeia*, Vol.2, 1988, HMSO, London, pp. A137-8.
6. Adams, R.P., 1995, *Identification of Essential Oil Components by Gas Chromatography-Mass Spectroscopy*, Allured Publ. Corp., IL.
7. Sefidkon, F., Dabiri, M. and Rahimi-Bidgoly, A., 1999, *Flav. Fragr. J.*, **14**, 405-8.
8. Baser, K.H.C., Kurkcuoglu, M., Ermin, N., Tumen G. and Malyer, H., 1999, *J. Essent. Oil Res.*, **11**, 86-8.
9. Loziene, K., Vaiciuniene J. and Venskutonis, P.R., 1998, *Planta Med.*, **64**, 772-3.
10. Tomei, P.E., Bertoli, A., Cioni, P.L., Flamini G. and Spinelli, G., 1998, *J. Essent. Oil Res.*, **10**, 667-9.
11. Kulevanova, S., Ristic, M., Stafilov, T. and Matevski, V., 1998, *J. Essent. Oil Res.*, **10**, 335-6.
12. Kulevanova, S., Ristic, M. and Stafilov, T., 1996, *Planta Med.*, **62**, 78-9.
13. Tumen, G. and Baser, K.H.C., 1994, *J. Essent. Oil Res.*, **6**, 663-4.
14. Iglesias, J., Vila, R., Canigueral, S., Bellakhdar, J. and Idrissi, A., 1991, *J. Essent. Oil Res.*, **3**, 43-4.

Volatile Constituents of the Leaves of *Ziziphus spina-christi* (L.) Willd. from Iran

ALIREZA ALIGHANADI (GHANNADI) and MOZHGAN MEHRI-ARDESTANI
Isfahan University of Medical Sciences,Faculty of Pharmacy and Pharmaceutical Sciences, Department of Pharmacognosy, Isfahan, Iran

1. INTRODUCTION

Ziziphus spina-christi (L.) Willd.(Rhamnaceae family) is a tree indigenous to the South of Iran. The leaves of this plant, which is locally known as "Sedr" and "Konar", have been used for washing both the hair and body. Plant leaves are also used in Iranian folk medicine as an antiseptic, antifungal and anti-inflammatory agent as well as for treating of skin diseases such as dermatitis[1-3]. A literature search has not revealed any previous work on the leaf oil of *Z. spina-christi*, but there were several reports on other chemical constituents and pharmacological properties of the plant[4-12].

2. PLANT MATERIAL

Z. spina-christi leaves were collected from Dashtestan in Bushehr province(close to Persian Gulf), Iran in March 1999. The plant was identified at the Herbarium Department of the Iranian Research Institute of Forests and Rangelands, Isfahan, Iran. A voucher specimen has been deposited in the Herbarium of the Pharmacognosy Department, Faculty of

Pharmacy and Pharmaceutical Sciences, Isfahan University of Medical Sciences, Isfahan, Iran. The oil was isolated by hydrodistillation of the powdered, dried leaves for 3h according to the method recommended in European Pharmacopoeia[13] to produce an oil in 0.1% yield. The oil was dried over anhydrous sodium sulphate and stored in a fridge.

3. GC AND GC-MS ANALYSIS

The oil was analyzed by GC and GC/MS. GC analysis was carried out on a Perkin-Elmer gas chromatograph Model 8500, equipped with a FID detector and a BP-1 capillary column (30 m x 0.25 mm, film thickness 0.25 µm). The operating conditions were as follows: carrier gas, helium with a flow rate of 2 mL/min; column temperature, 60°-275°C at 4°C /min; injector and detector temperatures, 280 °C; volume injected, 0.1µL of the oil; split ratio, 1:50.

GC/MS analysis was performed on a Hewlett Packard 6890 mass selective detector coupled with a Hewlett Packard 6890 gas chromatograph, equipped with a cross-linked 5% PH ME siloxane HP-5MS capillary column (30 m x 0.25 mm, film thickness 0.25 µm) and operating under the same conditions as described above. The MS operating parameters were as follows: ionization potential, 70ev; ionization current, 2 A; ion source temperature, 200°C; resolution, 1000. The oil constituents were identified by matching their mass spectra with Wiley 275. Lib. and reported data in literature[14-16]. The percentage composition of the oil was computed from GC peak areas without using correction factors.

4. CONCLUSION

Thirty- four constituents were identified which comprised 93.8% of the total oil. Geranyl acetone (14.1%), methyl hexadecanoate (10.0%), methyl octadecanoate (9.9%), farnesyl acetone C (9.9%), hexadecanol (9.7%), ethyl octadecanoate (8.0%), ethyl hexadecanoate (4.3%), *beta*-eudesmol (3.8%), methyl dodecanoate (3.2%), methyl tetradecanoate (2.6%), ethyl tetradecanoate (2.3%), tetradecanoic acid (1.6%), dodecanoic acid (1.4%), (E)-*beta*- ionone (1.4%), spathulenol (1.2%), terpinolene (1.2%), germacrene D (1.1%) and nerolidol (1.1%) were found to be the major constituents. Percentage composition of ethyl dodecanoate, *allo*- aromadendrene, *beta*-pinene, decane, dodecanol, hexadecane, tetradecanal, *beta*-caryophyllene, *alpha*-terpineol, *alpha*-pinene, 1,8-cineole, nerol, aromadendrene, *delta*-cadinene, *para*-cymene and limonene were less than 1 percent.

REFERENCES

1. Ghahreman, A., 1982, Flore de l'Iran en Couleur Natureile, Vol. 3. Institut des Recherches des Forets et des paturages, Tehran, p. 195.
2. Amin, G., 1991, Popular Medicinal Plants of Iran, Vol. 1. Ministry of Health Publications, Tehran, p. 67.
3. Nafisy, A., 1989, A Review of Traditional Medicine in Iran, Isfahan University Publications, Isfahan, p. 133.
4. Aynehchi, Y., and Mahmoodian, M., 1973, *Acta Pharm. Suec.*, **10**, 515-9.
5. Tschesche, R., Khokhar, I., Spilles, C., and Von Radloff, M., 1974, *Phytochemistry*, **13**, 1633.
6. Ikram, M., and Tomolinson, H., 1976, *Planta Med.*, **29**, 289-90.
7. Nawwar, M.A.M., Ishak, M.S., Michael, H.N., and Buddrus, J., 1984, *Phytochemistry*, **23**, 2110-11.
8. Shah, A.H., Ageel, A.M., Tariq, M., Messa, J.S., and Alyahya, M.A., 1986, *Fitoterapia*, **57**, 452-4.
9. Mahran, G.H., Glombitza, K.W., Mirhom, Y.W., Hartmann, R., and Michel, C.G., 1996, *Planta Med.*, **62**, 163-5.
10. Tanira, M.O.M., Ageel, A.M., Tariq, M., Mohsin, A., and Shah, A.H., 1988, *Int. J. Crude Drug Res.*, **26**, 56-60.
11. Glombitza, K.W., Mahran, G.H., Mirhom, Y.W., Michel, K.G., and Motawi, T.K., 1994, *Planta Med.*, **60**, 244-7.
12. Ali-Shtayeh, M.S., Yaghmour, R.M., Faidi, Y.R., Salem, K., and Al-Nuri, M.A., 1998, *J. Ethnopharmacol.*, **60**, 265-71.
13. European Pharmacopoeia, Vol.3, 1975, Maisonneuve SA, Sainte-Ruffine, p. 68.
14. Mclafferty, F.W., and Stauffer, D.B., 1991, The Important Peak Index of the Registry of Mass Spcetral Data, John Wiley & Sons, Inc., New York.
15. Adams, R.P., 1995, Identification of Essential Oil Components by Gas Chromatography / Mass Spectroscopy, Allured Publishing Co., Carol Stream.
16. Swigar, A.A., and Silverstein, R.M., 1981, Monoterpenes- Infrared, Mass, proton-NMR and carbon- NMR Spectra and Kovats Indices, Aldrich Chemical Co., Milwaukee.

Fatty Acid Composition of the Aerial Parts of *Urtica dioica* (Stinging nettle) L. (Urticaceae)

EYÜP BAĞCI
Fırat University, Faculty of Ars & Sciences, Department of Biology, Elazığ- Turkey

1. INTRODUCTION

There is growing interest in medicinal botanicals as part of complemantary medicine all over the world. In the United states, particularly both physicians and consumers are becoming aware of the use of herbals by native american societies[1]. Urtica dioica is one of these plants used for this purpose. In the last decade stinging nettle have been frequently used as a folk medicinal plant against cancer particularly in Turkey. It is reported that since these treatments are sometimes costly and have questionably efficacy and toxicity, proper scientific trials are needed to clarify whether such methods have a real role in cancer management[2].

The common wild plant nettle, especially *Urtica dioica*, is one of the most potent plants in producing direct irritation to the skin (urticaria)[3].

Urticaceae family has two genus, *Urtica* L. and *Parietaria* L.. *Urtica* genus is represented with 5 species growing naturally in Turkey according to Flora of Turkey. The various small species described as vicariads of this plant, as well as the numerous infraspesific taxa recorded from turkey and elsewhere, the genus need monographic treatment[4].

The folium Urticae has been using for a long time as medicinal plant. Baytop[5] reported that the leaves of *U.dioica* and *U. urens* have been using in Turkey and particularly *Urtica dioica* have been selling in the markets.

In the medical world there is a believe that this plant stimulates the metabolism in human body. It is used in tea as a medicine against rheumatism, gout, gall and liverproblems, spring and autumn cures.

Fatty acid containing lipids of vegatative plant parts consist mainly of glycolipids and phospholipids[6]. Somatic lipids are generally composed of commonplace fatty acids, but the acid of stored lipid sometimes contains unusual components which invite classification[6]. The fatty acid composition of whole leaves or their chloroplasts of plants show a remarkably consistent pattern over a wide variety of plant families, but the leaf cuticular waxes show variations with taxonomic possibilities[7].

It is reported that[8] the fatty acids with unusual structures often accumulate in seed oils. They are usually not present or present in much lower concentrations only, in the lipid of green leaves[9,10].

There are many study on the *Urtica dioica* chemicals[11-13]. In this study, it is aimed that to determine the fatty acid composition of the aerial parts of the *Urtica dioica* naturally growing in Turkey by using GC techniques. Thus the results obtained from the analysis were assessed in view of natuural product, medicinal plant and chemotaxonomy.

2. EXPERIMENTAL

In this research, the aerial parts of *Urtica dioica* were collected from natural habitats, Adana – Bahçe region, 1300 m.

2.1 Extraction of Lipids

5 gr. vegatative parts were homogenized with isopropanol. The latter procedure - extraction and purification - was made in according to Kates[14] method. Methylation was carried out according to Christie[15].

2.2 Gas Chromatography

In the GC. analysis of fatty acid methyl esters, the resultant mixture of fatty acid methyl ester was injected onto a UNICAM – 610 GC. FID detector, carrier gas (N, ml /min 2.5), capillary column (15 m X 0.32 mm) packed with 70% (BPX- 70) were used for analysis. The heat of column was 185 ^{0}C. 1 µl. was introduced onto the column. The quantity of fatty acid were calculated as % percentage (Tablo 1). Identification of the fatty acid components was carried out with help of authentic standart and retention time of components used in the fatty acid analysis. For all samples the procedure was performed in triplicate and the mean values were stated.

3. RESULTS AND DISCUSSION

Fatty acid composition of the aerial parts of *Urtica dioica* was shown in Table 1.

In the study, 18:3 (linolenic acid) was found as the major fatty acid component in the aerial parts of the nettle. 18:2 (linoleic acid -20.25) also is the second major fatty acid component relatively. In addition to these, palmitic acid (16:0 - 17.71) and (16:2 - 3.2) palmitoleic acid has large concentrations. (Table 1). The major fatty acid components of the *Urtica dioica* were oleic acid (8.83), lauric (3.47), myristic (1.79) and stearic acid (1.11).

In Antonopoulou et al.,[3] study on the fatty acid of *Urtica dioica* vegatative parts, four main classes of Phospholipids (i.e, posphatidyllinositol, phosphatidylethanolamine, phosphatidylcholine and lysophos-photidylcholine) were identified. They reported that a phospholipid that induced platelet aggregation was identified as platelet activating factor on the basis of biological, chemical and spectral methods.

Urtica dioica is one of the plant known as plant dermatits [16]. The other species within same genus, *Urtica pilulifera* takes place among the popular ethnobotanical plant in the Palestinian area, like in Turkey [17].

Table 1. Fatty acid composition of the aerial parts of *Urtica dioica*.

Fatty Acid components	GLC– Area %
10:0	0.064
12:0	3.47
14:0	1.79
15:0	0.42
16:0	17.71
16:1	3.2
16:2	4.4
18:0	1.11
18:1	8.83
18:2	20.25
18:3	31.4
20:0	1.21
20:1	0.30
20:2	0.08
20:4	0.85

It is reported that the fatty acid composition of the leaves of higher plants follow a generally consistent pattern in which only slight quantitative variations occur between most classes of plant, quantitatively the major fatty acids of leaves are α – linoleic and palmitic acids[6]. Oleic and palmitoleic acids are the most abundant of leaf monoenoic acids while a third monoenoic acid trans – 3 – hexadecanoic acid is represents in a minor proportion of the total fatty acids[6]. In Guerrero and Garcia[18] study, on the fatty acid and caroten components of some edible wild plants leaves, consist of stinging nettle, GC analysis showed the major fatty acid to be 18:3, 18:2 and 16:0 in the oil.

Fatty acid composition of the leaves of stinging nettle, was found very similar with the Nichols et al.[19] study results. Especially the high content of the linolenic and linoleic and palmitic acid concentrations.

The results showed that the analyzed plant species are rich sources of essential fatty acids, omega 6 and 3 fatty acid (18:2, 18:3). Particularly the fatty acid composition of sample studied here was found very different with the seed fatty acid composition of *Urtica dioica* (Aitzetmuller, 1994; unpublished results). The fatty acid composition of this species was consist from the usual fatty acid especially common fatty acid both vegatative and seed analysis. It will be necessary to enlarge the study with the whole family (Urticaceae) and genus specimen especially seed samples to evaulate the results in view of chemotaxonomy.

REFERENCES

1. Borchers, A.T., Keen, CL., Stern, JS., Gershwin, ME., 2000, *American J. of Clinical Nutrition*, **72** (2), 339-47.
2. Samur, M., Bozcuk, H.S., Kara, A., Savaş, B., 2001, *Supportive Care in Cancer*, **9** (6), 452-8.
3. Antonopoulou, S, Demopoulos, C.A., Andrikopoulos, N.K., 1996, *J. of Agricultural and Food Chemistry*, **44**(10), 3052-6.
4. Davis, P.H., 1982, Flora of Turkey and the East Aegean Islands, Vol. 7, Edinburgh University Press, Edinburgh.
5. Baytop, T., 1984, Türkiye'de Bitkilerle Tedavi, İstanbul Üniversitesi, Yay.No. 3255, İstanbul.
6. Hitchcock, C., Nichols, B.W., 1971, Plant Lipid Biochemistry, Academic Press, p.387.
7. Mazliak, P., 1968, In *Progress in Phytochemistry* (Reinhold L., Liwschitz, Y., eds.) Vol. 1, Interscience, New York. pp, 49-111.
8. Aitzetmuller, K., 1995, *Pl. Syst. Evol.*, **9**, 229-40.
9. Colombo, M.L., Tome, F., Bugatti, C., 1991, *Pl. Syst. Evol.*, **178**, 55-63.
10. Aitzetmuller, K., 1993, *J. High Resol.Chromatograph*, **16**, 488-90.
11. Kraus, R., Spiteller, G., 1991, *Liebigs Annalen Der Chemie*, **2**, 125-8.
12. Basaran, A.A., Akbay, P., Undeger, U., Basaran, N., 2001, *Toxicology*, **164** (1-3),171-2.

13. Karakaya, S., El SN., Tas, A.A., 2001, *International J.of Food Sciences and Nutrition*, **52**(6),501-8.
14. Kates, M., 1986, Techniques of Lipidology, Isolating analysis and identification of lipids, Elsevier, New York, p. 105.
15. Christie, W.W., 1989, Gas Chromatography and Lipids, The Oily Press.
16. Govern, TW., Barkley, TM., 1998, *Cutis*, **62** (2), 63-4. 11. Kraus, R., Spiteller, G., 1991. Ceramides From *Urtica dioica* roots. Liebigs Annalen Der Chemie. (2): 125-128.
17. Shtayeh, A. MS., Yaniv, Z., Mahajana, J., 2000, *J. of Ethnopharmocology*, **73** (1-2), 221-32.
18. Guerrero, G.L.J., Garcia, R.I., 1999, *European Food Research and Technology*, **209** (5), 313-6.
19. Nichols, B.,W., Stubbs, J. M. And James, A.T., 1967a, In *Biochemistry of Chloroplasts* (Goodwin, T.W., ed.), Vol. II, Academic Press, London and New York, pp. 677-90.

Fatty Acid Composition of *Aconitum orientale* Miller and *A. nasutum* Fisch. ex Reichb Seeds, A Chemotaxonomic Approach

EYÜP BAĞCI[1] and HASAN ÖZÇELİK[2]
[1] *Fırat University, Faculty of Arts & Sciences, Department of Biology, Elazığ, Turkey*
[2] *Süleyman Demirel., Faculty of Arts & Sciences, Department of Biology, Isparta, Turkey*

1. INTRODUCTION

Aconitum L. genus (Ranunculaceae) is represented with 3 species in Flora of Turkey[1]. They are *Aconitum orientale* Miller, *Aconitum nasutum* Fisch. ex Reichb. and *A. cochleare* Woroschin. In recent years, there are new addition to this number[2]. Genus *Aconitum* have been also represented in the Delphinideae Tribus with the *Consolida* and *Delphinium* genus, according to Tahtajan[3], Tamura[4] and Jensen & al[5]. Davis[1], reported that *Aconitum* L. is a very critical genus and has extremely toxic plants, many of whose species show more variability than they are often credited with. Utelli et al.[6], reported that the systematics of the yellow – flowered *Aconitum lycoctonum* species complex (Ranunculaceae) growing in Europa has long been considered difficult because of high morphological variability and hypothesized hybridization.

It has been demonstrated that the content and composition of fatty acids of seed lipids can serve as taxonomic markers in higher plants[7,8]. Fatty acids occur in high concentrations principally in plant seeds and fruit coats, where they may constitute up to 50 % of the dry weight. Variation in the fatty acids of seed fats has long been a point of interest to chemotaxonomists[8]. There are large differences between seed oil fatty acid patterns of various genera in the Ranunculaceae [9,10]. These may be caused by the presence and absence of chain – elongating enzyme systems (elongases) and by the presence and absence of several types desaturateses operating at the carbon atoms number five and six (or six and seven) of the fatty acid chain. Many members of Ranunculaceae contain unusual fatty acids in their seed oils. This leads to rather typical genus – spesific fatty acid patterns or fingerprints in these seed oils[11].

Since the early work of Kauffman & Barve[12], Bagby et al.[13], and Smith et al.[14], seed oils from representatives of the Ranunculaceae have attracted the attention of fatty acid researchers. This is from the wealth of different and unusual fatty acid structures that can be found at rather high levels in these seed oils[15].

Some chemicals, particularly terpenoid constituents of the *Aconitum* species have been studying for a long time by some researchers[16-20]. However, There are some systematic study on the genus[6,21]. Extracts of the plant *Aconitum* species are used in traditional chinese medicine predominantly as anti – inflammatory and analgesic agents and it is important with the aconitine alkaloids all over the world and in Turkey[18,22,23].

In this study, we aimed to determine the fatty acid composition of two *Aconitum* species naturally growing in Turkey, *A.orientale* and *A. nasutum*, by using GC techniques and to help to bringing out the chemotaxonomic relationships among the genera within the family Ranunculaceae.

2. EXPERIMENTAL

2.1 Seed samples

Aconitum nasutum seed samples were taken from Kars – Sarıkamış – Şenkaya (Erzurum) road, 20 km. to Sarıkamış. 2400 m. Özçelik 8626. *Aconitum orientale*: Gümüşhane - Artabel, alpine grass, 2000- 2150 m., Özçelik 8743.

2.2 Extraction of Lipids

5 gr. seed samples were homogenized with isopropanol. The latter procedure - extraction and purification - was made in according to Kates[24] method. Methylation were carried out according to Christie[25].

2.3 Gas Chromatography

The GC. analysis of fatty acid methyl esters (FAME), was carried out in F.Ü., Engineering Faculty, Chemistry Engineering Department. The resultant mixture of fatty acid methyl ester was injected onto a UNICAM – 610 GC. FID detector, carrier gas (N, ml /min- 2.5), capillary column (15 m x 0.32 mm) packed with 70% (BPX- 70) were used for analysis. The heat of column was 185 ^{0}C. 1 µl. was introduced onto the column. The quantity of fatty acid were calculated as % percentage (Table 1). Identification of the fatty acid components was carried out with help of authentic standart and retention time of components used in the fatty acid analysis. For all samples the procedure was performed in triplicate and the mean values were stated.

3. RESULTS AND DISCUSSION

The results of the seed oils fatty acid analysis of the *Aconitum nasutum* and *A. orientale* were shown in Table 1.

In the GC analysis, Linoleic acid (18:2) was found as major fatty acid component in the both *Aconitum* seed oil. Linoleic acid has comprised approximetely 50 % percentage of the oils. Oleic acid (18:1) was the second major fatty acid found in the oils. The quantity of Linoleic acid was 58.70 % in *A. orientale* and was 49.25 % in *A. nasutum*. In the same way, oleic acid was 27.77 % in *A. orientale* and was 33.03 % in *A. nasutum* oils.

The fatty acid composition of the *Aconitum* species studied in this study had in general usual fatty acid composition, they were not showed unusal fatty acid patterns. This is reported in the other *Aconitum* species [10,11].

The rest of the fatty acid components of *Aconitum* species studied here, had very low concentrations in the oils. While the palmitic acid was found as third major fatty acid component, myristic acid had lower concentrations in the both seed oil (Table 1). It is the same level with the *Aconitum napellus* and *A. turczaninovii*[11]. The others especially 20:0 groups poly-saturated and unsaturated- fatty acid components were not found or at the minimum concentrations determined. It is easy to say *Aconitum* seed oils studied has usual fatty acid composition and common fatty acids consists of the more of oil in the both species.

The fatty acid results obtained in the analysis were showed high congruence with the other *Aconitum* seed oils. In Aitzetmuller et al.[11] study, with different *Aconitum* species - *Aconitum barbatum, A. paniculatum, A. ranunculifolium, A. septentrionale, A. napellus*- oleic and linoleic acid components were found as the major fatty acid. 20:0 and unsaturated forms were determined in a low level or not found.

Utelli et al.[6] stated that there was high morphological variability within and among populations of the European *Aconitum* species and the morphological characters have no value as systematic characters. Systematic confusions on the *Aconitum*, have been reported in the Mucher[21] study with the chorology of *Aconitum* taxa in Europe.

Aitzetmuller et al.[11], reported that species of *Aconitum* do not contain fatty acids with 20 carbon atoms. On the other hand, same family genera, *Delphinium, Consolida, Nigella* and *Helleborus* species have contained C20 atoms. They found consistent C20 fatty acid pattern of 20:0 and the unsaturated forms of this fatty acid (20:1, 20: 2, 20:3) in the other genera of Ranunculaceae. They suggested that the chain elongation in the Ranunculaceae, 20:0 and long chain fatty acids had been lost only in the evolution of *Aconitum* (Aitzetmuller, et al.,[11]. Our results was supported this hypothesis.

It is known that a better knowledge of the seed fatty acid patterns could yield taxonomically useful results such as in the other plant family and genus [7,10].

The seed oil fatty acid patterns of Ranunculaceae were highly correlated with plant genera[7,10,26,27]. But the pattern diferences from genus to genus were very large, whereas those between different species of the same genus were often rather small, with only very few exceptions. Genus - to - genus differences in Ranunculaceae are larger than most family – to family differences elsewhere in the plant kingdom[10]. It is requires to enlarge the number of the studied species of this family and genus members to support this hypothesis. Our studies on the other plant genus in the Ranunculaceae and the others have been continued.

The fatty acid composition of the plants have contributed more knowledge on the phylogeny of the various genera and family. For example, From the point of fatty acid analysis, Aitzetmuller[10], reported that *Aquilegia* and *Thalictrum* were closely related because both contain highly unusual 18:3 trans fatty acid, Columbinic acid (18:3 δ5- trans, 9 cis, 12 cis). It is possible to find very strong clues to obtain some approachs to the phylogeny of plant species by using some chemicals like fatty acids. Aitzetmuller et al.[11], showed a chemotaxonomical relation of genera in the family Ranunculaceae, based solely on information obtained from seed oil fatty acid fingerprints.

Table 1. Fatty acid compositions of studied *Aconitum* species.

Fatty acid components	Percentage (%)	
	Aconitum orientale	*Aconitum nasutum*
Lauric acid 12:0	--	--
Myristic acid 14:0	3.10	3.37
Pentadecanoic acid 15:0	--	--
Palmitic acid 16:0	4.32	5.28
Palmitoleic acid 16:1	0.18	0.10
Margaric acid 17:0	--	--
Stearic acid 18:0	0.91	1.97
Oleic acid 18:1(α- 9)	27.77	33.03
Linoleic acid 18:2(α- 9,12)	58.70	49.25
Linolenic acid 18:3(α- 9,12,15)	2.21	4.75
Arachidic acid 20:0	0.11	0.09
Arachidik acid 20:1	0.08	0.10
Behenic acid 22:0	0.04	--

4. REFERENCES

1. Davis, P.H, 1982, Flora of Turkey and the East Aegean Islands, Vol. 7, Edinburgh University Press, Edinburgh.
2. Davis, P.H., 1988, Flora of Turkey and the East Aegean Islands, Vol. 10, Edinburgh University Press, Edinburgh.
3. Tahtajan, A., 1987, Systemea Magnoliophytorum, Leningrad, Nauka.
4. Tamura, M., 1993, Ranunculaceae, In *The Families and Genera of Vascular Plants* (Kubitzki, K.,ed.), Vol.2, Springer, Berlin.
5. Jensen, U., Hoot, S.B., Johannsen, J.T., Kosuge, K., 1995, *Pl. Syst. Evol. Suppl.*, **9**, 273-80.
6. Utelli, AB., Roy., BA., Baltisberger, M., 2000, *Plant Systematics and Evolution*, **224** (3-4),195-212.
7. Hegnauer, R, 1989, Chemotaxonomie der Pflanzen, Vol.VIII, Birkhuser, Basel, pp. 611-2.
8. Harborne, J.B., Turner, B.L., 1984, Plant Chemosystematics, Academic Press, London, pp. 180-91.
9. Aitzetmuller, K., Tsevegsüren, N., 1994, *J. Plant Physiol.*, **143**, 538-43.
10. Aitzetmuller, K., 1995, *Pl. Syst. Evol.*, **9**, 229- 40.
11. Aitzetmuller, K., - Tsevegsuren, N., Werner, G., 1999, *Pl. Syst. and Evol.*, **251**, 37-47.
12. Kauffman, H.P., Barve, J., 1965, *Fette, Seifen, Antrichmittel.*, **67**, 14-6.
13. Bagby, M.O., Smith, J.R., Mikolajczak, K.L., Wolff, I.A., 1962, *Biochemistry*, **1**, 632-9.
14. Smith, C.R., Kleiman, R., Wollf, I.A., 1968, *Lipids*, **3**, 37-42.
15. Aitzetmuller, K., 1996b, Seed fatty acids, chemotaxonomy and renewable sources, In *Oils – Fats- Lipids*, Proceedings of the 21st. World Congress of the International Society For fat Research, High Wycombe, Barnes, pp. 117-20.
16. Yue, J., Xu, J., Zhao, Q., Sun, H., Chen, Y., 1996, *J. Nat. Prod.*, **59** (3), 277-9.

17. Liu, H., Katz, A., 1996, *J. Nat. Prod.*, **59** (2), 135-8.
18. Ameri, A., 1998, *Archives of Pharmacology,* **357** (6), 585-92.
19. Fico, G., Braca, A., Bilia, A.R.,Tome, F., Morelli, I., 2000, *J. Nat. Prod.*, **63**(11),1563-5.
20. Rahman, A.-ur., Akhtar, F., Choudhary, M.I., Khalid, A., 2000, *J. Nat. Prod.*, **63** (10), 1393-5.
21. Mucher,W.,1993, *Annales Rei- Botanicae*, **33** (1), 51-76.
22. Baytop,T.,1984,Türkiye'de Bitkilerle Tedavi (Geçmişte ve Bugün),İstanbul Üniversitesi, Yay.No: 3255, İstanbul.
23. Gutser, U.T., Friese, J., Heubach, J.F., Matthiesen, T., Selve, N., Wilfert, B., Gleitz, J., 1997, *Archives of Pharmacology*, **357** (1), 39-48.
24. Kates, M., 1986, Techniques of Lipidology, Isolating analysis and identification of Lipids, Elsevier, New York, p. 105.
25. Christie, W.W., 1989, Gas Chromatography and Lipids, The Oily Press.
26. Aitzetmuller, K., Tsevegsüren, N., Ivanov, S.A., 1997, *Lipids,* **4**,385-8.
27. Aitzetmuller, K., 1993, *J. High Resol.Chromatograph*, **16**, 488-90.

New Peptide from a Bacterium Associated with Marine Sponge *Ircinia muscarum*

SALVATORE DE ROSA[1], MAYA MITOVA[2], SALVATORE DE CARO[1] and GIUSEPPINA TOMMONARO[1]
[1]*Istituto di Chimica Biomolecole CNR, via Campi Flegrei, 34, I-80078 Pozzuoli (Napoli) Italy;*
[2]*Institute of Organic Chemistry with Centre of Phytochemistry, Bulgarian Academy of Sciences, Sofia 1113, Bulgaria*

1. INTRODUCTION

Microorganisms are a rich source of new metabolites with a wide variety of biological activities, and some of them display significant practical applications[1,2]. Marine microorganisms are of considerable current interest as a new and promising source of biologically active compounds. They produce a variety of metabolites, some of which can be used for drug development[2,3]. It is also shown that some bioactive compounds, isolated from invertebrates originate from symbiotic microorganisms (e.g. tetrodotoxin, saxitoxin, okadaic acid, surugatoxins, etc.)[3-5]. Since the surfaces and internal spaces of Porifera are more nutrients rich than seawater and most sediment, they are a unique niche for the isolation of diverse microorganisms. In extreme case, bacteria occupy more than 40% of the tissue volume of a sponge[2,6]. It is therefore reasonable to believe that symbiotic microorganisms produce some sponge metabolites. Peptides, isolated from sponges, are also suspected to be of microbial origin from the presence of both d amino acids and unusual amino acids. Although that less 5% of the bacteria observed by microscopic method in the marine sponge are isolated and chemically characterised. This fact limits our possibility for explorations for new bioactive metabolites.

In the last two decades several peptides are isolated from marine organisms, because there are several reasons for progress in the chemistry of marine peptides: (1) development of reverse-phase HPLC enabled the isolation of peptides from a mixture of related compounds; (2) advances in spectroscopy, especially 2D-NMR and FAB mass spectrometry; (3) progress in chiral chromatography allowed the assignment of absolute stereochemistry of amino acids with small amounts of material.

Continuing our search for microorganisms associated with marine organisms, we have previously isolated a bacterium associated with the marine sponge *Dysidea fragilis*, from the Black Sea.[7,8] We now report the isolation of a bacterium, identified as *Pseudomonas* sp. (IM-1), from the marine sponge *Ircinia muscarum*, which was absent in the surrounding water. We succeeded in cultivating this bacterium and from the *n*-butanol extract of the culture, after removal of the bacteria, was isolated a new cyclic tetrapeptide (**1**).

2. RESULT AND DISCUSSION

A specimen of *Ircinia muscarum*, collected at 20 m dept in the gulf of Naples, was used for the cultivation of bacteria. A strain of bacterium *Pseudomonas* (IM-1) was isolated only from the sponge, but not in the surrounding water. The negligible changes in the bacterial composition through two years of cultivation without a sponge material is an indication that this bacterium is not a symbiont, but is only specifically associated with the tissues of *I. muscarum*. The bacterium was grown in Marine broth with a salt concentration of 35‰ at 37 °C, pH 7.5 for 48 h.

1

The n-butanol extract of culture (4 L), after removal of bacteria, was purified repeatedly by lobar C-18, followed by Sephadex LH-20 (MeOH) columns to give **1** as amorphous solid.

Table 1. NMR Spectra data of compound **1** in CD_3OD solution[a].

	^{13}C	^1H	HMBC(J_{C-H} = 10 Hz)
4-OH-Pro1			
αCH	58.7	4.56 dd (11.2, 6.5)	Hβ, Hγ, Hδ
βCH$_2$	38.1	2.32 ddd (13.4, 6.5, 2.5)	Hα, Hδ
		2.13 ddd (13.4, 11.2, 4.3)	
γCH	69.1	4.50 m	Hβ, Hδ
δ CH$_2$	55.2	3.49 ddd (12.2, 7.0, 5.2)	Hβ, Hγ
CO	173.1	--	Hα, Hβ
Leu1			
αCH	56.9	3.91 dd (9.6, 5.6)	Hβ, Hγ
βCH$_2$	43.3	1.72 ddd (14.6, 9.6, 5.5)	Hα, Hγ, Hδ
		1.63 ddd (14.6, 8.3, 5.6)	
γCH	25.6	1.82 m	Hα, Hβ, Hδ
δCH$_3$	23.2	0.99 d (6.6)	Hβ
δCH$_3$	21.9	1.03 d (6.6)	Hβ
CO	169.5	--	Hα, Hβ, Hδ (OH-Pro1)
4-OH-Pro2			
αCH	57.5	4.39 dd (8.1, 7.9)	Hβ, Hγ, Hδ
βCH$_2$	37.6	2.52 ddd (13.7, 8.1, 5.5)	Hα, Hδ
		2.27 ddd (13.7, 7.9, 2.4)	
γCH	68.8	4.46 m	Hβ, Hδ
δ CH$_2$	54.2	3.70 ddd (12.7, 4.3, 3.0)	Hβ, Hγ
CO	171.2	--	Hα, Hβ, Hα (Leu1)
Leu2			
αCH	54.6	4.21 dd (5.5, 5.3)	Hβ
βCH$_2$	39.3	1.94 m, 1.55 m	Hα, Hγ, Hδ
γCH	25.8	1.94 m	Hα, Hβ, Hδ
δCH$_3$	23.2	1.00 d (6.4)	Hβ
δCH$_3$	22.2	1.00 d (6.4)	Hβ
CO	169.0	--	Hα, Hβ, Hδ (OH-Pro2)

[a] Chemical shifts are referred to residual CD_3OD resonance. Multiplicities are indicated by usual symbols. Coupling constants (Hz) are in parentheses.

Compound **1** had $[α]_D$ − 18.1° (c = 0.011, MeOH) and showed a pseudomolecular ion peak at m/z 453.2714 (M + H$^+$, calculated 453.2704) in the HRFABMS (positive ion) spectrum, consistent with a molecular formula $C_{22}H_{36}N_4O_6$. The peptide nature of this compound was suggested by the molecular formula itself and from analysis of its ^{1}H- and ^{13}C-NMR spectra. Detailed analyses of the ^{1}H- and ^{13}C-NMR spectral data (Table 1) for **1**, with the aid of COSY-45, TOCSY, HMQC and HMBC spectra, established the presence of two leucine and two 4-OH-proline residues. This amino acid composition accounted for 6 out of the 7 degrees of unsaturation, requiring that **1** is a cyclic tetrapeptide. The amino acid composition was confirmed by HPLC (PICO-TAG) analysis of the acid hydrolysate of **1** after derivatization with phenylisothiocyanate (PITC) which revealed the presence of Leu and 4-OH-Pro.

The amino acid sequence of compound **1** was deduced by a detailed interpretation of HMBC spectrum that allowed us to assign the carbonyl signals and the sequence of the four amino acid units. The signals at lowest field (δ 173.1 and 171.2) were assigned to the carbonyl of the 4-OH-Pro units, while the remaining signals (δ 169.5 and 169.0) to the two Leu units. HMBC correlations observed between the δ methylene protons of 4-OH-Pro 1 (δ 3.49) and the carbonyl at δ 169.5 (Leu 1), α-proton of Leu 1 (δ 3.91) and the carbonyl of the second 4-OH-Pro unit (δ 171.2), and the δ methylene protons of 4-OH-Pro 2 (δ 3.70) and the carbonyl at δ 169.0 (Leu 2) defined the amino acid sequence.

The stereochemistry of the amino acid was determined by chiral HPLC analysis of the acid hydrolysate, which gave L-Leu, *trans*-4-OH-L-Pro and *cis*-4-OH-D-Pro.

The contemporary presence of both L-*trans*- and D-*cis*-4-OH-Pro, in the peptide **1**, showed the non-stereospecificity in the production of 4-OH-Pro by enzymatic pathway. Furthermore these results are further evidence that marine microorganisms produce peptide with unusual amino acid residues

3. EXPERIMENTAL SECTION

3.1 General Experimental Procedures

Optical rotation was measured on a JASCO DIP 370 polarimeter, using a 10-cm microcell. FABSMS were obtained on a VG-ZAB instrument equipped with a FAB source, using glycerol as a matrix. ^{1}H- and ^{13}C-NMR spectra were recorded at 500 and 125 MHz, respectively, on a Bruker AMX-500 spectrometer in CD_3OD, using the residual CD_3OD resonance at 3.48 ppm and 49.0 ppm as internal references, respectively.

3.2 Cultivation and Isolation

Ircinia muscarum (Dictyoceratida, Spongiidae) collected in the bay of Naples (Italy) at a depth of 20 m, was covered by seawater and transported immediately in the laboratory. Water samples were collected at the same time and place. The sponge was cut into small pieces and ground in a porcelain mortar. The homogenates obtained were diluted in Marine broth (2216 DIFCO) with a salt concentration of 35‰. Different dilutions were used: 10^{-1} (0.1 ml water or homogenate + 0.9 ml broth) 10^{-2}, 10^{-3} and 10^{-4}. We transferred 0.1 ml from every dilution in two Petri dishes containing marine broth agar with a final concentration of 0.35% sodium chloride. Incubation at 37°C for 24 h was performed and the bacterial isolates were repeatedly streaked to fresh nutrient agar plates in order to insure the purity of culture. Each strain obtained was placed in a separate flask and cultivated (37°C, marine broth, pH 7.5). We observed no qualitative differences in the bacterial composition in the interval 20-45°C. We worked at 37°C because the production of the bacterial biomass was more intensive at these conditions. IM-1 (a voucher specimen of the strain is maintained in the ICMIB-CNR collection) was identified as *Pseudomonas* sp.. The characterization was done by routine biochemical tests (API 20E, BioMérieux), modified for marine bacteria. Cultures (4 L) were incubated at 37 °C for 48 h, under shaking. The bacteria were isolated by centrifugation at 6000 rpm, for 15 min. The culture, after elimination of bacterium, was extracted first with EtOAc (300 ml x 3) and after that with *n*-butanol (300 ml x 3) and evaporated under reduced pressure. The *n*-butanol extract was chromatographed on a lobar C-18 column with H_2O/MeOH gradient. The fraction that eluted with 50% of MeOH was subjected to a Sephadex LH-20 column (4 x 100 cm) using MeOH (2 ml/min), to give the compound **1** (20 mg).

3.3 Compound 1

Compound **1**: white amorphous solid; $[\alpha]_D$ –18.1° (c = 0.011, MeOH); NMR data are reported in Table 1; HRFABMS *m/z* (%) 453.2714 (48) (M + H)$^+$ (calculated for $C_{22}H_{37}N_4O_6$, 453.2704), 341 (70), 227 (100).

3.4 Amino acid analysis

Compound **1** (200 µg) was hydrolysed with 6N HCl (500 µL) at 110 °C for 16 h. The acid hydrolysate, divided into two parts, was dried under N_2. A part, after derivatization with phenylisothiocyanate (PITC), was used for amino acid analysis by PICO-TAG (Waters). Leu and 4-OH-Pro were identified by co-injection with standard PTC amino acid derivatives. The second part was subjected to chiral HPLC (Phenomenex Chirex D-Penicillamine; 250 x 4.6 mm; 30% MeOH in 2 mM $CuSO_4$, 1 ml/min) detected at 254 nm. L-Leu, *trans*-4-OH-L-Pro and *cis*-4-OH-D-Pro were identified by co-injection with standard amino acids.

ACKNOWLEDGEMENT

This research was supported by CNR-Rome. One of the authors (M.M.) acknowledges an Italian Minister of Foreign Affairs fellowship.

REFERENCES

1. Pietra, F., 1997, *Nat. Prod. Report*, **14**, 453-64.
2. Fenical, W., 1993, *Chem. Rev.*, **93**, 1673-83.
3. Kobayashi, J., Ishibashi, M., 1993, *Chem. Rev.*, **93**, 1753-69.
4. Nakamura, H., Kishi, Y., Shimomura, O., Morse, D., Hastings, J.W., 1989, *J. Amer. Chem. Soc.*, **111**, 7607-11.
5. Schmitz, F.J., Prasad, R.S., Gopichand, Y., Hossain, M.B., Van der helm, D., Schmidt, P., 1981 *J. Amer. Chem. Soc.*, **103**, 2467-9.
6. Kobayashi, M., Kitagawa, I., 1994, *Pure & Appl. Chem.*, **66**, 819-26.
7. De Rosa, S., Milone, A., Kujumgiev, A., Stefanov, K., Nechev, I., Popov, S., 2000, *Comp. Biochem. Physiol.*, **126B**, 391-6.
8. De Rosa, S., De Giulio, A., Tommonaro, G., Popov, S., Kujumgiev, A., 2000, *J. Nat. Prod.*, **63**, 1454-5.

Sheep Brain Glutathione Reductase: Purification and Some Properties

N. LEYLA AÇAN and E. FERHAN TEZCAN
Hacettepe University, Faculty of Medicine, Department of Biochemistry, 06100 Ankara, Turkey

1. INTRODUCTION

Glutathione Reductase [GSSGR, NAD(P)H:Oxidized Glutathione Oxidoreductase, EC1.6.4.2] catalyzes the NADPH-dependent reduction of glutathione disulfide (GSSG). The enzyme was purified from several sources including yeast[1], mouse liver[2], sheep[3] and bovine brain[4], human erythrocytes[5] and jejunum[6]; and molecular and kinetic properties were elucidated.

GSSGR is responsible for maintaining glutathione in reduced state (GSH). By coupling with glutathione peroxidase reaction, it prevents cellular damage by oxidative stress. The importance of GSSGR in brain lies in the fact that brain is the organ with the highest risk of oxidative damage[7]. This article summarizes the purification and some molecular properties of sheep brain GSSGR.

2. MATERIAL AND METHOD

Fresh sheep brains were obtained from an Ankara slaughterhouse. The purification included ammonium sulfate fractionation (35-55%), heat denaturation (1 h at 65°C), 2',5'-ADP Sepharose 4B and Sephadex G-200 chromatography steps. GSSGR assays were carried out spectrophotometrically following the decrease in absorbance at 340 nm at 37°C.

A unit of activity is defined as the oxidation of 1 µmol of NADPH per min under the assay conditions[3].

3. RESULTS AND DISCUSSION

3.1 Purification and molecular properties

With the procedure described, about 11,000 fold of purification was achieved, with an overall yield of 40%[3]. Specific activity at the final step was 190 IU/mg and a single band was obtained at SDS-PAGE (Fig. 1). M_r of the purified enzyme was 116,000. The enzyme is a dimer of identical subunits, each containing a tightly bound FAD. The dissociation constant for FAD was estimated as 13 nM[8]. The molecular weights of the enzyme from other sources are within the range of 100-125 kDa[1-6]. From SDS-PAGE, subunit molecular weight was calculated as 64 kDa. This value is somehow greater than that of bovine brain, which was reported to be 55 kDa[4].

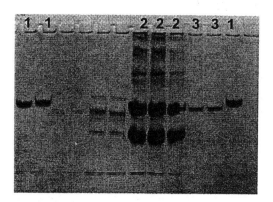

Figure 1. SDS-PAGE of purified GSSG: 1) Bovine albumin, 66,000; 2) Myosin, 205,000; β-Galactosidase, 116,000; Phosphorylase B, 97,000; Bovine albumin, 66,000; Egg albumin, 45,000; 3) Sheep brain GSSGR. (Stain: Coomassie Blue R250)

3.2 Kinetic properties

The optimum pH of the enzyme for NADPH and NADH are 6.8[3] and 5.2[9] respectively. The preferred substrate is NADPH. With NADH, specific activity reduced to 8 IU/mg[9]. The enzyme functions essentially irreversibly with a specific activity for the reverse reaction as 1 IU/mg when $NADP^+$ is the substrate[3].

GSSGR with its substrates NADPH and GSSG showed Ping-Pong kinetics with double substrate inhibition in the forward direction[10]. Km values for NADPH and GSSG were found to be 60.9 and 116.9 µM and Ki values were found to be 42.1 and 347.3 µM respectively. $NADP^+$ inhibition at low fixed concentration of NADPH was mixed type with a Ki of 281.5 µM and α of 0.048. It is concluded that the enzyme shows a hybrid Ping-Pong-ordered branched mechanism as proposed by Mannervik[11]. The reported Km values for NADPH range between 8-60.9 µM, and that for GSSG between 26-116.9 $µM^{1-6}$. For bovine brain those values are 8.1 and 36 µM respectively[4]. The differences in Km values are attributed to the differences in the ionic strength and the pH of the solutions used during measurements[5]. The kinetic behaviour of the enzyme with NADH was fitted into "ordered bi bi" mechanism with Km of 1011 µM and Ki of 207.8 µM for NADH[9]. Cadmium was found to be a competitive inhibitor with respect to GSSG and uncompetitive with respect to NADPH[12]. Similar results were reported for yeast GSSGR[13].

Although three dimensional structure of human erythrocyte GSSGR was elucidated[14] and a cDNA for mammalian enzyme has been cloned[15], amino acid sequence and the three dimensional structure of the enzyme from sheep brain remains to be elucidated.

REFERENCES

1. Massey, V., Williams Jr, C.H., 1965, *Biol. Chem.*, **40**, 4470-80.
2. Lopez-Barea, J., Lee, C.Y., 1979, *Eur. J. Biochem.*, **98**, 487-99.
3. Açan, N.L., Tezcan, E.F., 1989, *FEBS Lett.*, **250**, 72-4.
4. Gutterer, J.M., Dringen, R., Hirrlinger, J., Hamprecht, B., 1999, *J. Neurochem.*, **73**, 1422-30.
5. Worthington, D.J., Rosemeyer, M.A., 1975, *Eur. J. Biochem.*, **60**, 459-66.
6. Ogus, H., Ozer, N., 1991, *Biochem. Med. Metab. Biol.*, **45**, 65-73.
7. Dringen, R., 2000, *Prog. Neurobiol.*, **62**, 649-71.
8. Acan, N.L., Tezcan, E.F., 1991, In *Flavins and Flavoproteins* (Curti, B., Ronchi, S., Zanetti, G., eds.), Walter de Gruyter, Berlin, pp. 545-8.
9. Açan, N.L., Tezcan, E.F., 1994, *Dok. Eyl. Un. Tip Fak. Derg.*, **8**(2), 8-14.
10. Açan, N.L., Tezcan, E.F., 1991, *Enzyme,* **45**, 121-4.
11. Mannervik, B., 1973, *Biochem. Biophys. Res. Commun.*, **53**, 1151-8.
12. Açan, N.L., Tezcan, E.F., 1995, *Bioche. Mol Med.*, **54**, 33-7.
13. Serafini, M.T., Romeu, A., Arola, L.I., 1989, *Biochem. Int.*, **18**, 793-802.
14. Karplus, P.A., Schulz, G.E., 1987, *J. Mol. Biol.*, **195**, 701-29.
15. Tutic, M., Lu, X., Schirmer, R.H., Werner, D., 1990, *Eur. J. Biochem.*, **188**, 523-8.

Some Morphological and Phenological Characters of Tobacco (*Nicotiana tabacum* L.) Grown in Hatay Province of Turkey

DURMUŞ ALPASLAN KAYA and FİLİZ AYANOĞLU
Mustafa Kemal University, Faculty of Agriculture, Department of Field Crops, Antakya, Turkey

1. INTRODUCTION

In Hatay province the production of Turkish style tobacco is overall conducted with indigenous types. In the districts of Altınözü and Yayladağı where intense cultivation exists, mostly the types of Çakaldere and Küpeli as well as the types of Boynu Eğri, Kısa Bodur, Sarı Tütün and Kırmızı Çiçek are cultivated[1].

The aim of this study is to determine the morphological properties of the tobacco types, which are cultivated in the conditions of Hatay province. This will also be a step for production with qualified types and definition of the necessary materials for the betterment process will be enhanced.

2. MATERIALS AND METHODS

The experiment is conducted in the districts of Hatay province where Nicotiana tabacum L. species are cultivated in 1999 in total number of 126 cultivation fields and the distribution of the cultivation fields according to the districts is as follows: Yayladağı 54, Altınözü 32, İskenderun 16 and Antakya 24. The determination of the field, which will be subject to observation and evaluation, is done by taking account of the overall samples of the region as much as possible.

Habitus, the angle between the leaf and stalk, palm shape, leaf base, the appendages of the leaf base, leaf tip shape, status of inflorescence are examined and classified according to the CORESTA standards[2]. Besides, the features of the leaf surface, flower colour, plant height, stalk diameter, the length between internode, the number of the leaves, leaf area, leaf length, leaf width, diametrical ratio, ovality coefficient are determined in the 10 of the plants which are chosen randomly in each field.

3. RESULTS AND DISCUSSION

In the study 13 different types of tobacco are identified which are in different names and have quite different features, in terms of the features examined in the experiment. These ecotypes are Çakaldere, Küpeli Çakaldere, Küpeli, Uzun Bodur, Kısa Bodur, Boynu Eğri, Kırmızı Çiçek, Samsun, Sivri Çakaldere, Ziraat Tohumu, Çandır Tohumu, Boz Tütün and Kısa Çakaldere.

The ellipsoid habitus is found dominantly (94.4%) in Hatay province. It is followed by cylindrical (4.2%) and conical (3.2%) habitus with a low ratio. According to the angle between the leaf and the stalk, the following ratios are obtained: 48.4 % is curved-vertical (35-45°), 44.4% is vertical (<35°) and 7.1% is curved (45-65°). According to the form of the leaf palm, various kinds of leaves are found in the following ratios: round elliptical (53%), reverse oval ox-tongue (27.0%), ox-tongue (17.0%), elliptical (2.0%) and elliptical ox-tongue (1.0%). In respect to shape of the leaf base, most of the population are veiled (90.0 %) and, it's followed by wide winged base (4.0 %), wide winged narrowed base (3.0 %), slightly veiled (2.0 %) and without veiled (1.0 %), respectively. For the appendages of leaf base, there is a two pieces prolonged (49.0 %) on a large scale and, it's followed by asymmetric prolonged (29.0 %), symmetric two pieces prolonged (13.0 %), curly leaf tip prolonged (6.0 %), asymmetric two-pieces prolonged (2.0 %) and auricle shaped (1.0 %), respectively. For the shape of the leaf tip, most of the population have moderate sharp leaf tip (57.0 %), and it's followed by tobaccos which have sharp leaf tip (18.0 %), slightly sharp leaf tip (17.0 %), very sharp leaf tip (6.0 %) and long sharp leaf tip (2.0 %), respectively. In respect to status of inflorescence, most of the population are semi-spherical (58.0 %) and spherical (39.0 %) and on a low scale conical (1.0 %), inverted conical (1.0 %) and hexagon (1.0 %). Leaf surface characteristics which were observed were blistered leaf surface (36 %), waved leaf surface (35 %) and flat leaf surface (29 %). According to their flower colors, the dominant color was found to be pink with 49 %. It was followed by light pink (30 %), red (8 %), dark pink (7 %) and white (6 %) respectively.

It was observed that tobacco population of Hatay region had a broad range of variation according to their plant height, plant steam diameter, internode length, leaf number, leaf surface area, leaf length, leaf width, diametrical ratio and ovality coefficient (Table 1). The leaf width which is the most striking characteristic to realize the wealth of the region tobacco population. Such a variation of leaf width leads to a variability in the leaf shapes, which was rarely observed in other regions. The leading characteristic of tobacco of the region was their elliptical ox-tongue shape. Diametrical ratio, is a hereditary characteristic so does not vary even leaf dimensions vary with the environmental conditions and is a measure used in the betterment of tobacco and also in the determination of leaf form[3,4].

Table 1. Morphological characteristics of the tobacco cultivated in Hatay Province

Characteristics	Minimum	Maximum	Mean
Plant Height (cm)	51.0	240.0	121.9
Stalk Diameter (cm)	4.8	26.8	12.8
Internode Length (cm)	1.5	10.0	4.4
Leaf Number (number)	13.0	52.0	30.0
Leaf Surface Area (cm^2)	39.0	950.0	227.0
Leaf Length (cm)	10.0	54.5	27.2
Leaf Width (cm)	5.0	26.0	11.8
Diametrical Ratio	1.40	3.95	2.35
Ovality Coefficient	1.36	3.54	2.30

It was observed that the types obtained as a result of this research are different from each other regarding their morphological characteristics. However, the cause of this may be the applications such as fertilization, irrigation, etc. or the environmental circumstances. In the same line with this, there may be different types regarding their genotypic characteristics with the same phenotypic appearances. Thus, studies should be carried out to reveal genotypic differences in order to put forward differences.

REFERENCES

1. Ayanoğlu, F., Kaya, D. A., 1998, *M.K.Ü. Ziraat Fakültesi Dergisi*, 3(2), 71-8.
2. Anitia, N., Ioan, I., 1960, The Fundamental Morphological Charasteristics for The Botanical Description of Tobacco, CORESTA Information Bulletin No.4, 17-32.
3. Kostoff, D., 1945. Cytogenetics of The Genus Nicotiana, Sofia, pp. 203-4.
4. İncekara, F., 1979. Endüstri Bitkileri ve Islahı, Keyf Bitkileri ve Islahı, Ege Üniversitesi Matbaası, Ege Üniversitesi Ziraat Fakültesi Yayınları No: 84, p.180.

Genetic Diversity of Two Native Forest Tree Species in Turkey: *Pinus brutia* Ten. and *Cupressus sempervirens* L.

NURAY KAYA and KANİ IŞIK
Akdeniz University, Faculty of Arts and Sciences, Department of Biology, 07070 Antalya, Turkey

1. INTRODUCTION

Pinus brutia Ten. and *Cupressus sempervirens* L. are important forest tree species in Turkey for various economic and ecological reasons. *P. brutia* occupies 3.1 million ha. of forest land, which constitutes 36% of the forest areas in the country[1]. It grows from sea level up to 1500 meters on the Taurus Mountains[2,3]. This species is one of the most important forest trees in the region, providing both timber resources and amenity in Turkey; and it is preferable for afforestation of dry and degraded areas in the Mediterranean region and elsewhere, due to its wide tolerance range and drought resistance[4,5]. *C. sempervirens* L. also grows in several countries with Mediterranean climate[6,7]. It makes largest natural stands in Koprulu Kanyon National Park in Antalya, Turkey; making occasional mixed stands with *P. brutia*.

Information is slowly accumulating on the genetic diversity of these two species both in Turkey and abroad[4,8,9]. Genetic qualities could be determined by standard provenance and progeny tests[10]. In addition, genetic structure could be studied by using more novel techniques, one of which is isoenzyme analyses[11,12]. Genetic variation within- and among- populations can easily be determined at the molecular level by isoenzyme analysis[12]. Therefore, the aim of the present study is to extend our knowledge of genetic variation of the populations of these two species in Turkey, using isoenzyme analyses.

2. MATERIALS AND METHODS

Bulked seed material of *P. brutia* from nine natural stands in the region of Antalya, and single tree seed collections of *C. sempervirens* from three natural stands (Beskonak, Fethiye, Datca) in Southern and Western Turkey were obtained with the help of Ministry of Forestry of Turkey. For the analyses, the seeds were first germinated at 20 °C. Horizontal starch gel electrophoresis was used to analyze allele frequency patterns. Haploid tissue of seeds (megagametophytes) was analyzed according to Conkle et al.[13], with minor modifications[9,14]. Calculations of parameters of intra- and inter population genetic diversity (mean sample size per locus, mean number of alleles per locus, percentage of loci that were polymorphic, mean heterozygosity expected from Hardy-Weinberg proportions) were done with the BIOSYS-1 program[15].

3. RESULTS AND DISCUSSION

3.1 *Pinus Brutia*

A total of 23 loci and 57 alleles were observed in studying 14 enzyme systems. Of the 23 loci, seven (Got-3, Idh-1, Pgi-1, Pgm-1, Pgm-2 Sod-1 and Sod-2) were monomorphic and the remaining loci (Aco, Acp-2, Adh-2, Gdh, Got-1, Got-2, Mdh-1, Mdh-4, Mnr-1, Mnr-2, Mpi, Pgd-2, Pgd-3, Pgi-2, Sdh-1, Sdh-2) were polymorphic, considering 95% criterion for polymorphism. The average proportion of polymorphic loci per population ranged from 60.9% to 69.6%. The mean expected heterozygosity (H_{exp}) from Hardy-Weinberg proportions was 0.263[14]. The average number of alleles per locus per population was 2.1 (Table 1).

The mean expected heterozygosity was lowest in Duzlercami population (Table 1). The relatively low heterozygosity in Duzlercami population is probably associated with ecological conditions in this area, where the topography is relatively flat with homogeneous climatic and soil conditions. The other 8 populations exhibit considerable micro-site differences, possibly contributing their higher levels of genetic heterozygosity.

This study showed that the populations exhibit a rather high within population variation (H_S=0.263) and a moderate degree of among populations genetic variation (D_{ST}=0.015), which results in a high total variability within species (H_T=0.278).

Table 1. Genetic variability among populations of *P. brutia* based on isoenzyme analysis on 23 loci

Population name	Mean number of samples per locus	Mean number of alleles per locus	Percentage of polymorphic loci *	Hexp**
Kargi (Krg.)	129.5	2.0 (±0.2)	65.2	0.255 (±0.045)
Guzelbag (Gzl.)	128.1	2.1 (±0.2)	69.6	0.279 (±0.046)
Urlupelit (Url.)	126.8	2.2 (±0.2)	69.6	0.282 (±0.048)
Eskibag (Esk.)	128.8	2.1 (±0.2)	69.6	0.279 (±0.047)
Pinargozu (Pin.)	129.2	2.1 (±0.2)	65.2	0.251 (±0.047)
Duzlercami (Dzl)	123.2	2.0 (±0.2)	60.9	0.236 (±0.045)
Olimpos (Oli.)	127.6	2.1 (±0.2)	69.6	0.270 (±0.049)
Kumluca (Kum.)	129.2	2.1 (±0.2)	69.6	0.263 (±0.048)
Kargi (Kry.)	127.9	2.0 (±0.2)	69.6	0.271 (±0.046)
Mean (S.e.)	127.8	2.1		0.263

* at 95% criterion for polymorphism
** Hexp.: Expected mean heterozygosity based on Hardy-Weinberg equilirium (unbiased)

Of the total genetic variation, 5.3% was due to interpopulational differentiation (G_{ST}) and 94.7% was due to variation within populations (Table 2).

Table 2. Parameters of Genetic diversity in populations of *P. brutia* and *C. sempervirens*

Species	Number of populations	Hs *	Ht *	Dst *	Gst *
P. brutia	9	0.263 (0.045)	0.278 (0.048)	0.015	0.053
C. sempervirens	3	0.197 (0.047)	0.213 (0.055)	0.016	0.150

*Hs= Measure of genetic variation within populations, Ht= Measure of total genetic variation, Dst= Measure of genetic variation among populations, Gst= The relative degree of genetic differentiation = % Dst / Ht

This value is comparable with the results on some other coniferous species: For example, within population variation was reported to be 88% in *Pinus ponderosa*, 96% in *Pinus contorta*, 97% in *Pseudotsuga menziessii* and 97% in *Pinus rigida*[16]. Another study on various morphological traits of *Pinus brutia* also demonstrated that 96% of total genetic variation was due to within population variation[17]. Such high within population variation (or small variation among populations) are common among conifer populations

3.2 Cupressus sempervirens

A total of 22 loci and 46 alleles were observed in studying 13 enzyme systems. Of the 22 loci, 10 (Cat-1, Cat-2, Got-1, Got-2, Idh-1, Pgi-1, Pgm-2, Sdh, Sod-1 and Sod-2) were monomorphic and the remaining loci (Aco, Gdh, Got-3, G6pd, Idh-2, Mdh-3, Mdh-4, Mnr, Pgi-2, Pgm-1, 6pgd-1 and 6pgd-1) were polymorphic, considering 95% criterion for polymorphism. The average proportion of polymorphic loci per population was 45.5%. The mean expected heterozygosity (H_{exp}) from Hardy-Weinberg proportions was 0.202. The average number of alleles per locus per population was 1.8 (Table 3). The mean expected heterozygosity was highest in the Beskonak population, which is located in extremely steep terrains and canyons from 400 to 800 meters from the sea level[9].

Table 3. Genetic variability among populations of *C. sempervirens* based on isoenzyme analysis on 22 loci

Population name	Mean number of samples per locus	Mean number of alleles per locus	Percentage of polymorphic loci *	H_{exp}**
Beskonak	28.0	1.8 (0.2)	45.5	0.210 (0.054)
Datca	21.9	1.7 (0.2)	45.5	0.205 (0.051)
Fethiye	25.2	1.9 (0.2)	45.5	0.191 (0.052)
Mean	25.0	1.8		0.202

* at 95% criterion for polymorphism
** Hexp.: Expected mean heterozygosity based on Hardy-Weinberg equilirium (unbiased)

Within population variation (H_S) and among populations genetic variation (D_{ST}) were found to be 0.197 and 0.016, respectively. Total variability within species (H_T) was 0.213. Of the total genetic variation, 15.0% was due to interpopulational differentiation (G_{ST}) and 85.0% was due to variation within populations (Table 2)

4. CONCLUSION

Generally, within-species genetic diversity in forest tree species is higher than in other organisms. On the other hand, genetic differentiation among populations of a given forest tree species tends to be quite low[19]. Genetic variation and the structure of populations are formed by various evolutionary forces, especially breeding system, gene flow and natural selection[20,21]. If the genetic variation of a species is high, then the species has a higher ability to adapt to changing environmental conditions. The results of this study show that 1) both species are genetically variable and the populations differ in the kind and amount of variation they contain, 2) within population variation in *P. brutia* and *C. sempervirens* are quite high, indicating their wide adaptive abilities, and at the same time better opportunities for within population selection and breeding in tree improvement programs.

ACKNOWLEDGEMENT

The authors express their graduate to the Turkish Forest Service for providing the seed material, to UNESCO for supporting N. Kaya under a Biotechnology Research Fellowship for three months in Israel.

REFERENCES

1. Anonymous, 1980, Türkiye Orman Envanteri, Orman Bakanlığı, Orman Genel Müdürlüğü, Yayın No:13, Seri No: 630.
2. Critchfieldt, W. B., Little, E. L. Jr., 1966, Geographic distribution of the *Pinus* of the world, USDA Forest Service, Micellaneous Pub. 991, pp. 97.
3. Selik, 1958, *İst. Univ. Orm. Fak. Derg.*, **8a**, 161-98.
4. Oppenheimer, H. R., 1967, Mechanisims of Drought resistance in conifers of the Mediterrenean zone and the arid west of the USA. Part 1. Physiological and Anotomical Investigation. Final Report on Project No:A 10-Fs 7, Grant No: FG-Is 119, submitted to Israel, pp. 73.
5. Palmberg, C., 1976, *Silvae Genetica*, **24** (5-6), 150-60.
6. Emberger, L., Gaussen, H., Kassas, A., De Philipis, A., 1963, Bioclimatic map of the Mediterranean zone. Arid zone Research, UNESCO-FAO, Paris, pp.58.
7. Papageorgiou, A. C., Anetsos, P. K., Hattemer, H. H., 1994, *Forest Genetics*, **1**, 1-11.
8. Isık, K., Kara, N., 1997, *Silvae Genetica*, **46** (2-3), 113-20.
9. Korol, L., Kara, N., Isık, K., Schiller, G., 1997, *Silvae Genetica*, **46** (2-3), 151-5.
10. Isık, K., 1986, *Silvae Genetica*, **35**, 58-65.
11. Feret, P.P., Bergmann, F., 1976, Gel Electrophoresis of Proteins and Enzymes. In *Modern Methods in Forest Genetics* Miksche, J.P. (Ed), Springer-Verlag, New York, pp. 49-77.
12. Adams, W.T., 1993, Application of isozymes in tree breeding. In *Isozymes in Plant Genetics and Breeding*, Tanksley, S.D., Orton, T. J. (Eds), Part A. 1993, Elsevier Science Publishers B. V. Amsterdam, pp. 381-400.
13. Conkle, M.T., Hodgskiss, P.D., Nunnally, L.B., Hunter, S.C., 1982, Starch Gel Electrophoresis of Conifer Seeds: A Laboratory Manual, U.S.D.A. Gen. Techn. Rept. PSW- **64**, p.18.

14. Kara, N., Korol, L., Isık. K., Schiller, G., 1997, *Silvae Genetica,* **46** (2-3), 155-61.
15. Swofford, L. D., Selander, B. R., 1981, BIOSYS-1, A computer program for the analysis of allelic variation in genetics. User manual. Dept. of Genetics and Development, University of Illinois at Urbana-Champaign, Urbana, Il, USA.
16. Nicolic, D., Tucic, N., 1983, *Silvae Genetica,* **32**(3-4), 80-9.
17. Isık, K., Topak, M., Keskin, C., 1987, Kızılçamda (*Pinus brutia* Ten.) orijin denemeleri. Altı farklı populasyonun beş ayrı deneme alanında ilk altı yıldaki büyüme özellikleri. Orman Genel Müdürlüğü, Orman Ağaçları ve Tohumları islah Enstitüsü, Yayın No.3, p. 139.
18. Scaltsoyiannes, A., Rohr, R., Panetso, K.P., Tsaktsira, M., 1994, *Silvae Genetica*, **43**, 20-30.
19. Hamrick, J.L., 1989, Isozymes and the analysis of genetic structure in plant populations. In *Isozymes in Plant Biology* (Soltis, D.E., Soltis, P.S., eds.), Chapman and Hall., London, pp. 87-105.
20. Nevo, E., 1978, *Theor. Pop. Biol.,* **13**, 129-77.
21. Hamrick, J.L., Mitton, J.B., Linhart, Y.B., 1981, Levels of genetic variation in trees: Influence of life history characteristics. In *Proceedings of the Symposium on the isozymes of North American Forest Trees and insects* (Conkle, M.T., ed.), U.S.D.A. Gen. Techn. Rept. PSW-48, pp. 35-41.

The Vegetation Studies in the Pure Stands of Kürtün(Gümüşhane) Forests in Turkey

[1]MAHİR KÜÇÜK and [2]ÖMER EYÜBOĞLU
[1]*Eastern Black Sea Forestry Research Institute, 61040 Trabzon, Turkey*, [2]*Gazi University, faculty of Education, Department of Biology, Kırşehir, Turkey*

1. INTRODUCTION

The study area lies at the back of the Eastern Black Sea Region, and it belongs the Colchis Region of the Euro-Siberian Floral subsection in the Holarctic Floral Region.

The Harşit Valley is one of example of the region where the humid and temperate climate extends to the interior parts of Eastern Black Sea Region. This area is at the transition zone to the interior section of the Black Sea Region. Its climate is humid and it has local temperature. Therefore, when it is compared with the other forest areas, the richest part by the tree species and floristic composition is seen there.

According to Thorntwaite's classification, the climate is described as humid, mesothermal, only very slightly lacking water or not, and oceanic. The greater group of the soil is brownish colour and situated at high altitudes, in high mountains with alpine meadows, and grey brownish podzol soils[1].

2. MATERIAL AND METHOD

Before this study, in the year of 1992, the flora of region and the floristic composition of the pure stand types of the study area was investigated according to Braun-Blanquet method and plant associations and upper plant community were classified by taking the dominant-constant and characteristic species into consideration[1]. Also, in the year of 1998 with another study was investigated the ecological species or species groups of the same area[2].

At this study, the sub community of plant associations investigated the vegetation of the same area. The study area is about 2800 hectares.

At the investigation area 82 quadrats were selected to determine of the vegetation of the pure forest areas. Thirty-two from Oriental Spruce, 28 from Scotch pine, and 22 from Eastern Beech stands were selected. In these quadrats, the coverage degrees and sociability of the plants were put under record, with particular regard to the rules of plant sociology science. As a classifying method was used Braun-Blanquet's method too.

3. RESULTS AND DISCUSSION

There were mainly four distinct vegetation types in this region. These types are; forest, degraded forest, sub alpine, and alpine vegetation types. In the composition of the forest vegetation there are *Picea orientalis, Pinus sylvestris, Fagus orientalis, Quercus petraea* subsp. *iberica, Populus tremula* and *Abies nordmanniana* subsp. *nordmanniana. Picea orientalis* has got the widest distribution at this region and its altitudes are between 800 m. and 2000 meters. Whereas *Pinus sylvestris* is generally distributed at the vicinity of Kavraz and Söğüteli between 1400 m. and 2000 meters altitudes. The forests of *Fagus orientalis* lie around Örümcek and locally around Söğüteli, *Populus tremula* is generally found at ranges between 1000 m. and 1800 meters in a scattered way. *Abies nordmanniana* subsp. *nordmanniana* lies between 1000 m. and 1800 meters together with other tree species.

In this area are some species lie rarely like: Mountain Elm (*Ulmus glabra* Huds.), Maple species (*Acer trautvetteri* Medw., *Acer cappadocium* Gleditsch and *Acer campestre* L.), species of Mountain-ash (*Sorbus aucuparia* L., *Sorbus umbellata* (Desf.) Fritsch var *cretica* (Lind) Schneider and *Sorbus subfusca* (Ledep.) Boiss.), the Caucasus Linden (*Tilia rubra* DC. subsp. *caucasica* (Rubr.) V. Engler, also species of Oak (*Quercus hartwissiana* Steven., *Quercus macranthera* Fisch & Mey. subsp. *syspirensis* (C. Koch) Mentsky.), the Common Hornbean *(Carpinus betulus* L.) the Yew-tree (*Taxus* L.), the Anatolian Chestnut (*Castanea sativa* Miller), and Black Alder *(Alnus glutinosa* (L.) Gaerth. subsp. *barbata* (C. A. Mey.) Yalt.).

The alpine vegetation appears upper part of the forest vegetation in the Söğüteli part, at 1800 m. and 1900 meters, and in the Kavraz and Örümcek sections at 2000 m. to 2200 meters[3]. This vegetation covers even the highest peaks of the area (from 2500 m. to 2900 meters), and is mostly composed of grasses and much less of bushes. This greater part (around 18 000 hectares), extending to the highest peaks, serves as summer pastures and therefore suffers great damage, the structure of the vegetation becoming gradually ruined.

3.1 Definite Plant Associations within Forest Vegetation

As mentioned above, at the introduction section in the research area the previously selected the quadrats were classified according to the differential and characteristic species[4]. Thus, the following plant associations and sub communities were found (Table 1).

- *Piceetum orientali*
- *Vaccinio myrtilli-Pinetum sylvestri*
- *Rhododendro pontici-Fagetum orientali*

3.2 The Piceetum orientali association

This association firstly, is divided into the two sub associations, which are *Rhododendroetosum pontici* and Typicum sub associations. Then, *Rhododendroetosum pontici* sub associations are separated to the two variants again, which are *Abies nordmanniana ssp. nordmanniana-Vaccinium arctostaphylos* and typical variants.

Secondly sub association (Typicum sub association) is divided in to the three variants, which are Vaccinium arctostaphylos, Abies nordmanniana ssp. nordmanniana, and typical variants of the Piceetum orientali association (Table 1).

Three variants of the Piceetum orientali association is found as very widespread in the study area, which are *Vaccinium arctostaphylos, Abies nordmanniana ssp. nordmanniana-Vaccinium arctostaphylos* and typical variant (Typicum sub association).

The main species *Picea orientalis* is found naturally only in the Caucasus and North Eastern Anatolia. In Turkey, it is found in the region between the border to Georgia and the Ordu-Melet River. Here it occurs on mountain slopes facing the sea, either alone or together with other species.

The range of occurrence for *Picea orientalis* in the coastal facing area is situated between 1000 m. and 2000 meters, sometimes even reaching 2400 meters[5]. In this association, indicator plant species likely to be found are *Veronica officinalis* and *Oxalis acetosella*. It has been found that *Veronica officinalis* occurs in 66 and *Oxalis acetosella* in 56 percent of the quadrats investigated.

The general aspect of this association is composed of three layers, either with trees, bush, or grass. In the first layers, trees will appear to be 13 m. to 33 meters tall, with canopy density percentage of 25 to 90. In this part, *Abies nordmanniana subsp nordmanniana* is found together with *Picea orientalis*, and also, but much more rarely, *Populus tremula*.

The layer of bush holds *Rhododendron ponticum, Rhododendron luteum, Laurocerasus officinalis, Ilex colchica, Daphne pontica, Vaccinium arctostaphylos* and *Vaccinium myrtillus*.

The covering degree of the bush and grass layers varies from 5 to 90 percent. The amount of plants in the selected quadrats also varies between 3 and 15. The seedlings of *Picea orientalis* and *Abies nordmanniana* in the bush and grass layers were found between 35 and 47 percent.

3.2.1 The Vaccinio myrtilli - Pinetum sylvestri association

This association is represented with three variants in the study area. These are; *Rhododendron luteum, Quercus petraea ssp. iberica* and typical variants of the *Vaccinio myrtilli-Pinetum sylvestri* association. Final variant is found as very widespread in the study area. The variant of *Quercus petraea ssp. iberica* is found on the middle and low site quality class.

The differential species of this association are *Pinus sylvestris* and *Vaccinium myrtillus*. Within the forest-forming species *Pinus sylvestris* is one the most widely geographically distributed species. Occurring quite frequently in Europe, it is widespread in Siberia, the Caucasus, and Iran. In Turkey, it is found in Anatolia and Cappadocia. As for *Vaccinium myrtillus*, it is distributed in the Caucasus, Northern Asia, West, Middle, and Northern Europe, and Northern Anatolia[5]. In this association, *Vaccinium myrtillus* is found in over 65 percent of the areas investigated, and has a high coverage degree. In the places where it is found, it forms carpet-like layers and protects the soil. It prevents the seeds of *P. sylvestris* to reach on the ground, and it prevents the sunlight from the seedlings and increases competition of the roots.

The probable indicator species of this association are *Agrostis tenuis* and *Daphne pontica*. *A. tenuis* is found in over 79 percent of this areas surveyed, this species has a high coverage degree too. The secondly characteristic species for this association is *Daphne pontica*. It is found here and there within the groups of trees and it's occurrence lying in the vicinity of 57 percent.

Table 1 The plant associations of the pure stand in Kürtün (Örümcek, Kavraz and Söğüteli) Forests

A ss no.	Associations	Sub association	Variant	
1	Piceetum orientali	Rhododendroetosum Pontici	Abies nordmanniana ssp. nordmanniana Vaccinium arctostaphylos	1
			Typical variant	2
		Typicum	Vaccinium arctostaphylos	3
			Abies nordmanniana ssp.nord.	4
			Typical variant	5
2	Vaccinio myrtilli - Pinetum sylvestri		Rhododendron luteum	6
			Quercus petraea ssp.iberica	7
			Typical variant	8
3	Rhododendro pontici - Fagetum orientali	Piceetosum orientali	Vaccinium arctostaphylos	9
			Laurocerasus officinalis	10
			Typical variant	11

In this area, the association of *Pinus sylvestris - Vaccinium myrtillus* appears as trees, small trees, bush, and grass layers. In the forest-covered part, trees are 12 to 24 meters tall, and the percentages of canopy densities between 20 and 85. In this zone, at elevations between 1400 m. and 1500 meters, *Quercus petraea subsp. iberica* occurs infrequently.

In some parts of small trees layers, Juniperus oxycedrus subsp. oxycedrus are found together with *Quercus petraea* subsp. *iberica*.

In the bush layer, apart from *Vaccinium myrtillus* and *Daphne pontica* are found *Rhododendron luteum*, *Vaccinium arctostaphylos*, *Daphne glomerata* and *Bruckenthalia spicufolia*. Basically, this grass section displays a rather rich floristic composition. Its main component is plants from the family of the Gramineae, with a wide range of species. The coverage rates to the ground of the bush and grass layers change between from 10 to 90 percent.

3.2.2 The Rhododendro pontici - Fagetum orientali association

The Piceetosum orientali sub association in the study area represents this association. The *Piceetosum orientali* sub association has got three variants, which are Vaccinium arctostaphylos, Laurocerasus officinalis, and typical variant of the *Rhododendro pontici-Fagetum orientali*. The typical variant of *Rhododendro pontici-Fagetum orientali* association is found as the most widespread.

The differential species of the association, *Fagus orientalis* is found in Northern Anatolia, on the slopes of the mountains facing to the Black Sea. It is distributed frequently in the Directorates Forest Region of Bolu, Zonguldak, Kastamonu, Amasya and Giresun. Abroad, it occurs in the Crimea, the Caucasus, Northern Iran and Southeast Europe[5].

The component of this association is tree, bush, and grass layers. In the tree-grown part, tree heights were measured between 19 m. and 50 meters, and the canopy density was found between 20 and 95 percent. Apart from *Fagus orientalis, Picea orientalis, Acer cappadocicum var. cappadocicum, Carpinus betulus, Abies nordmanniana subsp. nordmanniana, Populus tremula* and *Ulmus glabra* occur but to a much less degree.

A great part of the bush section is covered with *Rhodonderon ponticum*. The other species belonging to this community are *Laurocerasus officinalis, Vaccinium arctostaphylos, Ilex colchica, Daphne pontica, Euonymus europaeus, Viburnum orientale, Corylus avellana, Euonymus latifolius* and *Sambucus nigra*. Since the bush section is so extensive, the grass layer is much hindered in its extension. This is why the floristic composition of the grass part is rather limited. The percentage to which the bush and grass surfaces are covered varies from 10 to 100 percent. In the analysed quadrats, the number of species goes from 4 to 12.

4. CONCLUSION

As a result of the lots of plant sociology studies certain characteristic species of plant associations are determined. If these studies are related with ecological data, ecological species groups can also be determined[2].

In this way, it can be determined whether forests and other natural habitat, are on the climax structure or not, and to be in the step of which succession (regressive or progressive succession) and where the potential of the soil, the balance of humidity, the soil acidity etc. are determined according to the characteristic species or ecological species groups.

Thus, the different types of forest and about groups of site units are obtained it can be gained very important information. Also it is the aim of a successful management. In our days, obtaining an inventory of site units represents a first step of management and silviculture in the countries concerned with modern management of the forest.

In this way, taking into account the lots of data of site units and the past history of the forest it can be planned the future of the forest more clearly and the faulty usage of the fields are prevented, thus its protect biological diversity.

As a result, after defining differential and characteristic species, which describe the plant associations, and after defining ecological species groups, which clarify the site (habitat), lots of forestry activities will become more effective, practical and economic.

REFERENCES

1. Küçük, M., 1998, The Flora of Kürtün (Gümüşhane) Örümcek Forest and The Floristic Composition of The Pure Stand Types, The Publications of Eastern Forest Research Institute, Technical Bulletin No. 5, Trabzon.
2. Küçük, M.; Altun, L., 1998, The Investigations on some Ecological Species Groups in Örümcek Forests, XIV. Biology Congress, 7-10 September 1998, Vol. I, Samsun.
3. Anonymous, 1984, General Directorate of Forest, Torul Forest Management, The Management Plan The Forests of Örümcek Region.
4. Davis, P. H., 1965-1982, Flora of Turkey and the East Aegean Islands, Vol.1-VI and VII, Edinburgh University Press, Edinburgh.
5. Quezel, P., Barbero, M., Akman, Y. 1980, *Phytocoenologia,* **8** (3/4), 236.

7th Year IPM Implementation: The Biodiversity of Pests and their Benefical Species in the Protected Vegetable in the Aegean Region of Turkey

NİLGÜN YAŞARAKINCI
Plant Protection Research Institute, 35040 Bornova, İzmir, Turkey

1. INTRODUCTION

The pests and their natural enemies were surveyed in 1991-1992 in İzmir Province of the Aegean Regin of Turkey[1,2]. The major pests were determined as: whiteflies (*Trialeurodes vaporariorum* Westw., *Bemisia tabaci* Gern.) on tomato; aphid (*Aphis gossypii* Glover) and spider mite (*Tetranychus cinnabarinus* Boiss.) on cucumber; spidermite (*T. urticea* Koch.) and aphids (*A. gossypii* and *Myzus persicae* Sulzer) on pepper. It were found that the pests populations couldn't be suppressed in spite of heavy wide spectrum pesticides treatments. The secondary pests were determined as; *Aculops lycopersici* (Massee) on tomato; *Thrips tabaci* Lindeman, whitefllies, leafminers on cucumber and pepper. It were recorded that the secondary pests were suppressed by the heavy treatments against the major pests.

2. RESULTS

The number of natural enemy species were less and their population densities were small throughout the season. The populations of the natural enemies began to increase at the end of the season after the pesticides treatments had ceased. The natural enemies species were recorded as; *Scolothrips longicornis* Priesner, *Chrysoperla carnea* Steph., *Episyrphus balteatus* DeGeer: *Metasyrphus corollae* Fabricius, *Chrysotoxum intermedium* Meigen; *Coccinella semptempunctata* L., *Adonia variegata* Goeze, *Scymnus marginalis* Rossi, *S. interruptus* Goeze, *S. subvilloaus* Goeze, *S. rubromaculatus* Goeze, *S. apetzi* Mulsant, *Propylaea quatuordecimpunctata* L.; *Macarolophus caliginosus* Wgn.; *Diglyphus isaea* Walker, *Hemiptarsenus zilahisebessi* Erdos, *Lysiphlebus fabarum* Marshall Walk, *Aphidius matricariae* Haliday and *Pronematus ubiguitus* McG..

IPM program has being carried out according to The National IPM Program since 1995. The pesticide treatments have been minimized by applying alternative control methods. The pesticide which has less side effects have been applied when necessary. The pesticide treatment have been reduced at the rate of %30-100. As a result of the IPM implementation, a sound natural balance has been achieved. The natural enemies have began to appear at the beginning of the season and their densities have increased throughout the season and suppressed some of the major pests populations. Additional to the former species *Typhlodromus pyri* Scheuten (Acarina:Phytoseidae);

Aphidoletes aphidimyza Rond. (Dip.:Cecidomyiidae); *Sphaerophoria rüppelli* Wiedeman, *S. scripta* L. (Dip.:Syrphidae); *Synharmonia conglobata* L., *Stethorus gilvirons* Muls., *Scymnus frontalis* Fabricius, *S. pallipediformis* Günther, *Henosepilacna elaterii* Rossi, *Psyllobora vigintiduopunctata* L., *Exochomus nigromaculatus* Goeze, *Coccinella undecimpunctata* L. (Col.:Coccinellidae); *Aelothrips collaris* Priesner (Thys.:Thripidae); *Nabis punctatus* C. (Nabidae), *Orius niger* W. (Anthocoridae), *Coranus aegyptius* F. (Deduviidae), *Piocoris erythrocephalus* P.S., *Geocoris lineola* Db. (Lygaeidae), *Deraeocoris serenus* D.Sc., *D. pallens* RT., *D. rutilus* H.S., *D. ruber* L., *D. schach* F. (Miridae) (Het.) were found as natural enemies.The population of whitefly and leafminer have been suppressed by their predator (*M. caliginosus*) and parasitoid (*D. isaea*), respectively, at the beginning of the season in first year of IPM implementation. The pests population densities have been reduced gradually year by year and become an unimportant pests. Exceptionally, whitefly population could be high if the conditions are favouroble in tomato plastic tunnels (high humidity, resource of the whitefly infestations and lack of predator). The aphid and spidermite populations also have been suppressed according to the conditions of the cucumber and pepper plastic tunnels, but mostly suppressed by the natural enemies. Spidermites have been suppressed by *S. gilvifrons*, *S. pallipediformis*, *S. ruppelli*, *C. carnea* on cucumber and *S. gilvifrons*, *S. rubromaculatus*, *S. apetzi*, *S. interruptus*, *S. frontalis*, *C. carnea*, *S. scripta*, *T. pyri* ve *S. longicornis* on pepper. Aphid suppressed by *C. carnea*, *A. aphidimyza*, *S. conglobata*, *A. variegata*, *C. septempunctata* on cucumber and *P. quatuordecimpunctata*, *A. variegata*, *S. scripta*, *A. aphidimyza*, *C.*

carnea, M. caliginosus on pepper. *T. tabaci* mostly suppressed by *O. niger* and *A. collaris* in the vegetable plastic tunnels.

In IPM implemented plastic tunnels when the side effect of the pesticides were reduced, the secondary or new pests namely *A. lycopersici* and *Cyrtopeltis tenius* on tomato, *Spodoptera littoralis, Nezera viridula* on tomato and cucumber, *Frankliniella occidentalis* Pergande and *Exolygus pratensis* L. on cucumber outbreak occurs.

REFERENCES

1. Yaşarakıncı, N., Hıncal, P., 1997, *Bitki Koruma Bülteni,* 1997, **37**(1-2), 79-89.
2. Yaşarakıncı, N., Fidan, Ü., Cınarlı, İ., Demir, G., Öz, S., Filiz, N., Koçer, H., Uçkan, A., Üstün, N., Hıncal, P., Altın, N., Taşdelen, P., Tokaç, A., Erdem, S., Aykut, N., Moroğlu, H., Ateş, N., Yalın, N., Saltabaş, M., Oktar, F., Erişen, İ., Yıldırım, B., Konak, N.,Ulusoy, F., Cengiz, M., Toker, A., 2000, The implementation of IPM on the protected vegetable, Türkiye 4. Entomoloji Kongresi, 12-15 Eylül 2000, Aydın, Turkey, pp. 23-32.

Phytopreparations from the Species of *Limonium* Mill

GALIYA ZHUSUPOVA, K. RACHIMOV, T. SHALAKHMETOVA and Z.H. ABILOV
al-Farabi Kazakh National University, Department of Chemistry, 95a Karasay-Batyr St., Almaty, Kazakhstan

1. INTRODUCTION

Our extensive long-term study of the species of *Limonium* Mill has shown that these herbs are medicinally valuable and that their derivatives demonstrate wide therapeutic effects. *Limonium* Mill is a member of *Plumbaginaceae* family. There are over 300 species of *Limonium Mill* in the world[1]. There are 18 species of *Limonium* Mill in Kazakhstan's saline soils; three of them are considered to be the most productive: *Limonium gmelinii* Wild ktze., *Limonium myrianthum* Shrenk ktze., *Limonium otolepis* Shrenk ktze[2]. These herbs are often of semi-subshrubbery type growing on strongly saline soils. A remarkable ability of all species of *Limonium* Mill is to move excess solutions of natrium and calcium salts from cell sap to the surface of an overgrown organ with the help of special glands[1,3].

2. PHARMACOLOGICAL ACTIVITIES OF PHYTOPREPARATIONS

Phytopreparations obtained on the basis of *Limonium* Mill possess antiviral, hepatoprotector, anti-inflammatory, anti-burn properties and consequently, can be used for medicinal purposes. The phytopreparations are harmless, nonallergenic, and do not accumulate in the body.

Experiments for the *Limonium* preparation influence were conducted on livers of white breedless rats at chronic impact of carbon tetrachloride. Fifty rats of reproductive age were divided into four experimental groups. In the first group, the rats inhaled carbon tetrachloride (CCl_4) for six months. In the second group, the rats received CCl_4 and 2 ml perorale of 2 % water solution of the preparation daily. In the third group, the rats received a 2 %-solution of the perorale during the same time. The fourth group was made up of intact rats.

After six months, the animals were put to sleep under ether anesthesia, and the liver parts were taken for histological analysis. Preparation of the fixed material was done in accordance with colloquial methods. The histological preparations dyed with hematoxilin-eozin and by Wan-Hyzon method were examined and photographed by light microscope MBI-15.

Analysis of the histological preparations has shown that the rats in both groups (intact and those which received the preparation) had not revealed any pathological changes in the liver. Chronic intoxication of the animals by CCl_4 caused progressive postnecrotic liver cirrhosis. Along with dystrophic and necrobiotic processes with prevailing fat degeneration of centerlobular hepatocytes and local colliquation necrosis, fibrous transformations were observed. There were widely revealed in the liver parenchyma the tissue-joining septa, both inside the liver lobules and on its periphery. Thus, so-called false lobules formed in the liver parenchyma of the rats intoxicated with CCl_4 for six months. The destructive processes were accompanied by chronic inflammation; infiltrates presented by polymorphous-nucleus leucocytes, lymphocytes and macrophages localized around the necrosis sites.

The animals from the second experimental group also developed liver cirrhosis, but with a different character of cirrhotic transformations. There were no false lobules; the fibrosis was actual in the central part of the liver lobule only. The hepatocytes were mostly subjected to hydropic dystrophy, and the fat degeneration of liver was not observed. Moreover, in comparison with the rats from the first experimental group, these rats did not have inflammation-cell infiltrates, which demonstrates anti-inflammation action of the investigated preparation. In the degeneration nodes there were observed hypertrophied hepatocytes with large hyperchromic nuclei, which demonstrated more active reparation processes in the animals' liver that received the phytopreparation. The obtained data demonstrates that the investigated preparation possesses some hepatoprotective action.

The antiviral activity of the *Limonium* preparation was studied on different cultures of influenza virus in vitro and in vivo. Experiments proved that the investigated drug in the amount of 2.5 mg per embryo demonstrates the maximum (100 %) level of virus inhibition activity with respect to all used viruses. A decrease in the preparation amount to 0.25 mg per embryo results in 100 % virusinhibition activity level with respect to some viruses and decreases to 80% for others. The antiviral activity of the *Limonium* preparation is comparable with that widely used in medicine preparation *Remantadine*, and at higher concentrations of 2.5 mg exceeds it. Since the viruses are in-cell parasites acting at a molecular-genetic level, neuraminidase activity was studied on a model influenza virus A/FPV/Rostock/34 and a Newcastle disease (culture La Sota) by standard methods in order to reveal the virusinhibition action of the preparation. Blocking the virus' fermentative activity was noticeable even at a minimal (0.005 %) concentration of the *Limonium* phytopreparation and reached its maximum at the 0.025 % concentration.

The technological schema of herb collection for use in phytopreporations is straightforward and environmentally safe. High positive bioactivity, commercial availability of raw materials, convenient location, and time of collection (vegetation) all support the idea of phytopreparation creation on the basis of *Limonium* Mill. Thus, the creation of phytopreparates with a wide spectrum of physiological activity on the basis of *Limonium* Mill is a very promising, tangible and commercially viable undertaking.

REFERENCES

1. Флора СССР, М., 1952, vol. XVIII, pp. 436-438
2. Флора Казахстана, Алма-Ата, Наука, 1961, vol. VII, pp.79-80
3. Faraday, C.D., Thomson W.W., 1986, *Journal of Experimental Botany*, **37**(177), 461-70.
4. Oxford, S., Lambkin, R., 1988, *DDT*, 3(10), 448

The Biological Activities of New Heterocylic Compounds Containing Nitrogen and Sulphur

FERAY AYDOĞAN, ZUHAL TURGUT, ÇİĞDEM YOLAÇAN and NÜKET ÖCAL
Yıldız Technical University, Department of Chemistry, Davutpaşa Campus, 34210- İstanbul, Turkey.

1. INTRODUCTION

Oxadiazole derivatives which belong to an important group of heterocyclic compounds have been subject of extensive study in the recent past. Numerous reports have appeared which highlight their chemistry and use[1-3]. Diverse biological activities such as antituberculostatic, antiinflamatory, analgesic, antipyretic, anticonvulsant, etc. have been found to be associated with oxadiazole derivatives[4,5]. Because of this reason we aimed to synthesize various 1,3,4-oxadiazole-2-thione derivatives to make remarkable contributions to this class of heterocyclic compounds. Here, we report the synthesis and characterization of some 5-alkyl substituted 1,3,4-oxadiazole-2-thiones (**3a-e**) and 5-alkyl substituted 3-(2,4-dimethylphenyl)-1,3,4-oxadiazole-2-thiones (**4a-d**) by using the synthetic procedure based on the ring closure reactions of appropriate acid hydrazides with carbon disulphide[6]. We also have synthesized some bis-Mannich bases by the reaction of **3a-e** with benzaldehyde and benzidine.

2. RESULTS

Except 3-(2,4-dimethylphenyl) substituted derivatives, all new compounds were examined for their antibacterial and antituberculostatic activity. The Mannich bases which contain benzidine were the most active, while the others had moderate activity. Compounds **3a-e** showed less antibacterial activity than their Mannich bases.

Antituberculosis activity The in vitro tuberculostatic activity of 1,3,4-oxadiazole-2-thione derivatives was studied against *Mycobacterium tuberculosis* using Lowenstein Jensen's egg medium by serial two fold dilution method and the retardation of the growth rate studied upto six weeks at 37 °C. The tuberculostatic concentration was 0.03 µg/ml.

Antibacterial activity Antibacterial activity of the compounds was tested by agar plate diffusion technique against *Staphylococcus aureus* using tetracycline as standard.

Compound Number	R_1	R_2	R_3
3a	OCH_3	Cl	H
3b	OC_2H_5	Cl	H
3c	CH_2COCH_3	OC_2H_5	H
3d	$CH_2COC_6H_5$	OC_2H_5	H
3e	$COOC_2H_5$	OC_2H_5	H
4a	CH_2Cl	OC_2H_5	2,4-dimethylphenyl
4b	OCH_3	Cl	2,4-dimethylphenyl
4c	CH_2COCH_3	OC_2H_5	2,4-dimethylphenyl
4d	C_6H_5	OCH_3	2,4-dimethylphenyl

The Biological Activities of New Heterocyclic Compounds 373

On the other hand, aldimines have been generally used as substrates in the formation of a large number of industrial compounds via cycloaddition, ring closure, replacement reactions, etc.[7,8] In addition, the aldimines of heterocyclic carbaldehydes, which are widely used in the production of pharmaceuticals, have taken an important place among the compounds of biological interest because of the conjugation and the groups that they contain within their molecules. Furthermore, most of the 4-thiazolidinones and their benzylidene derivatives display a large variety of activities such as antibiotic, diuretic, organoleptic, tuberculostatic, antileukemik and antiparasitical.[9,10] To our knowledge, little is known on fused thiazolidines known to possess these activities.[11] Moreover, little attention has been directed to the behaviour of this class of compounds toward phosphorus reagents.[12,13] Our continuous interest in the preparation and study of reactions of 4-thiazolidinones[14] prompted us to examine the behaviour of 5-benzylidene-3-aryl-2-(2-pyrrolyl)-4-thiazolidinones toward phosphonium ylide or phenylhydrazine under varied conditions to determine their cyclization reactions for possible biological activities of new compounds.[15]

R (Comp.no)= H (2a-e), CH_3 (3a-e)

REFERENCES

1. Potts, K., 1984, *Comprehensive Heterocyclic Chemistry*, (Katritzky, A. R., Rees, Ch.,eds.), Vol. 6, Pergamon Press, p. 427.
2. Kulkarni Y. D., Rowhani, A., 1989, *J. Indian Chem. Soc.*, **66**, 492-3.
3. Obi K., Kojima A., Fukuda, H., Hirai, K., 1995, *Bioorg. Med. Chem. Lett*, **5**, 2777- 82.
4. Mishra, L., Said, M. K., Itokawa, H., Takeya, K., 1995, *Bioorg. Med. Chem*, 3(9), 1241-5.
5. Suman, S. P., Bahel, S. C., 1979, *J. Indian Chem. Soc.*, **56**, 712.
6. Pramanik, S. S., Mukherjee, A., **1998**, *J. Indian Chem. Soc.*, **75**, 53-4.
7. Brown, A.D., Colvin, E.V., 1991, *Tetrahedron Lett.*, **32**, 5187.
8. Burwood, D.A., Gallucci, J., Hart, D.J., 1985, *J. Org. Chem.*, **50**, 5120.

9. Brown, F.C., 1961, *Chem. Rev.*, **61**, 463.
10. Raasch, M.S., *J. Heterocyclic Chem.*, **1974, 11**, 587.
11. Singh, V.P., Upadyay, G.S., Singh, H., 1992, *Asian J. Chem. Rev.*, **3**, 12.
12. Zhitar, B.E., Baranov, S.N., 1973, *Tezisy Dokl-Simp. Khim. Tekhnol. Geterotsikl Soedin., Goryuch. Iskop.* 2nd, 202; C.A. 1976, **85**, 192820.
13. Lugovkin, B.P., *Zh. Obshch. Khim.*, 1978, **48**, 1529; C.A. 1978, **89**,163673.
14. Kouznetsov, V., Öcal, N., Turgut, Z., Zubkov, F., Kaban, S., Varlamov, A.V., 1998, *Monatsh. Chem.*, **129**, 671.
15. Aydoğan F., Öcal N., Turgut Z., Yolaçan Ç., 2001, *Bulletin of Korean Chemical Society*, **22**(5), 476-9.

Synthetic Modification of Iridoids to Non-natural Indole Alkaloids

ÁKOS KOCSIS and PÉTER MÁTYUS
Semmelweis University, Department of Organic Chemistry, Budapest, Hungary

As a continuation of our studies on iridoids[1-4], we now report on some novel reactions of various aglycones of this class of compounds.

1: R = COOH, R^1 = H
2: R = H, R^1 = H

3: R = COOH
4: R = H

5: R = H, R^1 = [structure shown]

Loganin aglycone with L-tryptophane or tryptamine underwent a stereoselective reaction to give compounds **1** and **2**, respectively, whereas reaction of the same amino compounds with 5,9-epideoxyloganin also proceeded stereoselectively to afford compounds **3** and **4**, respectively. A condensation reaction of tryptamine with bisiridoid laciniatoside VII (the ester of deoxyloganic acid formed with loganin), isolated by us from *Dipsacus laciniatus*, led to the formation of compound **5**. The acidic deglycosylation of 3,4,7,8-tetrahydroasperuloside gave the tetracyclic aglycone **6**; when this reaction was carried out under mild conditions, **7** was obtained.

Interestingly, reaction of aglycon **7** proceeded in unusual manner with the participation of ring nitrogen of tryptamine to give compound **8**.

Reduction of secologanin with $NaBH_4$, followed by acidic deglycosylation gave aglycone **9**, which is also expected to be a suitable starting compound for synthesis of new alkaloide type compounds.

Structure of all novel compounds was unambiguously proved by spectroscopic methods. In compounds **1**, **2** and **5**, the β-steric position of 13b-H was supported by NOE observed between this hydrogen and 3-H$_\beta$ hydrogen.

On the other hand, steric position of 13b-H was proved to be α in compound **3**, based on the NOE interaction of this atom with hydrogen at 7-position. Configuration of the newly formed chiral centre in compounds **3** and **4** was found to be identical, since values of the proton-proton coupling constants in the two compounds are very similar.

Finally, formation of the C-N bond in compound **8** was supported by crosspeaks of 1-H of aglycone with C-2' and C-7a' of tryptamine moiety in the 2D HMBC spectrum.

Our experiments clearly demonstrate an efficient way for preparation of novel libraries of indole type alkaloids.

ACKNOWLEDGEMENTS

This work is dedicated to Prof. László F. Szabó on the occasion of this 70th birthday.

REFERENCES

1. Nagy, T., Kocsis, Á., Morvai, M., Szabó, L.F., Podányi, B., Gergely, A., Jerkovich, Gy., 1998, *Phytochemistry*, **47**(6), 1067-72.
2. Kocsis, Á., Szabó, L.F., Podányi, B., 1993, *J. Nat. Prod.*, **56**(9), 1486-99.
3. Böjte-Horváth, K., Hetényi, F., Kocsis, Á., Szabó, L.F., Varga-Balázs, M., Máthé, I., Tétényi, P., 1982, *Phytochemistry*, **21**(12), 2917-19.
4. Unpublished result

The Complexation of New 1,3-Dithiocalix[4]arene Containing Oxime Derivative

NACİYE YILMAZ COŞKUN[1], SABİHA MANAV YALÇIN[1], AYŞE GÜL GÜREK[2] and VEFA AHSEN[2,3]

[1]Yıldız Technical University, Art & Science Faculty, Department of Chemistry, 34210, Davutpaşa-Istanbul, Turkey. [2]TUBITAK, Marmara Research Center, Material & Chemistry, Technologies, Research Institue, P.O.Box 21, 41470, Gebze-Kocaeli, Turkey, [3]Department of Chemistry, Gebze Institue of Technoloy, 41400, Gebze-Kocaeli, Turkey.

1. INTRODUCTION

Calixarenes are three-dimentional cavity-containing macrocyclic compounds which are usually obtained by catalized condensation of p-alkylphenols and formaldehyde[1]. The calixarenes are widely used modules in supramolecular chemistry. In this field, most of the studies have been done on the smaller members of the family, calix[4]arenes and, more recently, calix[6]arene. Synthetic lipophilic, water-soluble, and ionophoric receptor molecules derivatived from calix[4]arenes have been used different technological applications. The easy accessibility and the selective functionalization at the phenolic hydroxy groups of calix[4]arenes have made this member of series increasingly attractive for the chemists involved in host-guest chemistry[2]. A host type, which is expected to fulfill demands for enzyme mimic is the well known calixarene. Calix[4]arenes can be specifically obtained in any of their four well defined conformational families (cone, partial cone, 1,3-alternate, 1,2-alternate). However, the small size of the cavity and the mobil nature of these compounds in solution leads to rapid inversion and the each other conformational changes at room temperature[3]. In particular, calixarenes have been used as sensors of metal ion, anion, neutral molecules and biological specials.[4-8] Molecular recognation is one of the most essential phenomena in biological process and it is a particular interest of life sciences to understand such processes. Thing et all. studied the selective modification of calixarenes at the diametrical 1,3-positions, either via removal of two phenolic groups or via replacement of two oxygen atoms by sulphur[9]. Also coordination chemistry of vic-dioxymates is in an intensive area of study and numerous transition metal complexes of this group of ligands have been investigated [10,11].

1.1 Results and Discussion

Herein, we report the synthesis of new dithioglyoxime ligands containing calix[4]arene. Firstly, 1,3-dimethoxy-2,4-dithiocalix[4]arene(**1**) was obtained according to the literature procedure[12]. The new oxime derivative(**2**) prepared by the reaction of 1,3-dimethoxy-2,4-dithiocalix[4]arene(**1**) metallic sodium and monochloroglyoxime in absolute ethanol. At this reaction, it was seen two compounds at TLC. These compounds have not been isolated with chromatographic procedures. We assumed that this product is conformer mixtures. Metal complexes of this product were prepared with metal salts in EtOH at 4.5-5 pH (scheme 1). It was observed two spots at TLC. In the ^{1}H NMR spectrum, compound **2** showed one methoxy signal, but not observed N-OH signal. However, in the ^{1}H NMR spectrum, metal complexes of compound **2** (compound **3**) observed N-OH signal at 12.59 ppm and hydrogen bridges signals at 16.6 and 17.3 ppm. Our researches are being continued.

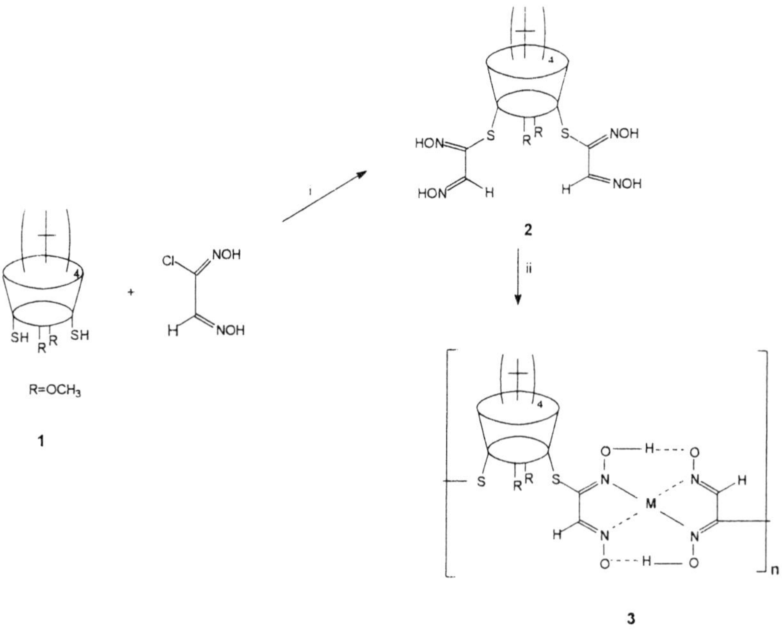

Scheme 1. (i) Abs. EtOH, metallic sodium ; (ii) EtOH, metal salts (Ni(II), Cu(II), Pd(II)), pH= 4.5-5.

REFERENCES

1. Gutsche, C.D., 1989, Calixarenes: Monographs. In *Supramolecular Chemistry* (Stoddart J.F., ed.), vol.1, RSC, Cambridge.
2. Vicens, J., Balmer, V., 1991, Calixarenes; A Versatile Class of Macrocyclic Compounds. In *Topics in Inclusion Science*, Kluwer Academic Press, Dordrect.
3. Gutsche, C.D., Dawan, B., Levin, J.A., No, K.H., Bauer, L.J.,1983, *Tetrahedron,* **39**(3), 409-26.
4. van der Veen, N.J., Egberink, R.J.M., Engbersen, J.F.J., van Veggel, F.J.M., Reinhoudt, 1999, *Chem. Commun.,* 681-2.
5. Leray, I., O'Reilly, F., Habib, J.L., Soumillion, J. –Ph., Valeur, B., 1999, *Chem. Commun.,* 795-6.
6. Chang, S.K., 1986, *J.Chem.Soc., Perkin Trans I,* 211.
7. Araki, K., Yanagi, A., Shinkai, S., 1993, *Tetrahedron,* **49**, 6763.
8. Lazzaratto,M., Sansone, F., Baldini, L., Casnati, A., Cozzini, P., Ungaro, R., 2001, *Eur. J.Org. Chem.,* 595-602.
9. Thing Y., Verboom, W., Groenen, van Loon, J.-D., Reinhoudt, 1990, *J.Cem.Soc., Chem. Commun.,* 1432-3.
10. Ahsen,V., Gürek, A.G., Gül A., Bekaroğlu, Ö., 1990, *J.Cem.Dalton Trans,* **5**.
11. Gürol, I, Ahsen, V., Bekaroğlu, Ö., 1992, *J.Chem.Soc., Dalton Trans,* 2283.
12. Delaigue,X., Hosseini,M.W., Kyritsakas,N., De Cian, A., Fischer, 1995, *J.Chem.Soc., Chem.Commun.,* 609-20.

Convenient Route to Quinoline-Tetrahydroquinolines from Quinoline-Carboxaldehydes

LEONOR Y. VARGAS MÉNDEZA[1], VLADIMIR V. KOUZNETSOV[1]
NÜKET ÖCAL[2], ZUHAL TURGUT[2] and ÇİĞDEM YOLAÇAN[2]
[1]*Industrial University of Santander, School of Chemistry Research Center for Biomolecules Laboratory of Fine Organic Synthesis, A.A. 678, Bucaramanga, Colombia,* [2] *Yıldız Technical University, Department of Chemistry, Davutpaşa Campus, 34210 Istanbul, Turkey.*

1. INTRODUCTION

Quinoline and tetrahydroquinoline structures are essential feature of many natural products. These heterocycles play a key role in heterocyclic and medicinal chemistry. Their synthesis by various methodologies has been published extensively[1-3]. However, in comparison to these systems, general synthetic methods of the preparation of diversely linked bisquinolines have been less developed. On the other hand, several N,N-(bisquinolin-4-yl)(hetero)alkanediamines may be useful agents against chloroquine-resistant malaria[4-6]. As a part of our research program on the chemistry of homoallylamines containing a (hetero)aromatic ring towards the synthesis of bioactive N-heterocycles[7,8], we are pursuing investigations on the synthesis of 1,2,3,4-tetrahydroquinolines containing quinoline nucleus. In the present paper, we report a simply and efficient two step synthesis of 2-(8'-quinolinyl)- and 2-(2'-quinolinyl)-1,2,3,4-tetrahydroquinolines using 6-exo-trig process where allyl group of accessible 4-N-arylamino-4-quinolinyl-1-butenes acts as an internal electrophilic C_3 synthon[8,9]. The proposed route is based on our experience in the construction of diverse heterocycles containing nitrogen via cationic intramolecular cyclisation reactions of similar N-phenyl substituted amino-1-butenes (homoallylamines) that are very versatile starting materials, possessing an π-electron rich aromatic ring and an allyl fragment[10-12].

2. RESULTS AND DISCUSSION

The N-8-quinolinylidenanilines **1a-c** and N-2-quinolinylidenaniline **1d** readily available from quinoline-8- and 6-methylquinoline-2-carboxaldehydes and the corresponding primary anilines were taken as basic precursors. These aldimines were transformed into the corresponding 4-N-arylamino-4-(8-quinolinyl)-1-butenes **2a-c**[8,9] and 4-N-arylamino-4-(2-quinolinyl)-1-butene **2d** through nucleophilic addition of the Grignard reagent using allyl magnesium bromide. This addition is one of key methods for preparing various homoallylamines,[13] which are of particular interest owing to the many possible transformations of the double bond of the allyl group. Thus, in continuation of our synthetic study, we used intermolecular alkylation reaction as a 6-exo-trig process where allyl group of compounds **2a-d** acts as an internal electrophilic C_3 synthon. Prepared 4-N-arylamino-4-(8-quinolinyl)-1-butenes **2a-c** and 4-N-arylamino-4-(2-quinolinyl)-1-butene **2d** were subjected to mediated-acid (85% H_2SO_4) intramolecular cyclisation giving the corresponding 2-(8'-quinolinyl)- and 2-(2'-quinolinyl)-1,2,3,4-tetrahydroquinolines **3a-d**, potentially bioactive heterocycles. The employed allyl electrophilic cyclisation of the aminobutenes **2a-d** afforded the desired products **3a-d** in 35-91% yields (scheme 1, table 1).

Table 1. Properties of the Componds 3a-d.

Compounds 3	R	R_1	Yields, %	Physical aspect
a	CH_3	CH_3	76	Orange crystals, m.p. 80-81 °C
b	CH_3O	H	91	Yellow oil
c	CH_3CH_2O	H	35	Red oil
d	CH_3	CH_3	89	Yellow crystals, m.p. 201-202°C

NMR analysis of compounds **3a-c** revealed that they possess a cis-configuration of the methyl group and the quinoline ring disposed both equatorially at C-4 ($J_{4a,3a}$ = 11.2-12.1 Hz) and C-2 ($J_{2a,3a}$ = 11.1-11.3 Hz), respectively. Based on this ^{1}H NMR analysis and comparison with those observed previously for the related compounds,[14] we assumed that the major isomer (t_R 44.01 min) of mixture **3d** has the cis-configuration.

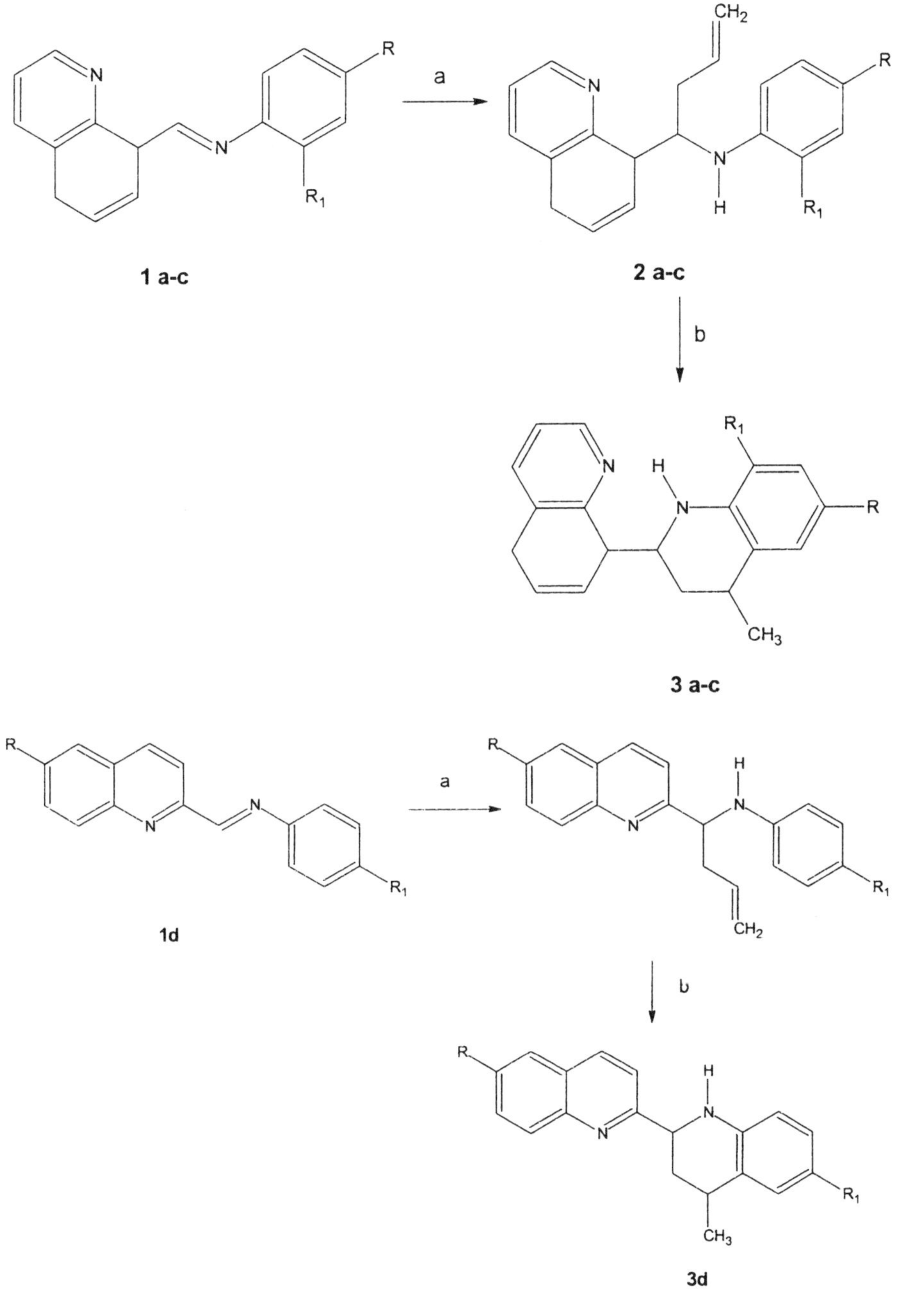

Conditions. a. $CH_2=CH-CH_2MgBr/Et_2O$; b. 85% $H_2SO_4/CHCl_3$

Scheme 1. Synthesis of Quinoline-Tetrahydroquinolines

REFERENCES

1. Jones, G., 1984, Pyridines and Their Benzo Derivatives: (v). Synthesis. In *Comprehensive Heterocyclic Chemistry* (Katritzky A.R., ed.), Vol.2, Pergamon Press, Oxford, p. 395.
2. Katrizky, A.R., Rachwal, S., Rachwal, B., 1996, *Tetrahedron*, **52**, 15031.
3. Kouznetsov, V.V., Palma, A., Ewert, C., Varlamov, A., J., 1998, *Heterocycl. Chem.*, **35**, 761.
4. Vennerstrom, J.L., Ellis, W.Y., Ager, A.L., Andersen, S.L., Gerena, L., Milhous, W.K., 1992, *J. Med. Chem.*, **35**, 2129.
5. Raynes, K., Galatis, D., Cowman, A.F., Tilley, L., Deady, L.W., 1995, *J. Med. Chem.*, **38**, 204.
6. Raynes, K., 1999, *Int. J. Parasitol.*, **29**, 367.
7. Urbina, J.M., Cortés, J.C., Palma, A., López, S.N., Zacchino, S.A., Enriz, D.R., Ribas, J.C., Kouznetsov, V. V., 2000, *Bioorg. Med. Chem.*, **8**, 691.
8. Öcal, N., Yolaçan, Ç., Kaban, Ş., Vargas Méndez, L.Y., Kouznetsov, V.V., 2001, *J. Heterocycl. Chem.*, **38**, 233.
9. Vargas Méndez, L.Y., Kouznetsov, V.V., Poveda, J.C., Yolaçan, Ç., Öcal, N., Aydoğan, F., 2001, *Heterocycl. Commun.*, **7**, 129.
10. Kouznetsov, V.V., Palma, A., Rozo, W., Stashenko, E., Bahsas, A., Amaro-Luis, J., 2000, *Tetrahedron Lett.*, **41**, 6985.
11. Vargas Méndez, L.Y, Rozo, R., Kouznetsov, V., 2000, *Heterocycles*, **53**, 785.
12. Kouznetsov, V.V., Palma A.R., Aliev A.E., 1998, *Anales de Quimica. Int. Ed.,* **94**, 132 .
13. Bloch R., 1998, Chem. Rev., **98**, 1407.
14. Kuznetsov, V.V., Aliev, A.E., Prostakov, N.S., 1994, *Khim. Geterotsikl. Soedin.*, 73 (1994); Chem. Abst., 1994, **121**, 300.738.

Quantum Chemical Research of Quercetin, Myricetin, their Bromo- and Sulpho Derivatives

GALIYA ZHUSUPOVA and V. GAPDRAKIPOV
al-Farabi Kazakh National University, Department of Chemistry, 95a Karasay-Batyr St., Almaty, Kazakhstan

1. INTRODUCTION

Structural diversity of flavonoids due to the rate of heterocyclic ring reduction, aromatic rings hydroxylation (A and B) and their wide presence in nature creates presuppositions for scientific application of these compounds for correction of more than 40 types of biochemical processes in living organisms[1].

The most common flavonols in plants are quercetin and myricetin characterised by wide spectrum of therapeutic action combined with low toxicity and absence of allergic and cumulative properties[2-4].

Quercetin
$\Delta H_f = -224.27$ kcal/mol

Myricetin
$\Delta H_f = -268.06$ kcal/mol

2. RESULTS

It is known that flavonoids normalize peroxide oxidation of membrane lipids[5]. In this connection it is of interest to study the interaction processes of hydrogen atoms of quercetin hydroxy-groups with OH radical and the reaction ability determination for mono- and disubstituted quercetin bromine derivatives with respect to hydroxyl radical[6]. The quantum-chemical calculations were done usinng the PM3 method[7]. Table 1 presents the energies of the OH-bond homolytic dissociation in hydroxyl groups in positions 3, 4', 7 in quercetin and its bromine derivatives.

Table 1. The energies of homolytic and heterolytic dissociation O-H - bonds in quercetin and its Br-derivative. (kcal/mol)

№	Br'[a]	Br''[a]	-H[b]	h[c]	Δh[d]	h(A⁻)[e]	Δh[f]
1	-	-	-	-224.74	-		
2	-	-	3	-208.26	16.48	-265.39	394.23
3	-	-	4	-207.90	16.84	-269.28	398.12
4	-	-	7	-177.06	47.68	-269.88	398.72
5	5	-	-	-215.29	-		
6	5	-	3	-199.12	16.17	-259.49	397.78
7	5	-	4	-197.48	17.81	-264.13	402.42
8	5	-	7	-186.89	28.40	-260.96	399.25
9	6	-	-	-213.90	-		
10	6	-	3	-197.69	16.21	-256.50	396.18
11	6	-	4	-196.80	17.10	-259.83	399.51
12	6	-	7	-184.05	29.85	-261.76	401.44
13	8	-	-	-214.79	-		
14	8	-	3	-198.69	16.10	-256.98	395.77
15	8	-	4	-197.50	17.29	-260.07	398.86
16	8	-	7	-184.47	30.32	-261.89	400.68
17	6	8	-	-203.36	-		
18	6	8	3	-187.67	15.69	-247.86	398.08
19	6	8	4	-186.48	16.88	-250.23	400.45
20	6	8	7	-161.29	42.07	-254.64	404.86

Heat of [a] position of Br; [b] Position of eliminating atom H; [c] Heat of formation of the radicals into reactions $A \rightarrow A^{\cdot} + H^{\cdot}$; [d] Affinities to hydrogen atom; [e] formation of the anions into reactions $A \rightarrow A^- + H^+$; [f] Affinities to proton.

As one can see from the presented data, the OH-bond dissociation energies are very low in comparison with hydroxyl groups in alcohols and carbon acids of the aliphatic series. This is explained by strong stabilization of forming aromatic quercetin radical or its bromine derivatives[8]. Data presented in the table demonstrate that the least energies consumption correspond to the OH- bond homolytic disintegration of hydroxygroups in positions 3 and 4'. Thus, in terms of energetic practicability, the reaction should pass through these exact positions. The weaker activity of the hydroxygroup hydrogen atom in the 7th position is conditioned by the presence of carbonyl group in the 4th position, which, in this case, impedes the formation of the quinoid structure stabilizing the radical.

Introduction of one more bromine atom into the quercetin structure (positions 6 or 8, or 5') has no significant effect on the role of the hydroxigroup hydrogen atom in 3, 4 positions, but increases the activity in the position 7. Introduction of 2 bromine atoms in the quercetin positions 6, 8 makes the activity of the hydroxygroup hydrogen in the 7th position close to the activity of unsubstituted quercetin.

Quercetin and myricetin can be convenient syntones for introduction into their molecules of various atoms and groups of atoms for establishing the relationship between the structure of these compounds, their reaction ability and biological activity. Because of the orientation influence of the hydroxygroups, the reaction-possible positions for the reactions of electrophilic displacement such as bromation and sulfonation in quercetin are the positions 6, 8, 2', 5', 6' and in myricetin – 6, 8, 2', 6' correspondingly. Electronic-acceptor influence of the additional hydroxyl group in myricetin, which increases conjugation in the molecule, is both in its reaction ability and bioactivity. Myricetin is less active in reactions as quercetin, but is more biologically active.

Quantum chemical calculations for reactions of electrophilic displacement showed the following:

- Reactivity of quercetinin in reactions of bromization and sulphuration is higher than of myricetin.
- Bromization of both quercetin and myricetin is more favorable into the position 8 than 6. It is in good agreement with the experimental data for myricetin with methanol as a solvent. However, it is not so for quercetin – position 6 with dioxan as a solvent. It is likely caused by a nature of the solvent, which influences the process of electrophyl substitution.
- Sulphogroups are to be substituted into various positions of quercetin in the following sequence: 5'>6>8. There really was 5'-monosulphoacid as the main product of monosulphuration, two isomeric 6, 5'- and 8, 5'- disulphoacids – of disulphuration in the experiment.

REFERENCES

1. P.J. Macander Plant Flavonoids in Biology and Medicine. New York, 1986, p. 489-792.
2. Г.Е. Жусупова, М.С. Ержанова, К.Д. Рахимов, С.М. Верменичев *О противоопухолевой активности природных оксифлавонов и их производных.* Всесоюзный симпозиум по фенольным соединениям, Таллин, 1987, pp.42-43
3. S.R. Husalne, J. Cillard, P. Cillard // Phytochem, 1987, vol. 26, N 9, P. 2489-2491
4. Растительные лекарственные средства./ род ред. Н.П. Максютиной, Киев: Здоровье, 1985, p. 99-101
5. J. Pincemall, Deby., Lion, // Proc. 7th Hunng. Bioflavanoid Symp. Budapest, 1986, p. 423-436.
6. Г.Е. Жусупова, А.Д. Нагимова, М.С. Ержанова, ХПС 1995, № 4, p. 709-711
7. J.J.P. Stewart, J. Comp. Chem., 1989, 10, 221.
8. В.И. Веденеев, Л.В. Гуревич, В.Н. Кондратьев, В.А. Медведев, Е.Л. Франкевич *Энергии разрыва химических связей. Потенциалы ионизации и сродство к электрону.* Справочник. АН СССР, М., 1962, 216 с.

Destruction and Conservation of Turkish Orchids

EKREM SEZİK
Gazi University, Faculty of Pharmacy, Department of Pharmacognosy, 06330 Ankara, Turkey

1. INTRODUCTION

Turkey is a rich country of terrestrial orchids. The first list of Turkish orchids was prepared by myself in 1976[1,2] after "Flora Orientalis". Later, different investigations had also been made on the orchids of Turkey[3,12]. Orchidaceae part was prepared by Renz and Taubenheim in "Flora of Turkey"[13] in 1984. In the same year, my book "Orkidelerimiz -Türkiye'nin Orkideleri (Our Orchids - Orchids of Turkey)" was published[14]. After 1984, many new orchid species are found in especially South and South-West Anatolia and published[15,20]. In 1998, J.Kreutz published his book "Die Orchideen der Turkei" on Turkish orchids[21]. Volume 11 of "Flora of Turkey" has been published in 2001 and Orchidaceae part of the flora prepared by J.Kreutz[22].

According to all these papers and books, there are approximately 150 orchid taxa in Turkey.

1.1 Orchids growing in Turkey

Terrestrial orchids have creeping, much reduced, fibrous or fleshy rhizomes or tuber - like roots. Table 1 shows the orchids have rhizomes and roots growing in Turkey.

Most of the orchids in Turkey are tuberous (Table 2). On the other hand, tuberous orchids have been used to obtain salep since centuries.

Table 1. Turkish Orchids have rhizom and roots

Genera	Taxa
Cephalanthera	6
Corallorhiza	1
Epipactis	9
Epipogium	1
Goodyera	1
Limodorum	1
Listera	2
Neottia	1
Total	23

Table 2. Turkish Orchids have tubers

Genera	Taxa	Genera	Taxa
Aceras	1	*Anacamptis*	1
Barlia	1	*Coeloglossum*	1
Comperia	1	*Dactylorhiza*	12
Gymnadenia	1	*Himantoglossum*	3
Neotinea	11	*Ophrys*	60
Orchis	31	*Platanthera*	3
Serapias	6	*Steveniella*	1
Traunsteinera	1	*Spiranthes*	1
		Total	125

2. WHAT IS SALEP?

Tuberous orchids have 2 tubers. One is the old tuber and it is wrinkled, shrunken, and brown in colour. The other one has a light colour, spherical to ovoid shape. It is daughter tuber and called "iyisi "in Turkish (means the good one).

2.1 Collecting

During the flowering season, the whole plant is dug out together with its tubers. The wrinkled old tuber is thrown away and the daughter (young) one is collected. Tubers are carefully washed to get rid of the soil on the surface of tubers.

After washing, they are boiled in water, buttermilk, ayran (diluted yougurt), and on occasion in whole milk. The boiling medium differs according to the salep producing areas. After boiling, tubers are rinsed with cold water and dried in the sun. They are spread on the floor or threaded and the strings are hung in a sunny place .The tuber are very hard when they are dried properly[2,14,23].

Salep can be stored in tuber form for years without any decomposition of its content.

2.2 Collecting areas

There are 5 major collecting areas in Anatolia[2,14,23]. These are showed below. Turkish names of the commercial saleps are shown in italics and collecting areas are given in the second line.

Salep collecting areas and commercial saleps

- **North Anatolia**
    Kastamonu Salep(Kastamonu salep)[24]
    Kastamonu ,Tokat(Maden) Provinces
- **South -West Anatolia**
    Muğla salep (Muğla salep)[2,28,29]
    Muğla Province (from Söke to Fethiye)
- **South Anatolia**
    Antalya Salep (Antalya Salep)
    Antalya Province (from Elmalı to Gazipaşa)
    Silifke Salep (Silifke Salep)
    İçel Province (Around Silifke)
- **South-East Anatolia**
    Maraş Salep (Maraş Salep)[25,27]
    Kahraman Maraş, Adıyaman, Malatya Provinces
- **East Anatolia**
    Van Salep (Van Salep)
    Van,Muş, Bitlis Provinces

2.3 Usage of Salep

Salep powder, in Turkey, is used to prepare either ice cream or a hot drink prepared with milk. Salep tubers have polyholosides. The main polyholoside is a glucomannan.

Salep is used as a binder in the production of ice cream because of its glucomannan content. Salep also retards melting of ice-cream. The famous ice cream prepared by using salep is named as " Maraş Dondurması" (ice cream from Kahraman Maraş)[25,27]. Maraş ice-cream is very hard. It is not easy to cut in pieces.

2.4 Orchids used to obtain Salep

Around 120 taxa belonging the genera Ophrys, Orchis, Himantoglossum, Serapias, Anacamptis, Comperia, Barlia, Dactylorhiza, Aceras, Neotinea are used to obtain salep in Anatolia[2,14,23,26].

Table 3. Orchids used to obtain Salep

Genera	Taxa
Aceras	1
Anacamptis	1
Barlia	1
Comperia	1
Dactylorhiza	12
Himantoglossum	3
Neotinea	1
Ophrys	60
Orchis	31
Serapias	6
Total	**117**

2.5 Salep Production

In some years 15 tons of salep may be exported from Turkey. 10 tons of salep is exported yearly in avarage. Domestic use of salep has been twice the export amount at least. That means, appr. 20 tons salep is obtained in Turkey every year.

Let us compare the different sorts of salep.

Table 4. Mean weights of different commercial saleps

Sort of Salep	Mean weight of tubers in g	Mean no of tubers per kg salep
Muğla Salep	0.23	4.348
Kastamonu Salep	0.50	2.000
Silifke Salep	0.35	2.857
Antalya Salep	0.21	4.762
Maraş Salep	1.60	625
Van Salep	1.00	1.000

Salep is collected mostly by women, children and shepherds in poor rural areas.

The tubers are sold to the dealers either fresh or dried. The orchids are rarely found in nature and the tubers are small. A peasant can collect only one kg of tubers in 1-2 days.

During the drying process fresh orchid tubers lose 70-90 % of their weight. The loss of water depends on the season of collection .40 million of tuberous orchids are destroyed in order to produce salep. Orchid destruction has reached unbelievable level.

The earning of a collector per day is about 2 million TL (appr. 1.5 $). The salary of a peasant working in Forestry Department is appr. 15-20 million TL (10-15 $). On the other hand, salep collection takes place only during 2 (at least 3) weeks in the flowering season of orchids[14,26].

Table 5. Marketing of Salep Price and the profits

	PRICE (Tl) per Kg	PROFIT (Tl) per Kg
Peasant	3-5.000.000.-	3-5.000.000.-
Merchants(1.)	18-25.000.000.-	15-20.000.000.-
Merchants (2.)	25-35.000.000.-	7-10.000.000.-
Market	35-45.000.000.-	10.000.000.-

Annual income is only marginal importance even for the poor villager. The merchants do not only sell salep. They also sell crude wool, leather, furs, mushroom, herbs, etc. That means, prohibition of collecting salep will not affect so much, the economical situation of the peasants and merchants.

2.5 Substitutes of Salep

On the other hand, the use of salep substitutes has significantly increased in recent years. CMC, soluble starch, guar gum, rice starch are the substitutes of salep.

The officially recorded Turkish export of salep was 75 tons in 1993. It is impossible, because it makes appr. 150 million of orchids. That value clearly shows, salep powder is adulterated with CMC, starch etc.

3. PROTECTION OF ORCHIDS

Which way will we stop the destruction of orchids?

I tried to inform the people, government departments, bureaucrats, scientists even collectors. I organised seminars, prepared post-cards, radio and TV programs. My articles on Turkish orchids and their destroying are published in popular science journals[30,31]. I also prepared documentaries for TRT (Turkish Radio Television).

Unfortunately the destroying of Turkish orchids still goes on.

3.1 A Project proposal for Muğla Region

I prepared a project proposal for Muğla Region. I chose Muğla. Because this region is the richest area of Turkey for orchids[14,21]. Most of the endemic species are grown in this area and surroundings. On the other hand, there is a huge destroying of orchids in Muğla Region (Table 6). This project can be a model for the other salep producing areas.

Table 6. Orchids used to obtain salep in Muğla

	Number of Taxa
Aceras antropophorum	1
Anacamptis pyramidalis	1
Barlia robertiana	1
Comperia comperiana	1
Dactylorhiza romana	1
Neotinea maculata	1
Ophrys species	31
Orchis species	18
Serapias species	4
Total	**59**

3.2 Conservation of Orchids of Muğla (CORMU)

Project Steps

EDUCATION
Educating pogramme for
Local University & schools
Local officials of Ministry of Forestry
Muhtars (Headmen of villages)
Traders, vendors & ice-cream producers
Tourism agencies
INFORMATION
Media
Local newspapers
TV & radio channels
Printed Materia
Brochures
Books
Postcards
RESEARCH
Collaborative projects between
Gazi & Muğla University
Ministry of Forestry
Ministry of Environment

SOCIAL & SCIENTIFIC ACTIVITIES
Public Conferences
Scientific Meetings
Orchid Festival
Orchid Exursions

Let me explain shortly this project model.
First step is to collect orchid tubers will be forbidden for 2 years
Forbidding is not enough. I can summarise the project as follows:

3.2.1 Education

Local University & Schools
Students having education at different levels will be given lectures on topics of salep-orchid relationship, and destruction of orchids. The youngster should be taken as the future administrators, vendors, businessmen, etc.;

therefore educating, informing them on this subject will bring long-term benefits on protection of orchids.

Muhtars (Headmen of villages)

Muhtar is the elected alderman (headman) of a village. Because muhtar has considerable personal communication with the peasants, he (or she) has key importance on informing people about the destroying of orchids.

Ministry of Forestry

Officials need to be convinced on the subject as they constitute the authority on collection of plants, of course orchids.

Salep vendors

Salep vendors get the most financial benefit, so convincing these people may harbour certain difficulties. These people must inform about on salep-orchid-ice cream relationship and alternatives to salep.

Tourism agencies

Representatives from companies active in the field of tourism were given information on the subject. So they may have expected to organise "orchid excursions" especially on out - of -season months like April.

3.2.2 Information

Local press, TV and radios

The local media has an important function as to inform the people about the destruction of orchids in Muğla. News, articles on orchids and destruction of them appearing in local newspapers will have an important impact .A documentary on the subject will be provided to local TV channels for broadcasting. Also, interviews covering the topics on radio and TV channels will be beneficial especially on young people and housewives, who are regular listeners and watchers.

Printed materials

Printed materials include postcards, brochures and booklets for sale to tourists mainly.

3.2.3 Research

A joint-project between Gazi University, Faculty of Pharmacy and Muğla University, Dept. of Biology of 2 years duration about the distribution of orchids in the Muğla Region, localities of the endemic species, and the degree of their destruction will be carried out. The project includes the training of 3 taxonomists from Muğla University on orchids. Educational slides and other materials about the orchids found in the Region will be prepared as well.

3.2.4 Social Activities

Orchid festival will be organised around 15 of April (the flowering time most of the orchids). The program will include scientific activities (a conference, symposium,etc.), contests (different topics on photography, clothing ,etc..), tea-parties, balls etc.

An international symposium on orchids may be hosted by the governor of the town or rector of Muğla University. This symposia could become a traditional event.

4. CONCLUSION AND REMARKS

Orchids in Turkey have been gathered for centuries to obtain "Salep. In addition, over urbanisation and environmental pollution accelerated the destruction in the last 30 years. I am drawing attention to this destruction in my lectures, papers etc. Recently some of European scientists have also expressed concerned on this subject. My project proposal aiming to decrease and stop the destruction of orchids is presented in this paper. I believe that destruction will be slowed down in the mentioned region where extinction of some species is a potential danger. In case the project is realised in achievement, others may follow throughout the country.

REFERENCES

1. Boissier, E., 1884, Flora Orientalis 5 (Orchidaceae),Genevae.
2. Sezik, E., 1967, Türkiye'nin Salepgilleri, Ticari Salep Çeşitleri ve Özellikle Muğla Salebi Üzerinde Araştırmalar (Researces on the orchids of Turkey, Commercial saleps and especially on Muğla Salep),Ph.D. Thesis, İstanbul University,Faculty of Pharmacy, İstanbul.
3. Sezik, E., 1969, Türkiye'nin Orkideleri (Orchidaceae Family in Turkey), TÜBİTAK II. Bilim Kongresi (II. Science Congress of Turkish Research Council), Ankara,17-19 November.
4. Sezik, E., 1982, Türkiye'de Orchidaceae Familyası (Orchidaceae Family in Turkey), IV. Bitkisel İlaç Hammaddeleri Toplantısı Kitabı (Proceedings of the Symposium on Plant Originated Crude Drugs), pp.77-83, 153.
5. Sundermann, H., 1969, *Die Orchidee*, **20**, 309.
6. Sundermann, H.,Taubenheim,G., 1978, *Die Orchidee*, **29**, 172.
7. Sundermann, H.,Taubenheim,G., 1981, *Die Orchidee* , **29**, 202.
8. Sundermann, H.,Taubenheim,G., 1981, *Die Orchidee*, **32**, 214.

9. Sundermann,H.,Taubenheim,G.,1982, *Die Orchidee*, **33**, 222.
10. Taubenheim,G., 1977, *J.Ber.Nat.V.Wupp.*, **29**, 78.
11. Taubenheim,G., 1979, *Die Orchidee*, **30**, 223.
12. Taubenheim,G., 1980, *Die Orchidee*, **31**, 5.
13. Renz,J.,Taubenheim,G.,1984, Orchidaceae. In *Flora of Turkey and the East Egean Islands* (Davis, P.H,, Mill,R., Tan,K. eds.), Vol.8, Edinburgh University Press, Edinburgh, p.632.
14. Sezik,E., Orkidelerimiz - Türkiye' nin Orkideleri (Our Orchids. - Orchids of Turkey), Güzel Sanatlar Matbaası , İstanbul , pp.166.
15. Baumann,B. et H., 1991, Hybridogene Populationen zwischen *Orchis anatolica* Boiss. und *O.quadripunctata* Cyr. ex Ten. In der Ost-mediterraneis, Mitt. Bl.Arbeitskr.Heim.Orch. Baden-Württ., **23**, 203-242.
16. Kreutz,C.A.J., 1997, *Euorchis*, **9**, 48-76.
17. Kreutz,C.A.J., 1997, *J.Eur.Orch.*, **29** ,653-98.
18. Kreutz,C.A.J., Peter,R.,1998, *J.Eur.Orch* , **30** ,81-156.
19. Robatsch ,K., 1991, *Ber.Arbeitskrs.Heim.Orchid.*, **8**, 61-3.
20. Rückbrodt D. Et U., 1996, *J.Eur.Orc.*, **28**, 391-403.
21. Kreutz,C.A.J., 1998, Die Orchideen der Türkei, CIP-Gegevens Koninklijke Bibliotheek, Den Haag.
22. Kreutz,C.A.J., 2000, Orchidaceae. In *Flora of Turkey and the East Aegean Islands* (Güner,A., Özhatay,N., Ekim,T., Başer,K.H.C., eds.), Vol.11, Edinburgh University Press, Edinburgh.
23. Baytop,T., Sezik, E., 1968, *J.Fac.Pharm.İstanbul*, **4**, 61-8.
24. Sezik,E., Özer,Y.B., 1983, Kastamonu Salebinin Menşei ve Kastamonu Civarının Orkideleri (The Origin of Kastamonu Salep and the Orchids of Kastamonu), Project of Turkish Research Council (TÜBİTAK), No:TBAG 424, Ankara.
25. Sezik,E.,1969, Muğla Salebinin Menşei ve Kalitesi (The Origin and Quality of Muğla Salep),TÜBİTAK II. Bilim Kongresi November 17-19 (II. Science Congress of Turkish Research Council), Ankara.
26. Sezik, E., 1969, *J.Fac.Pharm. İstanbul*, **5**, 77-9.
27. Sezik, E.,Baykal,T., 1991 : Maraş Salebinin Menşei ve Maraş Civarının Orkideleri (The origin of Maraş Salep and the Orchids of Maraş),Project of Turkish Research Council (TÜBİTAK), No:TBAG 664, Ankara.
28. Sezik,E.,Baykal,T., 1988, *Doğa Türk Eczacılık Dergisi*, **1**, 10-6.
29. Sezik, E., 1989, Turkish Orchids and Salep,Orchidees Botanique du Monde Entier Colloque 11, Paris, pp.181-9,
30. Sezik,E., 1990, *Image*,**12**,11-5.
31. Sezik,E. ,1990, *Bilim ve Teknik*, **23**, 5-8.
32. Kasparek, M., Grimm, U., 1999, *Economic Botany* , **53**, 396-406.
33. Ertuğ,F., 2000, *Economic Botany* , **54**, 421-2.

Annex

International Union of Pure and Applied Chemistry, Organic and Biomolecular Chemistry Division in conjunction with The Society of Biological Diversity-Ankara-Turkey

Plenary lectures presented at the 3rd IUPAC International Conference on Biodiversity (ICOB-3) held in Antalya, Turkey 3-8 November, 2001 published in "Pure and Applied Chemistry, Vol. 74, No.4, pp. 511-584, 2002.

Bilge Şener	iv	Preface
Atta-ur-Rahman and M. Iqbal Choudhary	511	Biodiversity – A wonderful source of exciting new pharmacophores. Further to new theory of memory.
Ronald J. Quinn, Priscila de Almeida Leone, Gordon Guymer and John N. A. Hooper	519	Australian biodiversity via its plants and marine organisms. A high – throughput screening approach to drug discovery.
K. Hüsnü Can Başer	527	Aromatic biodiversity among the flowering plant taxa of Turkey.
Neriman Özhatay	547	Diversity of bulbous monocots in Turkey with special reference. Chromosome numbers
Günay Sarıyar	557	Biodiversity in the Alkaloids of Turkish *papaver* species
Ya-ping Zhang, Xiao-xia Wang, Oliver A. Ryder, Hai-peng Li, He-ming Zhang, Yange Yong and Peng-yan Wang	575	Genetic Diversity and conservation of endangered animal species.

Author Index

Abegaz, B.M., 71
Abilov, Z.H., 367
Abou-Gazar, H., 257
Abou-Gharbia, M., 63
Achmad, S.A., 91
Açan, N.L., 341
Afshan, F., 109
Aftab, K., 279
Ahsen, V., 379
Akalın, E., 309
Aktay, G., 261
Alighanadi, A., 319
Amer, W.M., 197
Arısan, M., 273
Ayanoğlu, F., 227, 345
Aydoğan, F., 371
Ayub, Q., 35
Bağcı, E., 323, 329
Balasubramanian, K., 253
Başer, K.H.C., 309
Bedir, E., 257
Begum, S., 109
Benkhaled, M., 303
Benkiki, N., 303
Bingöl, F., 261
Bowers, W.S., 49
Bruneau, C., 303
Cavallisforza, L.L., 35
Çalış, İ., 137, 257
Dagne, E., 79
De Caro, S., 335
De Rosa, S., 335

Demirci, B., 309
Ekizoğlu, M., 273
Ercil, D., 273
Erdemoğlu, N., 291, 297
Eryılmaz, B., 261
Eyüboğlu, Ö., 355
Faizi, S., 109
Flores, F.A., 85
Foo, L.Y., 265
Fujimoto, H., 85
Gapdrakipov, V., 387
Garson, J.M., 179
Grishin, E.V., 161
Gül Gürek, A., 379
Hakim, E.H., 91
Hameed, A., 35
Hatano, T., 265
Hegazy, A.K., 197
Heywood, V.H., 13
Higa, T., 169
Hussein, A.K.M., 59
Ishibashi, M., 85
Ismail, M., 35
Işık, K., 349
Ito, H., 265
Jain, S.C., 101
Juliawaty, L.D., 91
Kabouche, Z., 303
Kaloga, M., 269
Kardar, N., 109
Kaushik, N., 283, 287
Kaya, D.A., 345

Kaya, N., 349
Kayser, O., 265
Khaliq, S., 35
Khan, I.A., 257
Kırıcı, S., 227
Kiderlen, A.F., 265
Kleinschmit, J., 1
Kleinschmit, J.R.G., 1
Kocsis, Á., 375
Kolodziej, H., 265, 269, 273
Korolkova, Yu.V., 161
Kouznetsov, V.V., 383
Koyuncu, M., 247
Küçük, M., 355
Lu, B.-R., 23
Makmur, L., 91
Manav Yalçın, S., 379
Mansoor, A., 35
Mátyus, P., 375
Mazhar, K., 35
Mehdi, S.Q., 35
Mehri-Ardestani, M., 319
Mitova, M., 335
Mohyuddin, A., 35
Mujahidin, D., 91
Mutanyatta, J., 71
Nagamatsu, C., 85
Ohtani, I.I., 169
Ok, K., 217
Okuyama, E., 85
Olsen, C.E., 101
Öcal, N., 371, 383
Özalp, M., 273
Özçelik, H., 329
Özgüven, M., 227
Papaioannou, M., 35
Parmar, V.S., 101
Pluzhnikov, K.A., 161
Prasad, A.K., 101
Priyadarshini, R., 253
Qamar, R., 35
Rachimov, K., 367

Radha, A., 253
Radtke, O.A., 269
Rahman, S., 35
Raj, H.G., 101
Raman, N., 253
Rasheed, M., 109
Roy, M.C., 169
Rughunathan, R., 253
Ruiz, J., 85
Sajjadi, S.E., 315
Sakar, M.K., 273
Satake, M., 85
Sekita, S., 85
Sezik, E., 391
Shalakhmetova, T., 367
Shimamura, K., 85
Shiroto, O., 85
Siddiqui, B.S., 109
Siddiqui, S., 35
Singh, B.G., 283
Syah, Y.M., 91
Şener, B., 291, 297
Tan, N., 269
Tan, R.-X., 151
Tanaka, J., 169
Taşdemir, D., 187
Tezcan, E.F., 341
Tommonaro, G., 335
Turgut, Z., 371, 383
Tyler-Smith, C., 35
Vargas Méndeza, L.Y., 383
Volkova, T.M., 161
Watterson, A.C., 101
Yaşarakıncı, N., 363
Yeşilada, E., 119
Yılmaz Coşkun, N., 379
Yolaçan, Ç., 371, 383
Yoshida, T., 265
Yuenyongsawad, S., 85
Zhusupova, G., 367, 387
Zou, R.-X., 151

Subject Index

Aceraceae, 154
Aceras, 392
Acetaminophen toxicity, 261, 263
Acetoxycoumarins, 105
Aconitum nasutum, 329
Aconitum orientale, 329
Acorn worms, 172
Aedes aegyptii, 113
Aegean region, 363
Aegilops, 28, 31
AFLP, 15
Africa, 35, 71
African flora, 79
African populations, 43
Agro-ecosystem, 59
Agropyron, 28, 30
Ailments, 122
Aiolopus, 61
Algae, 169
Allelochemicals, 179, 180
Allelopathic chemicals, 157
Allelopathic effects, 179
Allium schoenoprassum, 247
Allium tuncelianum, 247
Allium vineale, 247
Allylation, 383
Altitudinal diversity, 197
Amaryllidaceae, 248
Amino acid analysis, 339
Anabasis articulata, 59
Anacamptis, 392
Anacardiaceae, 154
Anatolia, 247, 309
Anatomical characteristics, 311, 312

Angiosperms, 152, 154
Animal bite, 129, 130
Animal diseases, 131
Antibacterial activity, 273, 275, 276, 372
Anticancer agent, 174
Anticancer compounds, 80
Anti-cancer metabolites, 187
Anti-fouling paints, 184
Antifungal activity, 273, 275, 276
Anti-HIV compounds, 81
Antiinflammatory activity, 146
Anti-leishmania compounds, 82
Antileishmanial activity, 265, 269, 271
Anti-malarial compounds, 83
Antimicrobial activity, 145, 253-255
Antimicrobial metabolites, 156
Antioxidant activity, 261
Antioxidant, 260
Anti-oxidative effect, 85, 106, 146
Antirepellent1, 29, 130
Antituberculosis activity, 372
Antitumor activity, 174
Antitumor agents, 157
Antiviral activity, 368
Aphis, 61
Apocynaceae, 82
Aquaculture, 184
Arachnida venoms, 162
Araucariaceae, 152
Aromatic biodiversity, 315
Aromatic plants, 13
Arthropod toxins, 161, 165
Arthropods, 61
Artocarpus sp., 91, 92

Artocarpus altilis, 94
Artocarpus bracteata, 96
Artocarpus champeden, 92
Artocarpus lanceifolius, 93
Artocarpus maingayii, 94
Artocarpus rotunda, 95
Artocarpus scortechinii, 96
Artocarpus teysmannii, 96
Ascidians, 174
Asclepiadaceae, 83
Asphodelaceae, 84
Asphodelus microcarpus, 59
Asteronotus cespitosus, 181
Australia, 181
Autosomal markers, 35
Autosomal microsatellite analysis, 38
Autosomes, 36
Autosomol microsatellites, 42, 44
Azadirachta indica, 111, 114, 116, 283
Azadirachtin, 283, 285

Barlia, 392
Berberidaceae, 154, 248
Betula pendula, 8
Betulaceae, 154
Biallelic polymorphism, 37
Bignoniaceae, 83
Bioactive compounds 71, 79, 85, 253
Biodiversity 23, 59, 109, 110, 119, 137, 151, 222, 224, 253, 363
Biogenesis, 97
Biological activity, 74, 97, 105, 109, 143, 371, 379, 383
Bioprospecting, 20
Birch, 8
Biregional, 208
Blood pressure lowering activity, 279
Blood samples, 37
Brachycerus spinicollis, 60
Bromization, 389
Bulbine, 72, 73, 75
Burseraceae, 80

Cachyrs, 309
Calicheamicin conjugates, 65
Calix[4]arene, 379
Capsicum annuum, 261
Capsodes, 60
Cardio-vascular complaints, 129
Carnivore, 60
CBD, 20, 223
CCI-779, 63, 65, 67
CD33 antibody, 63

Celastraceae, 85
Cell cycle, 188
Central Nervous System diseases, 126, 127
Characteristic species, 355
Chemical diversity, 13, 113, 169
Chemical prospecting, 283
Chemical signals, 179
Chemical structures, 155
Chemical variation, 14, 15
Chemistry, 74
Chemodiversity, 91, 151, 155
Chemosystematic markers, 147
Chemotaxonomic study, 309
Chemotaxonomy, 313, 323, 329
Chemotypes, 14
Chenopodiaceae, 154
Chikusichloa, 29
Chilli pepper, 261
China, 26
Chorological analysis, 212
Chuchuhuasi, 85
Classical field, 4
Climate, 200
Clonal archives, 8
Coastal vegetation, 214
Coeloglossum, 392
Collection, 4
Combretaceae, 80
Combretum erythrophyllum, 80
Command mechanism, 217
Commercial saleps, 393
Commiphora erlangeriana, 80
Comperia, 392
Complexation, 379
Compositae, 154
Conservation approaches, 17
Conservation, 1, 3, 13, 23, 24, 391, 397
Construction, 184
Convention on Biological Diversity 20, 223
Coral reefs, 169, 176
Corals, 173
CORMU, 397
Corynanthe pachyceras, 82
Cosmopolitan, 208
Coumarin, 53, 101
Croton tiglium, 81
Cruciferae, 154
Cryptolepis sanguinoşenta, 83
Cupressaceae, 152
Cupressus sempervirens, 349, 352
Cyclamen hederifolium, 247
Cynodon dactylon, 253

Cytostatic activity, 143
Cytotoxic activity, 143
Cytotoxic compounds, 80
Cytotoxicity, 91, 175

Dactylorhiza, 247
Dactylorhiza, 392
DAMC, 102
Data analysis, 38
Data Bank, 120
DEET, 53
Dermatological disorders, 124
Destruction, 391
Deteritivivore, 60
Determination, 227
Dicots1, 54, 202, 203
Dicotyledonae, 154
Dictyotaceae, 169
Dipsacus laciniatus, 375
Dipterocarpaceae, 81
Distribution of data, 122
Diterpenoids, 115, 269
Diversity, 154
DNA analysis, 35
DNA based markers, 36
Documentation, 119
Domestication, 227
Dorstenia, 72, 73, 76, 77
DPPH, 86, 260
Drasterius biomaculatus, 62
Drug discovery, 63, 64
Dysidea fragilis, 336

East Asian populations, 35
Ecdysone, 54
Ecdysteroids, 55
Ecogeographical sampling, 16
Ecosystems, 59
Ecotypes, 228, 233
Edible plants, 101
Education, 397
Egypt, 59
Elymus, 28, 30, 31
Elytrigia, 28, 31
Endangered plants, 71
Endemic, 208, 315
Endemism, 205, 206, 211
Endocrine diseases, 127
Endophyte, 151, 155
Endophyte-harboring plants, 152
Endophytes producing taxol, 158
Endophytic bacteria, 152
Enteropneusta, 172

Enzyme, 101
Eremirus spectabilis, 247
Ericaceae, 154
Erymopyrum, 31
Essential oil, 230, 231, 310, 313
Ethnic groups, 36
Ethnobotanical uses, 131
Ethnomedical use, 72
Eucalyptus, 287, 289
EUFORGEN, 2, 9
Euphorbiaceae, 81
European populations, 35
European yew, 291
Evaluation, 4
Ex situ conservation, 4, 7, 17, 25

Fabaceae, 154
Fagaceae, 154
Fatty acids, 323, 325, 329, 333
Fauna, 60
Fenoxycarb, 51
Fenugreek, 101
Ferulago, 309, 310
Flavonoids, 97, 310, 313, 387
Floristic relations, 206
Floristic richness, 202, 207, 209
Folk illnesses, 128, 129
Folk medicine, 119
Folk remedies, 122
Food, 131
Forest area, 5
Forest tree, 349
Forest vegetation, 357
Forests, 355
Free market environmentalism, 224
Free market mechanism, 217
Free radical scavenging, 257
Fritillaria imperialis, 247
Fritillaria persica, 247
FRLHT, 17
Fumariaceae, 248
Functional metabolites, 156
Furocoumarins, 303

Galanthus elwesii, 247
GAR-936, 63, 67
Garlic, 247
Gastro-intestinal disorders, 123, 124
GATT, 222
GC analysis, 325, 331
GC-MS analysis, 316, 320
Genetic conservation, 8, 14
Genetic data, 42

Genetic diversity, 8, 13, 23, 109, 349, 352
Genetic erosion, 24
Genetic markers, 4
Genetic relationships, 35
Genetic variation, 15
Genital system problems, 127, 128
Genomics, 64
Genotype, 109
Gentianiaceae, 154
Geology, 200
Geophytes, 247
Geraniaceae, 248
Germany, 8
Glaucosciadium, 309
Globulariaceae, 138, 144
Glutathione reductase, 341
Glycine gracilis, 30, 32
Glycine max, 30, 32
Glycine soja, 30, 32,
Glycine tabacina, 32
Glycine tomentella, 32
Glycine, 23
Glycylcyclines, 69
Gramineae, 152, 253
Groups of phenylethanoids, 138-143
Gümüþhane, 355
Gymnadenia, 392
Gymnocarpus decandrum, 59
Gymnosperms, 152

Haematopoietic system, 131
Haloragaceae, 154
Haplogroup frequencies, 41
Hatay, 345
Helicoverpa armigera, 287, 289
Heraclenol, 303, 307
Herbal teas, 138
Herbivore, 60
Herbivore-toxic metabolites, 157
Heterocyclic compounds, 371
Heterodera schachtii, 55
Heterogamia syriaca, 60
Himantoglossum, 392
Homo sapiens, 36
Homoisoflavonoids, 74
Hordeum vulgare, 28
Hordeum, 30
Host plants, 155
Human evolution, 35
Human genom diversity, 35
Humiles, 309
Hygroryza, 29
Hypolepidaceae, 152

Immune system diseases, 127
Implementation, 363
In planta cultivation, 155
In situ conservation, 4, 6, 17, 27
In vitro cultivation, 155
India, 283
Indian biodiversity, 253
Indol alkaloids, 375
Indonesian plants, 91
Indonesian sponge, 176
Influenza virus, 369
Information, 398
Insect growth regulators, 50
Insect predators, 59
Insecticidal metabolites, 157
Insects, 54
Integration, 6
Interactions, 181
Inventory, 3
IPM program, 363, 364
IPR, 20
Iran, 315, 319
Ircinia muscarum, 335
Iridaceae, 248, 249
Iridoids, 375
Iris spuria ssp. *musulmanica*, 247
Isocyanide sponges, 182
Isofuranonaphthoquinones, 75
Isolariciresinol, 291
Isopimpinellin, 303, 307
IUFRO, 2

Juvabione, 51
Juvocimene II, 51

Kigelia pinnata, 83
Kinetic properties, 342
Kniphofia foliosa, 84
Kürtün, 355

Labidura reparia, 61
Lamiaceae, 138, 144
Lamioideae, 147
Lariciresinol, 297
Latitudinal diversity, 197
Latrotoxins, 161, 164
Lauraceae, 154
Lawsonia alba, 112
Ledebouria, 72, 73
Leersia, 29
Leguminosae, 154
Leishmania major, 83
Leishmania sp., 265, 271

Subject Index

Leishmaniasis, 269
Leishmanicidal activity, 266
Lenfoid system, 131
Leucojum aestivum, 247
Leymus chinensis, 28
Leymus secalinus, 28
Leymus, 30
Lignans, 291, 297
Liliaceae, 249, 250
Limonium, 367
Linalool, 53
Lipid peroxidation, 261
Lipids, 324, 325, 331
Liver microsomes, 102
Local participation, 20
Loganiaceae, 83
Loganin, 375
Loranthaceae, 154
Lung ailments, 124, 125

Malvaceae, 154
Managed forests, 5
Marine animals, 179
Marine organisms, 169
Marine sponges, 187, 335
Market mechanism, 217
Maytenus amazonica, 85
Medicinal plants, 13, 79, 101, 109, 110, 323
Medicinal uses, 123
Medon ochracen, 61
Menispermaceae, 82
Mentha aquatica, 227
Mentha arvensis, 227
Mentha longifolia, 227
Mentha piperita, 227
Mentha pulegium, 227
Mentha sp., 227
Mentha spicata ssp. *spicata*, 227
Mentha suaveolens, 227
Messor, 60
Metabolic diseases, 127
Metabolites, 151, 169
Methylcoumarins, 101
Microbial infections, 126
Microorganisms, 335
Microsatellites, 35
Mixed economies, 217
Mode of action, 161
Mollusks, 171
Monocots, 152, 202, 203
Monocotyledonae, 152
Monomorrium, 61

Monotes africanus, 81
Moraceae, 91
Morphological characteristics 310, 311, 345, 347
Mylotarg^{TM} 63, 64
Myricetin, 387
Myrtaceae, 154, 287

N,N-Diethyl-m-toluamide, 53
NAPRALERT, 253
Narcissus poeticus, 247
Native, 349
Natural forests, 5
Natural juvenile hormone, 51
Natural products leads, 63
Natural products, 91
Neem fruits, 117
Neem, 283
Nematodes, 54
Neoplastic diseases, 127
Neotinea, 392
Nepetoideae, 147
Network, 2
Neurotoxins, 161
New leads, 110
Nicotiana tabacum, 345
Noble hardwood species, 8
Non-African populations, 43
Non-natural, 375
Nudibranch, 181

Oak, 8
Ocimum basilicum, 279
Octocorallia, 173
Okinawan sponges, 175
Oleaceae, 138, 144, 154
Ophrys, 247, 392
Ophtalmologic problems, 130
Opisthobranchia, 171
Oral diseases, 130
Oral hygiene, 130
Orchards, 8
Orchidaceae, 153, 250, 391
Orchids, 394
Orchis, 247, 392
Ornamental, 247
Ornitogalum narbonense, 247
Oryza, 23, 25, 29
Oryza granulata, 27
Oryza nivara, 23
Oryza officinalis, 27
Oryza rufipogon, 23, 24, 27
Oryza sativa, 26

Oryzeae, 26
Otolologic problems, 130
Overharvesting, 19

Pacific pest control, 49
Pakistan, 35
Pakistani ethnic groups, 39
Pakistani populations, 42
Paleotropic, 208
Palmaceae, 153
Pantropic, 208
Paprika, 261
Parasitic infections, 126
Parietaria, 323
PC analysis, 42
PCR, 15
Peptide, 335
Peptides, 161
Perfect competition, 219
Peruvian herbal medicine, 85
Pesticidal activity, 287
Pests, 363
Phenotype, 109
Phenyl anthraquinones, 75
Phenylethanoid glycosides, 137, 144, 257
Pheromones, 52
Philanthus, 61
Phyllidia ocellata, 183
Phyllidid nudibranchs, 182
Phylogenetic analysis, 42
Phytochemical repellents, 53
Phytoecdysteroids, 54
Phytogeographic region, 208
Phytogeographical groups, 208
Phytojuvenoids, 51
Phytopreparations, 367, 368
Piceetum orientali, 357
Pinaceae, 152
Pinus brutia, 349, 350, 351
Piper nigrum, 112
Piperitone, 53
Piperonal, 53
Plant associations, 359
Plant hormones, 157
Plant protection strategies, 54
Plantago albicans, 59
Plantation, 5
Platanthera, 392
Plumbaginaceae, 367
Pluriregional, 208
Podophyllotoxin-type, 80
Poisonous plants, 329
Polymorphism, 35, 36

Population genetics, 35
Populations, 37
Porteresia, 29
Prangos, 309
Prenylated flavonoids, 76
Proanthocyanidins, 265
Prostaglandin inhibition, 85
Protection, 396
Protein transacetylase, 105
Prunus avium, 8
Psathyrostachys, 30
Pseudomonas, 336
Pseudoroegneria, 30
Pssamobius porcicollis, 61
Pteridophytes, 152
Pure stands, 355, 359
Pyrus pyraster, 8

QSAR, 104
Quality aspects, 227
Quantum, 387
Quercetin, 387
Quercus robur ssp. *petraea*, 8
Quercus robur, 8
Quinoline, 383
Quinolinecarboxaldehydes, 383

RAFI, 20
Rank-order list of plants, 132
Ranunculaceae, 250
Ranunculaceae, 329
Rapamycin, 65-67
Rapateaceae, 153
RAPD, 15
Recombination, 36
Red Sea, 197, 199, 213
Repellents, 52, 129, 130
Research, 398
RFLP, 15
Rhabdastrella globostellata, 187, 191
Rhamnaceae, 154
Rhamnaceae, 319
Rhododendro pontici-Fagetum orientali 357, 360
Rosaceae, 154
Rosmarinic acid, 147
Rubiaceae, 154
Ruta montana, 303
Rutaceae, 154

Sahlep, 247, 391
Salep collecting areas, 393
Salep production, 394

Subject Index

Salep, 392
Salicaceae, 154
Salvia cilicica, 269
Sampling, 15
Sapotaceae, 154
Scilla, 72
Scorpion venoms, 162
Scrophulariaceae, 138, 144
Secale cornutum, 28
Secale, 28, 30
Sedum sartorianum ssp. *sartorianum*, 273
Semi-desert ecosystem, 59
Serapias, 392
Sheep brain, 341
Skeleto-muscular problems, 125
Social activities, 399
Soil fauna, 59
Soja, 30
Solanaceae, 154
Sorbus torminalis, 8
South-Asian region, 110
Soybean, 32
Species diversity, 204, 205, 210
Spider venoms, 161
Spinach, 55
Spinacia oleracea, 55
Spiranthes, 392
Sponge diet, 181
Sponge metabolites, 184
Sponges, 174, 179, 181
Stephania dinklagei, 82
Sternbergia clusiana, 247
Steveniella, 392
Stinging nettle, 323
Structural diversity, 387
Structure analysis, 42
Structure, 161, 163
Strychnos icaja, 83
Stylissa massa, 187, 189
Stylotella aurantium, 189
Sub community, 355
Sulphuration, 389
Sumilarv, 51
Sustainable utilization, 23
Synthetic modification, 375

TAase, 102
Tamil Nadu State, 283
Tariffs, 222
Taxa of endophytes, 55
Taxaceae, 152, 291, 297
Taxiresinol, 297, 298
Taxodiaceae, 152

Taxol, 157, 158
Taxus baccata, 291, 297
Taxus, 158
Temperate hardwood species, 1, 3
Terrestrial orchids, 391
Tetracyclines, 68
Tetrahydroquinoline, 383
Threats to genetic conservation, 17
Thymeleae hirsute, 59
Thymus, 315
Tobacco, 345
Toxicity, 287
Toxins, 161
Trade liberalization, 222
Traditional medicine, 119
Transacetylase, 102
Transacetylate, 102
Transport industry, 184
Traunsteinera, 392
Trigonella foenum-graecum, 101
TRIPs, 223
Triterpenoids, 114, 116
Triticeae, 23, 24, 29
Triticum aestivum, 28
Triticum, 30
Tropinota squalida, 61
TUHIB, 120, 121, 133
Tulipa sylvestris, 247
Turkey, 119, 227, 247, 324, 345, 349, 355, 363, 391
Turkish flora, 119
Turkish Orchids, 391, 392
Turkish traditional medicine, 120

Ubiquitous distribution, 152
Unirgional, 208
Upper respiratory track ailements 124, 125
Urological complaints, 125, 126
Urtica dioica, 323
Urticaceae, 323
Utilization, 1, 3, 5

Vaccinio myrtilli-Pinetum sylvestri 357, 358
Vegetation, 355
Verbascum wiedemannianum, 257
Vic-dioxime, 379
Vitaceae, 154
Vitamin C, 261
Voacanga africana, 19
Volatile constituents, 319

Wild Cherry, 8
Wild harvesting, 19
Wild Pear, 8
Wild relatives of crop, 23
Wild relatives, 25
Wild rice, 26
Wild service tree, 8
Wild soybean, 30
Wild, 227
World Bank report, 19
WRCS, 23

Xanthotoxin, 307
Xestospongia sp., 187, 193
Y Chromosome, 35, 36
Yield, 227, 229

Zingiberaceae, 153
Zizania, 29
Ziziphus spina-christi, 319
Zophosis, 61